FUZZY RELATION EQUATIONS AND
THEIR APPLICATIONS TO KNOWLEDGE ENGINEERING

**THEORY AND DECISION LIBRARY**

General Editors: W. Leinfellner and G. Eberlein

Series A: Philosophy and Methodology of the Social Sciences
Editors: W. Leinfellner (Technical Universtiy of Vienna)
G. Eberlein (Technical University of Munich)

Series B: Mathematical and Statistical Methods
Editor: H. Skala (University of Paderborn)

Series C: Game Theory, Mathematical Programming and Operations Research
Editor: S. H. Tijs (University of Nijmegen)

Series D: System Theory, Knowledge Engineering and Problem Solving
Editor: W. Janko (University of Economics, Vienna)

---

SERIES D: **SYSTEM THEORY, KNOWLEDGE ENGINEERING AND PROBLEM SOLVING**

Volume 3

---

Editor: W. Janko (Vienna)



**Scope**

This series focuses on the design and description of organisations and systems with application to the social sciences. Formal treatment of the subjects is encouraged. Systems theory, information systems, system analysis, interrelated structures, program systems and expert systems are considered to be a theme within the series. The fundamental basics of such concepts including computational and algorithmic aspects and the investigation of the empirical behaviour of systems and organisations will be an essential part of this library. The study of problems related to the interface of systems and organisations to their environment is supported. Interdisciplinary considerations are welcome. The publication of recent and original results will be favoured.

For a list of titles published in this series, see final page.

# FUZZY RELATION EQUATIONS AND THEIR APPLICATIONS TO KNOWLEDGE ENGINEERING

*by*

ANTONIO DI NOLA and SALVATORE SESSA
*Università di Napoli, Facoltà di Architettura, Napoli, Italy*

WITOLD PEDRYCZ
*University of Manitoba,*
*Department of Electrical Engineering, Winnipeg, Canada*

and

ELIE SANCHEZ
*Université Aix-Marseille II,*
*Faculté de Médecine, Marseille, France*

*Foreword by*

LOTFI A. ZADEH
*University of California, Berkeley, U.S.A.*

SPRINGER-SCIENCE+BUSINESS MEDIA, B.V.

Library of Congress Cataloging in Publication Data

Fuzzy relation equations and their applications to knowledge engineering / Antonio Di Nola ... [et al.].
p. cm. -- (Theory and decision library. Series D, System theory, knowledge engineering, and problem solving)
Includes bibliographies and indexes.
ISBN 978-90-481-4050-3 ISBN 978-94-017-1650-5 (eBook)
DOI 10.1007/978-94-017-1650-5
1. Expert systems (Computer science) 2. Fuzzy systems. I. Di Nola, Antonio. II. Series.
QA76.76.E95F886 1989
006.3'3'01511322--dc20 89-34246

ISBN 978-90-481-4050-3

---

*printed on acid free paper*

**Foreword**

It took many decades for Peirce's concept of a relation to find its way into the microelectronic innards of control systems of cement kilns, subway trains, and tunnel-digging machinery. But what is amazing is that the more we learn about the basically simple concept of a relation, the more aware we become of its fundamental importance and wide ranging ramifications. The work by Di Nola, Pedrycz, Sanchez, and Sessa takes us a long distance in this direction by opening new vistas on both the theory and applications of fuzzy relations – relations which serve to model the imprecise concepts which pervade the real world.

Di Nola, Pedrycz, Sanchez, and Sessa focus their attention on a central problem in the theory of fuzzy relations, namely the solution of fuzzy relational equations. The theory of such equations was initiated by Sanchez in 1976, in a seminal paper dealing with the resolution of composite fuzzy relational equations. Since then, hundreds of papers have been written on this and related topics, with major contributions originating in France, Italy, Spain, Germany, Poland, Japan, China, the Soviet Union, India, and other countries. The bibliography included in this volume highlights the widespread interest in the theory of fuzzy relational equations and the broad spectrum of its applications.

In the context of applications, the importance of the theory of fuzzy relational equations derives from the fact that human knowledge may be viewed as a collection of facts and rules, each of which may be represented as the assignment of a fuzzy relation to the unconditional or conditional possibility distribution of a variable. What this implies is that knowledge may be viewed as a system of fuzzy relational equations. In this perspective, then, inference from a body of knowledge reduces to the solution of a system of fuzzy relational equations. This basic idea underlies the theory of approximate reasoning based on fuzzy logic as well as various versions of fuzzy Prolog and, in particular, Professor Baldwin's language FRIL, which is a Prolog-based language for inference from fuzzy relations.

The work of Di Nola, Pedrycz, Sanchez, and Sessa has a dual purpose: first, to present an authoritative and up-to-date account of the theory in a rigorous, thorough, and complete fashion; and second, to describe its applications, especially in the realm of knowledge-based systems.

The theoretical part addresses the major issues, among them: fuzzy relational equations in residuated lattices, the lower solutions of max-min equations, the measures of fuzziness of solutions, the max-min decomposition, and fuzzy relational equations with triangular norms. In a transition to applications, the authors consider an issue which is of high intrinsic importance, namely, the approximate solution of fuzzy relational equations. In this and other chapters, the authors make the reading easier for the non-mathematician by describing solution algorithms and applying them to well-chosen examples.

The second part, which deals with applications, develops a systematic approach to knowledge representation and inference based on the theory developed in the earlier chapters. In addition to the applications to knowledge-based systems, the authors present a lucid account of the basic ideas underlying the analysis and design of fuzzy logic controllers. Such controllers have proved to be highly successful in a variety of applications ranging from industrial process control and robotics to medical diagnosis and traffic control.

An important issue which is addressed in the chapters dealing with knowledge-based systems is that of the validation of production rules and the related problems of reduction and reconstruction. In these chapters, there is a great deal that is new in the application of the theory of fuzzy relational equations to the problem of inference.

The authors of this volume have played a leading role in the development of the theory of fuzzy relational equations and its applications. Not surprisingly, the book reflects their high expertise and expository skills. Much of the material is new; the writing is lucid and well-motivated; and the references are a model of thoroughness and organization. Di Nola, Pedrycz, Sanchez, and Sessa deserve our thanks and congratulations for authoring an outstanding text which is certain to become an important landmark in the development of the theory of fuzzy sets and its applications.

Lotfi A. Zadeh
Berkeley, California

## Table of Contents

## Preface

The growing interest to knowledge engineering has forced the burning need to search novel tools which mimic human processes of perception, decision-making, object recognition, etc.

Fuzzy set theory is an approach studied extensively in this area. On the other hand, the notion of fuzzy set, i.e. a set with not sharply defined boundaries, is very natural for a human being. Moreover, this approach is very convenient since the user-friendly man-system communication is performed in a linguistic (not numerical) fashion.

The aim of this monograph is to provide the reader with fundamentals of fuzzy relations indicating clearly their applications to knowledge engineering as, e.g., verification of a knowledge base (in the sense of its consistency and relevancy), designing of inference mechanisms, reduction of a knowledge base, propagation of uncertainty in different reasoning schemes.

This book is organized in 15 Chapters. Chapters 1÷9 contain theoretical backgrounds of the theory of fuzzy relation equations. In these Chapters, the fuzzy sets are defined and studied on several types of lattice: of course, this in the spirit of the symbolic computations, characteristic for Artificial Intelligence. In Chapters 10÷14, containing the above mentioned applications to knowledge engineering, the fuzzy sets are expressed in the real unit interval in order to process the vague information coming from the expert.

Chapter 15 contains two useful and extensive bibliographies of papers on fuzzy equations, fuzzy relations and related topics. Each Chapter has its appropriate references.

The following diagram illustrates some main routes for studying the Chapters of this book. Of course, the reader should consider this diagram only as indicative.

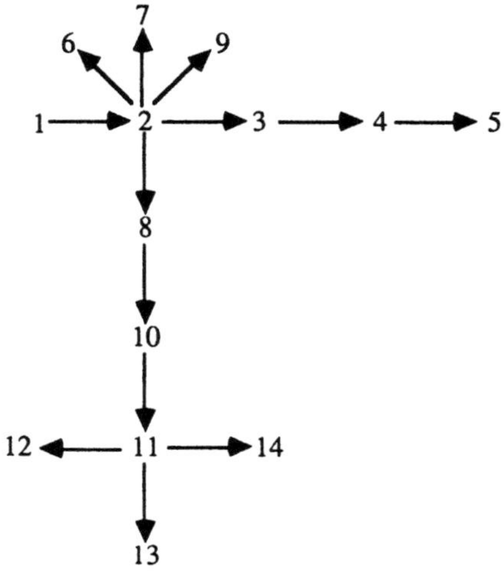

## Acknowledgements

*Thanks are due to* N. Blanchard, Z.Q. Cao, D. Dubois, S. Gottwald, H. Hashimoto, M. Higashi, G. Hirsch, K.H. Kim, G.J. Klir, W. Kolodziejczyk, V. Novak, A. Ohsato, H. Prade, R.M. Tong, P.Z. Wang, D. Willaeys, R.R. Yager *for providing preprints and reprints of their works and* S.Z.Guo *for providing a list of Chinese papers on fuzzy relation equations, partially contained in the bibliographies of Ch.15.*

## List of Abbreviations

| | |
|---|---|
| Ch(s). | Chapter(s) |
| Corol. | Corollary |
| Def(s). | Definition(s) |
| Eq(s). | Equation(s) |
| Ex(s). | Example(s) |
| Fig(s). | Figure(s) |
| KB | Knowledge Base |
| Prop(s). | Proposition(s) |
| Sec(s). | Section(s) |
| Thm(s). | Theorem(s) |
| [x, Ch.y] | Reference x of Chapter y |

# CHAPTER 1

## INTRODUCTORY REMARKS ON FUZZY SETS



### 1.1. Remarks on Fuzzy Sets

This Ch. can be regarded as an introduction to this book that originates from the primordial need to collect results from several papers on fuzzy relation equations. First of all we would like to emphasize some considerations of a general character in fuzzy set theory. This will enable the reader to get a certain perspective on this field of research. Moreover these remarks could suggest a philosophy behind the methodology of fuzzy sets.

A fundamental method in any science is the classification of the states of an individual physical process by means of groups of observations that admit generalization to physical laws. One of the major obstacles in this scheme is the separation of processes into individual observables. In fact, the Gestalt school [5] and the Heisenberg uncertainty principle suggest this is quite impossible.

In real life, one is commonly faced with many classes of nonrandom objects. Consider, for instance, such categories as *tall* man, *old* town, *high* inflation rate. All of them convey a semantic meaning, significant for a certain community; however, the concept of membership of an object to such a class is not obvious.

Therefore, application of methods based on formal two-valued logics to describe real world facts may be crucially limited or at least viewed as artificial. Remember Russell [7]:

*"All traditional logic habitually assumes that precise symbols are being employed. It is therefore not applicable to this terrestrial life, but only to an imagined celestial existence."*

Following Zadeh [10], the founder of fuzzy set theory:

*"In classical two-valued systems, all classes are assumed to have sharply defined boundaries. So either an object is a member of a class or it is not a member of a class. Now, this is okay if you are talking about something like mortal or not mortal, dead or alive, male or female, and so forth. These are examples of classes that have sharp boundaries.*

*But most classes in the real world do not have sharp boundaries. For example, if you consider characteristics or properties like tall, intelligent, tired, sick, and so forth, all of these characteristics lack sharp boundaries. Classical two-valued logic is not designed to deal with properties that are a matter of degree. This is the first point.*

*Now, there is, of course, a generalization of two-valued logic. And these generalized logical systems are called multivalued logics. So in multivalued logical systems, a property can be possessed to a degree".*

These sentences point to the discrepancy between a precise mathematical formalism and the ill-defined vague relationships or phenomena it is used to handle. A gap between existing theory and interpretation of its results has been discussed by Black [3]. He suggests that *"with the provision of an adequate symbolism the need is removed for regarding vagueness as a defect of language"*.

He strongly argues that vagueness should not be equated with subjectivity. For instance, a color accepted as "blue" is vague in itself, but not subjective since its sensation among all human beings is roughly similar.

Even in different branches of applied mathematics one can find an extensive list of concepts that are not clearly defined. Take, for instance, some examples in numerical analysis: *sparse* matrix, *linear* approximation of a function in a *small* neighbourhood of a point.

It is worth remembering the problem stated by Borel [4] (cfr. also [6]) who discusses the ancient Greek sophism of the pile of seeds (cfr. Fig.1.1.):

*"One seed does not constitute a pile nor two, nor three ... from the other side everybody will agree that 100 million seeds constitutes a pile. What therefore is the appropriate limit? Can we say that 325647 seeds don't constitute a pile but 325648 do?"*

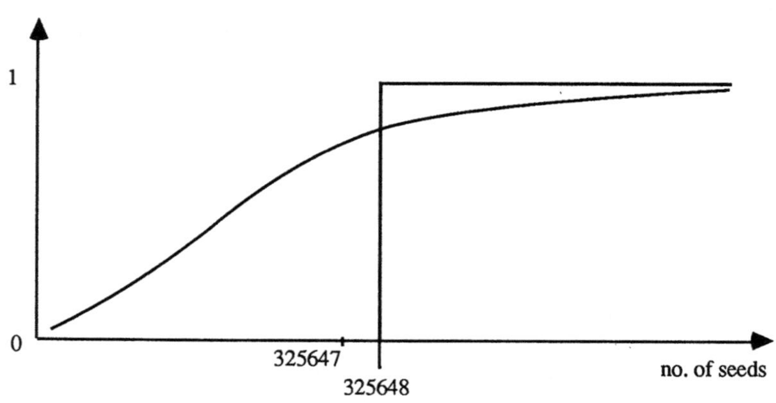


Fig. 1.1

Notion of the pile illustrated as a fuzzy set and a crisp set

The dramatic need appeared to construct a convenient flexible and efficient tool for handling and processing fuzzy categories with soft borderlines. It is created by man-machine communication systems, e.g., various types of knowledge-based systems, where the need for formal mechanism of data handling and processing comes originally from natural language.

The idea of a fuzzy set proposed by Zadeh in 1965 [8] is a step in this direction. This concept arose originally in the areas of pattern classification and information processing [1]. He defines a fuzzy set (class) A in **X** by a membership function A(x) which associates with each element of **X**, say x, a real number in [0,1]. A(x) represents the grade of membership of x to A. For A(x)=1, one has a total belonging, while the value of the membership function A(x) equalling 0 excludes x from the class A.

Following Zadeh [10], we remember:

*"... tallness becomes a matter of degree, as does intelligence, tiredness, and so forth. Usually you have degrees between zero and one. So you can say, for example, that a person is tall to the degree 0.9...*

*So something as simple as "John is tall" would require multivalued logic unless you arbitrarily establish a threshold by saying "somebody over 6 feet tall is tall and those who are less than 6 feet tall are not tall". In other words, unless you artificially introduce some sort of a threshold like that, you will need multivalued logic.*

*But even though these multivalued logical systems have been available for some time, they have not been used to any significant extent in linguistics, in psychology and in other fields where human cognition plays an important role... And this is where fuzzy logic enters the picture.*

*What differentiates fuzzy logic from multivalued logic is that in fuzzy logic you can deal with fuzzy quantifiers, like "most", "few", "many" and "several".*

*Fuzzy quantifiers have something to do with enumeration, that is, with counting... In multivalued logic you have only two quantifiers, "all" and "some", whereas in fuzzy logic you have all the fuzzy quantifiers.*

*This is one of the important differences".*

In [9] Zadeh also declares:

*"the key elements in human thinking are not numbers, but labels of fuzzy sets, that is, classes of objects in which the transition from membership to non-membership is gradual rather than abrupt. Indeed the pervasiveness of fuzziness in human thought processes suggests that much of the logic behind human reasoning is not the traditional two-valued or even multivalued logic, but a logic with fuzzy truths, fuzzy connectives and fuzzy rules of inference. In our view, it is this fuzzy, and as yet not well understood, logic that plays a basic role in what may well be one of the most important facts of human thinking, namely, the ability to summarize information, to extract from the collections of masses of data impinging upon the human brain those and only those subcollections which*

*relevant to the performance of the task at hand*".

This statement corresponds to the so-called principle of incompatibility [9]. It points out that there is a contradiction between precision and significant relevance, i.e. we are unable to formulate precise and yet significant statements.

Nowadays fuzzy set theory is in its twenties. It is not surprising that the membership function is sometimes considered as a probability function. In [2], one can find the following self-evident explanation:

*"randomness has to do with uncertainty concerning membership or non-membership of an object in a non-fuzzy set, while fuzziness has to do with classes in which there may be grades of membership intermediate between full membership and non-membership*".

We can easily point to a philosophical distinction between probability and fuzziness. To accomplish this, we put the value of the membership function of A in x to be equal to a, i.e. $A(x)=a$, and the probability that x belongs to A to be equal to a, i.e. $P\{x\in A\}=a$, $a\in[0,1]$. Upon observation of x, the a priori probability $P\{x\in A\}=a$ becomes a posteriori probability, i.e. either $P\{x\in A\}=1$ or $P\{x\in A\}=0$. But A(x), a measure of the extent to which x belongs to the given category, remains the same, in other words the randomness disappears, but the fuzziness remains.

## 1.2. Some Useful Bibliographies

To the best of our knowledge, for useful information, we give here some general bibliographies on fuzzy sets. For detailed recent and regularly updated literature, we refer the reader to the specific Secs of the international journals *FUZZY SETS AND SYSTEMS* and *BUSEFAL*, edited by the well known authors D.Dubois and H.Prade.

J. De Kerf, A Bibliography on Fuzzy Sets, *J. Comp. Applied Math.* 1 (1975), 205-212.

B.R. Gaines and L.J. Kohout, The Fuzzy Decade: a Bibliography on Fuzzy Systems and Closely Related Topics, *Internat. J. Man Machine Studies* 9 (1977), 1-68.

A.Kandel, Bibliography: A Compilation of Approximately 1000 Important References (Fuzzy 1000) on Fuzzy Set Theory and Its Applications, in: *Fuzzy Mathematical Techniques with Applications*, Addison-Wesley Publ. Co., Reading, Mass. (1986), pp.225-266.

A.Kandel and W.J. Byatt, Fuzzy sets, Fuzzy Algebra and Fuzzy Statistics, *Proceedings of the IEEE*, vol. 66, n.12 (1978), 1619-1639.

A. Kandel and R.R. Yager, A 1979 Bibliography of Fuzzy Sets and Their Applications, in: *Advances in Fuzzy Set Theory and Application* (M.M. Gupta, R.K. Ragade and R.R. Yager, Eds.), North-Holland, Amsterdam (1979), pp.621-744.

J.Maiers and Y.S. Sherif, Application of Fuzzy Set Theory, *IEEE Trans. Syst. Man. Cybern.*, vol. SMC-15, no.1 (1985), 175-189.
R.R. Yager, An Useful Bibliography for Fuzzy Sets, in: *Recent Advances in Fuzzy Set and Possibility Theory* (R.R.Yager, Ed.), Pergamon Press, Elmsford, N.Y., 1982.
R.R. Yager, Fuzzy Sets: A Bibliography, *Intersystems Publications*, Seaside, CA, 1983.

## References

[1] R.E. Bellmann, R. Kalaba and L.A. Zadeh, Abstraction and pattern classification, *J. Math. Anal. Appl.* 13 (1966), 1-7.
[2] R.E. Bellmann and L.A. Zadeh, Decision-making in a fuzzy environment, *Management Sciences* 17 (1970), 141-164.
[3] M. Black, Vagueness: an exercise in logical analysis, *Philosophy Science* 4 (1937), 427-455.
[4] E. Borel, *Probabilité et Certitude*, Press Univ. de France, Paris, 1950.
[5] C.Corge, *Elements d'Informatique*, Larousse, Paris, 1975.
[6] R.C. Godal and T.J. Goodman, Fuzzy Sets and Borel, *IEEE Trans. Syst. Man. Cybern.*, vol. SMC-10, no. 10 (1980), 637.
[7] B. Russell, Vagueness, *Austral. J. Philosophy* 1 (1923), 84-92.
[8] L.A. Zadeh, Fuzzy Sets, *Inform. and Control* 8 (1965), 338-353.
[9] L.A. Zadeh, Outline of a new approach to the analysis of complex systems and decision processes, *IEEE Trans. Syst. Man. Cybern.*, vol. SMC-2 (1973), 28-44.
[10] L.A. Zadeh, Coping with the imprecision of the real world: an interview with L.A. Zadeh, *Communications of the ACM* 27, no.4 (1984), 304-311.

# CHAPTER 2

## FUZZY RELATION EQUATIONS IN RESIDUATED LATTICES



*We would like to underline the following statement of Goguen [5]:*
*"The importance of relations is almost self-evident. Science is, in a sense, the discovery of relations between observables... Difficulties arise in the so-called "soft" sciences because the relations involved do not appear to be "hard", as they are, say, in classical physics. A thoroughgoing application of probability theory has relieved many difficulties, but it is clear that others remain. We suggest that further difficulties might be cleared up through a systematic exploitation of fuzziness".*

*This statement has induced us to dwell upon the world of fuzzy relations, which, as is well-known, constitute a wide generalization of the classical concept of Boolean relation (or matrix) equations, e.g., Rudeanu [7].*

*In [14] Zadeh and Desoer have stressed the equivalence between the study of relations and general system theory, pointing out the importance of studying fuzzy relation equations: if we think of a fuzzy relation as a system, the composition of a fuzzy relation with a fuzzy set (input) yields a fuzzy set (output).*

*We begin this Ch. by recalling basic definitions of the theory of fuzzy relation equations. In Sec.1, we assume that the equations are defined over complete right-residuated lattices. This hypothesis is maintained in Sec.2 which contains the fundamental theorems of the theory. In Secs.3 and 4, the results are established in the context of complete Brouwerian lattices, whereas dual equations are presented in Sec.5 in the setting of complete dually Brouwerian lattices.*

*For the terminology on lattices used in this book, we refer the reader to Birkhoff's book [1].*

### 2.1. Basic Definitions

Zadeh [8, Ch.1] defines a fuzzy set A of a nonempty set **X** as a real function, denoted also by A, with membership values in the unit interval [0,1]. Goguen [5] enlarges the definition of A, substituting [0,1] by a suitable partially ordered set (poset) **L**, whose lattice structure reflects an analogous one in the set **F(X)**={A:**X**→**L**} of all fuzzy sets of **X**.

Note that if **L**={0,1}, then A becomes the characteristic function which defines a subset A of **X**.

We recall some classical concepts of lattice theory [1]. A *lattice* **L** is a poset with two internal composition laws "$\wedge$" and "$\vee$", $a\wedge b$ and $a\vee b$ being the greatest lower bound (g.l.b.) and the least upper bound (l.u.b.), respectively, of any two elements $a,b\in \mathbf{L}$. A lattice **L** is *complete* if any subset of **L** has g.l.b. and l.u.b. The symbol "$\leq$" stands for the partial ordering of **L**; 0 and 1 denote the universal bounds of **L**. A lattice **L** is *totally ordered* (or *linear*) if $a\wedge b\in \{a,b\}$ for any two elements $a,b\in \mathbf{L}$. A *po-groupoid* **L** is a poset with a binary multiplication "·" (sometimes the elements linked by this operation will be juxtaposed) satisfying the *isotonicity condition*:

$$a \leq b \quad \text{implies} \quad xa \leq xb \quad \text{and} \quad ax \leq bx \tag{2.1}$$

for all $a,b\in \mathbf{L}$. We say that a po-groupoid **L** is *right* (resp. *left*)*-residuated* if for all $a,b\in \mathbf{L}$ there exists a largest $x\in \mathbf{L}$, denoted by $a\rightarrow b$ (resp. $a\leftarrow b$) such that $ax\leq b$ (resp. $xa\leq b$).

A right (resp.left)-residuated lattice **L** is a lattice with a binary multiplication with respect to which **L** is a right (resp. left)-residuated po-groupoid. **L** is a *residuated* lattice if it is a right and a left-residuated lattice and it is *Brouwerian* if we define $a\cdot b=a\wedge b$ for all $a,b\in \mathbf{L}$.

PROPOSITION 2.1. *In any complete right (resp. left)-residuated lattice* **L**, *we have:*

$$a\Big(\bigvee_{i\in \mathbf{I}} x_i\Big) = \bigvee_{i\in \mathbf{I}} (a\,x_i) \qquad \Big[\text{resp. } \Big(\bigvee_{i\in \mathbf{I}} x_i\Big)a = \bigvee_{i\in \mathbf{I}} (x_i\,a)\Big], \tag{2.2}$$

***I*** *being an arbitrary set of indices.*

PROOF. Let $x=\bigvee_{i\in \mathbf{I}} (a\,x_i)$, then $ax_i\leq x$, i.e. $x_i\leq(a\rightarrow x)$ for all $i\in \mathbf{I}$, so $\bigvee_{i\in \mathbf{I}} x_i\leq(a\rightarrow x)$. From the condition of isotonicity (2.1), we deduce that

$$a\Big(\bigvee_{i\in \mathbf{I}} x_i\Big) \leq a\cdot(a\rightarrow x) \leq x. \tag{2.3}$$

On the other hand, $ax_i\leq a\big(\bigvee_{i\in \mathbf{I}} x_i\big)$ for all $i\in \mathbf{I}$, and then

$$x = \bigvee_{i\in \mathbf{I}} (a\, x_i) \leq a \cdot ( \bigvee_{i\in \mathbf{I}} x_i) . \tag{2.4}$$

From (2.3) and (2.4), (2.2) follows. ■

This Proposition is given in [4]. Note that if **L** is a linear lattice, then $a\to b=1$ if $a\leq b$ and $a\to b=b$ if $a>b$. From now on, we will write $a\alpha b$ instead of $a\to b$ in accordance with the symbolism in [8] and $a\alpha' b=a\leftarrow b$. If the operation "·" is commutative, of course we have $a\alpha b=a\alpha' b$. It is easy to verify the following properties in any Brouwerian lattice **L**:

(i) $a \wedge (a\, \alpha\, b) \leq b$,

(ii) $a\, \alpha\, (b \vee c) = (a\, \alpha\, b) \vee (a\, \alpha\, c)$,

(iii) $a\, \alpha\, (a \wedge b) \geq b$,

where $a,b,c\in \mathbf{L}$.

We now recall usual definitions of fuzzy set theory [8, Ch.1].

Let **L** be a complete lattice with universal bounds 0 and 1.

DEFINITION 2.1. *Let* $A,B\in \mathbf{F}(\mathbf{X})$. *We say that* B *contains or includes* A, *in symbols* $B\geq A$, *iff* $B(x)\geq A(x)$ *for any* $x\in \mathbf{X}$ *and we say that* A *is equal to* B, *in symbols* $A=B$, *iff* $A(x)=B(x)$ *for any* $x\in \mathbf{X}$. *The fuzzy union and intersection of* A *and* B *are pointwise defined as* $(A\vee B)(x)=A(x)\vee B(x)$ *and* $(A\wedge B)(x)=A(x)\wedge B(x)$ *for any* $x\in \mathbf{X}$.

DEFINITION 2.2. *Let* **Y** *be a nonempty set. A fuzzy relation* R *between* **X** *and* **Y** *is a fuzzy set of the Cartesian product* $\mathbf{X}\times\mathbf{Y}$, *i.e. an element of* $\mathbf{F}(\mathbf{X}\times\mathbf{Y})$.

Since no misunderstanding can arise, we also identify fuzzy relations and fuzzy sets with their membership functions.

DEFINITION 2.3. *We call the inverse or the transpose of* $R\in \mathbf{F}(\mathbf{X}\times\mathbf{Y})$, *the fuzzy relation* $R^{-1}\in \mathbf{F}(\mathbf{Y}\times\mathbf{X})$ *defined as* $R^{-1}(y,x)=R(x,y)$ *for all* $x\in \mathbf{X}$, $y\in \mathbf{Y}$.

If we assume, additionally, that **L** is a po-groupoid, then we can give the following definitions:

DEFINITION 2.4. *Let* $A\in \mathbf{F}(\mathbf{X})$ *and* $R\in \mathbf{F}(\mathbf{X}\times\mathbf{Y})$. *We define the max-*$\odot$ *composition of* R *and* A *to be the fuzzy set* $B\in \mathbf{F}(\mathbf{Y})$, *in symbols* $B=R\odot A$, *given by*

$$B(y) = (R\odot A)\,(y) = \bigvee_{x\in \mathbf{X}} [A(x) \cdot R(x, y)] \tag{2.5}$$

*for any* $y \in \mathbf{Y}$.

DEFINITION 2.5. *Let* $\mathbf{Z}$ *be a nonempty set,* $Q \in \mathbf{F}(\mathbf{X}\times\mathbf{Y})$ *and* $S \in \mathbf{F}(\mathbf{Y}\times\mathbf{Z})$. *We define max-*$\odot$ *composition of* Q *and* S *to be the fuzzy relation* $T \in \mathbf{F}(\mathbf{X}\times\mathbf{Z})$, *in symbols* $T=S\odot Q$, *given by*

$$T(x, z) = (S\odot Q)(x, z) = \bigvee_{y\in \mathbf{Y}} [Q(x, y) \cdot S(y, z)] \tag{2.6}$$

*for all* $x \in \mathbf{X}$, $z \in \mathbf{Z}$.

The formula (2.6) is clearly an extension of (2.5) and when **L** is a complete Boolean lattice (where $\wedge$ coincides with "$\cdot$"), (2.6) becomes the classical Boolean matrix product, e.g., Rudeanu [7]. Using the isotonicity condition (2.1), one can immediately prove that if $S_1 \le S_2$, with $S_1, S_2 \in \mathbf{F}(\mathbf{Y}\times\mathbf{Z})$, then $S_1\odot Q \le S_2\odot Q$ for any $Q \in \mathbf{F}(\mathbf{X}\times\mathbf{Y})$ and $R\odot S_1 \le R\odot S_2$ for any $R \in \mathbf{F}(\mathbf{Z}\times\mathbf{X})$.

From now on, we also assume that **L** is a complete residuated lattice.

DEFINITION 2.6. *Let* $A \in \mathbf{F}(\mathbf{X})$ *and* $B \in \mathbf{F}(\mathbf{Y})$. *We define the* $\alpha$*-composition of* A *and* B *to be the fuzzy relation* $(A\alpha B) \in \mathbf{F}(\mathbf{X}\times\mathbf{Y})$ *defined for all* $x \in \mathbf{X}$, $y \in \mathbf{Y}$, *as*

$$(A\alpha B)(x, y) = A(x)\, \alpha\, B(y) \tag{2.7}$$

DEFINITION 2.7. *Let* $Q \in \mathbf{F}(\mathbf{X}\times\mathbf{Y})$ (*resp.* $S \in \mathbf{F}(\mathbf{Y}\times\mathbf{Z})$) *and* $T \in \mathbf{F}(\mathbf{X}\times\mathbf{Z})$. *We define the* $\alpha$- (*resp.* $\alpha'$-)*composition of* $Q^{-1}$ (*resp.* S) *and* T (*resp.* $T^{-1}$)*to be the fuzzy relation* $(Q^{-1}\alpha T) \in \mathbf{F}(\mathbf{Y}\times\mathbf{Z})$ (*resp* $(S\alpha' T^{-1}) \in \mathbf{F}(\mathbf{Y}\times\mathbf{X})$*defined for all* $y \in \mathbf{Y}$, $z \in \mathbf{Z}$ (*resp.* $x \in \mathbf{X}$), *as*

$$(Q^{-1}\, \alpha\, T)(y, z) = \bigwedge_{x\in \mathbf{X}} [Q^{-1}(y, x)\, \alpha\, T(x, z)], \tag{2.8}$$

*resp.*

$$(S\, \alpha'\, T^{-1})(y, x) = \bigwedge_{z\in \mathbf{Z}} [S(y, z)\, \alpha'\, T^{-1}(z, x)]. \tag{2.8'}$$

For brevity, we put $(Q^{-1}\alpha T)=S^*$.

## 2.2. Max-$\odot$ Composite Fuzzy Equations

Usually, we call Eq. (2.6), a max-$\odot$ composite fuzzy equation and, as already said above, it generalizes the classical Boolean equation (e.g. [2], [6], [7]).

We present an extension of the results of Luce [6], which solves the Eq. $R\odot A=B$, where A and B are Boolean matrices and R is unknown. In [7], Rudeanu stressed that analogous results can be established in a Brouwerian lattice. In this case, the greatest solution (in the sense of the fuzzy inclusion) is found by Sanchez [8].

We propose two problems:

(q1) *given Q and T, determine S such that Eq. (2.6) holds,*

(q2) *given S and T, determine Q such that Eq. (2.6) holds.*

In order to give the resolution of the problems (q1) and (q2), we define the following sets.

$$\mathcal{L} = \mathcal{L}\,(Q, T) = \{S\in \mathbf{F}(\mathbf{Y}\times\mathbf{Z}) \colon S\odot Q \leq T\},$$

$$\mathcal{S} = \mathcal{S}\,(Q, T) = \{S\in \mathbf{F}(\mathbf{Y}\times\mathbf{Z}) \colon S\odot Q = T\},$$

$$\mathbf{Q} = \mathbf{Q}\,(S, T) = \{Q\in \mathbf{F}(\mathbf{X}\times\mathbf{Y}) \colon S\odot Q = T\}.$$

LEMMA 2.2. $S^*\in\mathcal{L}$ *and further* $S^*\geq S$ *for any* $S\in\mathcal{L}$ .

PROOF. From (2.8), we have $S^*(y,z)\leq[Q(x,y)\alpha T(x,z)]$ for all $x\in \mathbf{X}$, $y\in \mathbf{Y}$, $z\in \mathbf{Z}$. From the isotonicity condition (2.1), we deduce that

$$(S^*\odot Q)\,(x, z) = \bigvee_{y\in\mathbf{Y}} [Q(x, y)\cdot S^*(y, z)] \leq \bigvee_{y\in\mathbf{Y}} \{Q(x, y)\cdot [Q(x, y)\alpha T(x,z)]\} \leq$$

$$\bigvee_{y\in\mathbf{Y}} T(x, z) = T(x, z)$$

for all $x\in \mathbf{X}$, $z\in \mathbf{Z}$, i.e. $S^*\in\mathcal{L}$.

Let $S\in\mathcal{L}$, then $[Q(x,y)\cdot S(y,z)]\leq T(x,z)$, i.e. $S(y,z)\leq[Q(x,y)\alpha T(x,z)]$ for all $x\in \mathbf{X}$, $y\in \mathbf{Y}$, $z\in \mathbf{Z}$. Thus

$$S(y, z) \leq \bigwedge_{x\in\mathbf{X}} [Q(x, y)\,\alpha\, T(x, z)] = S^*(y, z)$$

for all $y\in \mathbf{Y}$, $z\in \mathbf{Z}$. ■

THEOREM 2.3. $\mathcal{S}\neq\emptyset$ *iff* $S^*\in\mathcal{S}$. *Further,* $S^*\geq S$ *for any* $S\in\mathcal{S}$.

PROOF. Let $\mathcal{S}\neq\emptyset$ and $S\in\mathcal{S}\leq\mathcal{L}$. Then $S\leq S^*$ by Lemma 2.2 and $T=S\odot Q\leq S^*\odot Q\leq T$. Thus $S^*\in\mathcal{S}$ and the reverse is obvious. ■

Hence Thm.2.3 solves problem (q1). Analogously, using (2.8'), the problem (q2) is solved by the following result:

THEOREM 2.4. $\mathbf{Q}\neq\emptyset$ *iff* $(S\alpha' T^{-1})^{-1}\in\mathbf{Q}$. *Furthermore,* $(S\alpha' T^{-1})^{-1}\geq Q$ *for any* $Q\in\mathbf{Q}$.

Lemma 2.2. and Thm. 2.3 are given in [4]. For complete right-residuated lattices. Also Turunen [13] independently established Thms. 2.3 and 2.4, which we illustrate in the example below, where the max-$\odot$ composition is the max-min composition and **L** is a complete Brouwerian lattice (non totally ordered).

EXAMPLE 2.1. Let **N** be the set of nonnegative integers. In accordance with Szàsz [12, p.84], we define a∧b=l.c.m.{a,b},a∨b=g.c.d.{a,b},a≤b iff a is multiple of b, where a,b∈**N** and l.c.m. (resp. g.c.d.) stands for the smallest (resp. greatest) common multiple (resp. divisor) between a and b. Note that 0 and 1 are the universal bounds of this lattice structure **L**=(**N**,∧,∨,≤), which is complete and Brouwerian and with operator "α" given by

$$a\ \alpha\ b = \text{g.c.d.}\ \{x\in\mathbf{N}\colon a\wedge x\leq b\}$$

for any a,b∈**N**.

This lattice will be also used in other examples. For motives which we will illustrate in Sec.3.6, we assume, for simplicity, **Z** singleton, i.e. **Z**={z}. Further, $\mathbf{X}=\{x_1,x_2,x_3\}$, $\mathbf{Y}=\{y_1,y_2,y_3\}$ and $Q\in\mathbf{F}(\mathbf{X}\times\mathbf{Y})$, $S\in\mathbf{F}(\mathbf{Y}\times\mathbf{Z})$ are defined as

$$Q=\begin{Vmatrix} 3 & 6 & 8 \\ 4 & 7 & 9 \\ 2 & 5 & 3 \end{Vmatrix}, \qquad S=\begin{Vmatrix} 3 \\ 2 \\ 6 \end{Vmatrix}.$$

We have (since α≡α'):

$$T = S\odot Q=\begin{Vmatrix} 3 \\ 2 \\ 2 \end{Vmatrix}, \quad S^* = Q^{-1}\alpha T = \begin{Vmatrix} 1 \\ 2 \\ 6 \end{Vmatrix}, \quad (S\alpha T^{-1})^{-1} = \begin{Vmatrix} 1 & 3 & 1 \\ 2 & 1 & 1 \\ 2 & 1 & 1 \end{Vmatrix}.$$

We see, verifying Thm.2.3,that $S^*\odot Q=T$ and $S^*\geq S$ while, verifying Thm.2.4,we have $S\odot(S\alpha T^{-1})^{-1}=T$ and $(S\alpha T^{-1})^{-1}\geq Q$.

In the above example, the universes of discourse are finite. In the case of infinite sets, we quote from [8] a more sophisticated example:

EXAMPLE 2.2. Let **X**, **Y**, **Z** be the set of positive real numbers, L=[0,1], ∧≡min, ∨≡max, $Q\in\mathbf{F}(\mathbf{X}\times\mathbf{Y})$ and $S\in\mathbf{F}(\mathbf{Y}\times\mathbf{Z})$ given by $Q(x,y)=\exp[-k(x-y)^2]$ for all $x\in\mathbf{X}$, $y\in\mathbf{Y}$,

$S(y,z)=\exp.[-k(y-z)^2]$ for all $y\in \mathbf{Y}$, $z\in \mathbf{Z}$, where $k\geq 1$ is a constant. The max-min composition gives $T=S\odot Q$ defined by $T(x,z)=\exp[-k(x-z)^2/4]$ for all $x\in \mathbf{X}$, $z\in \mathbf{Z}$. If we suppose Q and T to be assigned, we find that $S^*(y,z)=\exp[-kz^2/4]$ if $y\leq z/2$ and $S^*(y,z)=\exp[-k(y-z)^2]$ if $y\geq z/2$. It is seen that $S^*\odot Q=T$ and $S^*\geq S$.

If $\mathbf{L}$ is a complete Brouwerian lattice, Thms.2.3 and 2.4 become Thms.2.5 and 2.6 of [8], respectively. Henceforth, whenever that $\mathbf{L}=[0,1]$, we regard it as in Ex.2.2, i.e. as a linear lattice with respect to its usual operations min and max.

We shall often use the following Lemma in the sequel.

LEMMA 2.5. *Let* $S_1,S_2\in \mathbf{S}$ *and* $S\in \mathbf{F}(\mathbf{Y}\times\mathbf{Z})$ *such that* $S_1\leq S\leq S_2$. *Then* $S\in \mathbf{S}$.

PROOF. It is trivial since $T=S_1\odot Q\leq S\odot Q\leq S_2\odot Q=T$. ■

THEOREM 2.6. $\mathbf{S}$ *is an upper semilattice, i.e. for all* $S_1,S_2\in \mathbf{S}$, *the fuzzy union* $S_1\vee S_2$ *belongs to* $\mathbf{S}$.

PROOF. By Prop. 2.1, we have:

$$[(S_1\vee S_2)\odot Q]\,(x, z) = \bigvee_{y\in \mathbf{Y}} \{Q(x, y)\cdot [S_1(y, z) \vee S_2(y, z)]\} =$$

$$\bigvee_{y\in \mathbf{Y}} \{[Q(x, y)\cdot S_1(y, z)] \vee [Q(x, y)\cdot S_2(y, z)]\} =$$

$$\{\bigvee_{y\in \mathbf{Y}} [Q(x, y)\cdot S_1(y, z)]\} \vee \{\bigvee_{y\in \mathbf{Y}} [Q(x, y)\cdot S_2(y, z)]\} =$$

$$T(x,z) \vee T(x,z) = T(x,z)$$

for all $x\in \mathbf{X}$, $z\in \mathbf{Z}$. ■

This result appears in [4], but already Sanchez in [10 and 11] pointed out this result in complete Brouwerian lattices, since the max-min composition of fuzzy relations is distributive on unions but not on the intersection of fuzzy relations, i.e. $S_1\wedge S_2$ does not belong generally to $\mathbf{S}$ if $S_1, S_2\in \mathbf{S}$, as is shown in the following example [8]:

EXAMPLE 2.3. Let $\mathbf{X}=\{x_1,x_2,x_3\}$, $\mathbf{Y}=\{y_1,y_2,y_3,y_4\}$, $\mathbf{Z}=\{z_1,z_2,z_3\}$, $\mathbf{L}=[0,1]$ and $Q\in \mathbf{F}(\mathbf{X}\times\mathbf{Y})$, $T\in \mathbf{F}(\mathbf{X}\times\mathbf{Z})$, $S_1,S_2\in \mathbf{S}$ so defined:

$$Q=\begin{Vmatrix} 0.2 & 0.0 & 0.8 & 1.0 \\ 0.4 & 0.3 & 0.0 & 0.7 \\ 0.5 & 0.9 & 0.2 & 0.0 \end{Vmatrix}, \quad T=\begin{Vmatrix} 0.7 & 0.3 & 1.0 \\ 0.6 & 0.4 & 0.7 \\ 0.8 & 0.9 & 0.2 \end{Vmatrix},$$

$$S_1=\begin{Vmatrix} 0.3 & 0.5 & 0.2 \\ 0.8 & 1.0 & 0.0 \\ 0.7 & 0.0 & 0.5 \\ 0.6 & 0.3 & 1.0 \end{Vmatrix}, \quad S_2=\begin{Vmatrix} 0.1 & 0.4 & 0.0 \\ 0.8 & 0.9 & 0.4 \\ 0.7 & 0.3 & 0.4 \\ 0.6 & 0.0 & 1.0 \end{Vmatrix}.$$

Clearly $S_1 \odot Q = S_2 \odot Q = T$ and

$$\begin{Vmatrix} 0.1 & 0.4 & 0.0 \\ 0.8 & 0.9 & 0.0 \\ 0.7 & 0.0 & 0.4 \\ 0.6 & 0.0 & 1.0 \end{Vmatrix} = S_1 \wedge S_2 \leq S^* = \begin{Vmatrix} 1.0 & 1.0 & 0.2 \\ 0.8 & 1.0 & 0.2 \\ 0.7 & 0.3 & 1.0 \\ 0.6 & 0.3 & 1.0 \end{Vmatrix},$$

but

$$[(S_1 \wedge S_2) \odot Q]\,(x_1, z_2) = \bigvee_{j=1}^{4} \,[Q(x_1, y_j) \wedge (S_1 \wedge S_2)\,(y_j, z_2)] = 0.2 < 0.3 = T\,(x_1, z_2)$$

and hence $S_1 \wedge S_2 \notin \mathbf{S}$.

Later we shall illustrate some sufficient conditions which specify when $S_1 \wedge S_2$ lies in $\mathbf{S}$.

REMARK 2.1. If the operation "·" is commutative [13], any result established in the set $\mathbf{S}$ can be reformulated in the set $\mathbf{Q}$. Indeed, from Defs.2.3 and 2.5, one deduces that $(S \odot Q)^{-1} = Q^{-1} \odot S^{-1}$. Then Eq.(2.6), i.e. $S \odot Q = T$, is equivalent to Eq. $Q^{-1} \odot S^{-1} = T^{-1}$. As stressed by Sanchez in [8] and [10], one can derive Thm.2.3 from Thm.2.4 changing $Q^{-1}$ into S and $T^{-1}$ into T and vice versa.

## 2.3. Another Characterization Theorem

The results of this Sec. have been established in [3]. Although the fundamental Thm.2.3 assumes the existence of the greatest element of $\mathbf{S}$, if $\mathbf{S} \neq \emptyset$, when Eq.(2.6) is assigned on finite sets or infinite sets, here we determine other elements of $\mathbf{S}$, different from $S^*$. This problem is studied, in the context of the complete Brouwerian lattices, for max-min fuzzy equations and we present another characterization theorem, different from Thm.2.3, whose proof suggests how to build other elements of $\mathbf{S}$.

For any $z \in \mathbf{Z}$, let $P_z \in \mathbf{F}(\mathbf{X}\times\mathbf{Y})$ be a fuzzy relation ($P_z \leq Q$) such that

$$\bigvee_{y\in \mathbf{Y}} P_z(x,y) = T(x,z) \tag{2.9}$$

for any $x \in \mathbf{X}$. Further, we define the sets:

$$\mathcal{P}(y,z) = \{a(y,z) = a \in \mathbf{L} : Q(x,y) \wedge a = P_z(x,y),\ \ x \in \mathbf{X}\}$$

for all $y \in \mathbf{Y}$, $z \in \mathbf{Z}$. For any $z \in \mathbf{Z}$, $\mathcal{P}(z)$ denotes the set of fuzzy relations $P_z \in \mathbf{F}(\mathbf{X}\times\mathbf{Y})$ for which (2.9) holds and such that $\mathcal{P}(y,z) \neq \emptyset$ for any $y \in \mathbf{Y}$. Then the following characterization holds:

THEOREM 2.7. $\mathbf{S} \neq \emptyset$ *iff* $\mathcal{P}(z) \neq \emptyset$ *for any* $z \in \mathbf{Z}$.

PROOF. If $\mathbf{S} \neq \emptyset$, let $S \in \mathbf{S}$ and then

$$\bigvee_{y\in \mathbf{Y}} [Q(x,y) \wedge S(y,z)] = T(x,z). \tag{2.10}$$

Further, we define the fuzzy relations $P_z \in \mathbf{F}(\mathbf{X}\times\mathbf{Y})$ by

$$P_z(x,y) = Q(x,y) \wedge S(y,z) \tag{2.11}$$

for all $x \in \mathbf{X}$, $y \in \mathbf{Y}$, $z \in \mathbf{Z}$. Thus (2.9) can be deduced from (2.10) and, because of (2.11), $S(y,z)$ lies in $\mathcal{P}(y,z)$ for all $y \in \mathbf{Y}$, $z \in \mathbf{Z}$ and then $P_z \in \mathcal{P}(z)$ for any $z \in \mathbf{Z}$.

Vice versa, let $\mathcal{P}(z) \neq \emptyset$ for any $z \in \mathbf{Z}$. Let $P_z \in \mathcal{P}(z)$, then $\mathcal{P}(y,z) \neq \emptyset$ for all $y \in \mathbf{Y}$, $z \in \mathbf{Z}$ and choose an arbitrary element $a(y,z) \in \mathcal{P}(y,z)$ for which we have:

$$Q(x,y) \wedge a(y,z) = P_z(x,y).$$

Hence

$$\bigvee_{y\in \mathbf{Y}} [Q(x,y) \wedge a(y,z)] = \bigvee_{y\in \mathbf{Y}} P_z(x,y) = T(x,y),$$

since (2.9) holds. This implies that the fuzzy relation $S \in \mathbf{F}(\mathbf{Y}\times\mathbf{Z})$ pointwise defined by $S(y,z)=a(y,z)$ for all $y \in \mathbf{Y}$, $z \in \mathbf{Z}$, is an element of $\mathbf{S}$, i.e. $\mathbf{S} \neq \emptyset$. ■

We illustrate this theorem by a suitable example borrowed from [3]:

EXAMPLE 2.4. Let $\mathbf{L}$, $\mathbf{X}$ and $\mathbf{Y}$ be as in Ex.2.1, $\mathbf{Z}=\{z_1,z_2\}$ and $Q \in \mathbf{F}(\mathbf{X}\times\mathbf{Y})$,

$T \in \mathbf{F}(\mathbf{X}\times\mathbf{Z})$ be defined as

$$Q = \begin{Vmatrix} 3 & 6 & 8 \\ 4 & 7 & 9 \\ 2 & 5 & 3 \end{Vmatrix}, \qquad T = \begin{Vmatrix} 3 & 1 \\ 2 & 2 \\ 2 & 2 \end{Vmatrix}.$$

We have

$$S^* = \begin{Vmatrix} 1 & 1 \\ 2 & 2 \\ 6 & 2 \end{Vmatrix}$$

and $\mathbf{S}\neq\emptyset$ since $S^*\odot Q=T$. We put:

$$P_1 = P_{z_1} = \begin{Vmatrix} 3 & 12 & 24 \\ 12 & 28 & 18 \\ 6 & 20 & 6 \end{Vmatrix}, \qquad P_2 = P_{z_2} = \begin{Vmatrix} 15 & 6 & 8 \\ 20 & 14 & 36 \\ 10 & 10 & 12 \end{Vmatrix}.$$

We easily see that $P_1$ and $P_2$ verify condition (2.9) and further we have:

$$\mathbf{P}(y_1,z_1) = \{3\},\ \mathbf{P}(y_2,z_1) = \{4\},\ \mathbf{P}(y_3,z_1) = \{6\},$$

$$\mathbf{P}(y_1,z_2) = \{5\},\ \mathbf{P}(y_2,z_2) = \{2\},\ \mathbf{P}(y_3,z_2) = \{4\}.$$

Hence $P_1\in\mathbf{P}(z_1)$ and $P_2\in\mathbf{P}(z_2)$; i.e. $\mathbf{P}(z_i)\neq\emptyset$ for i=1,2.

The proof of Thm.2.7 guarantees the existence of an element $S\in\mathbf{S}$ defined by $S(y,z)=a(y,z)$, where $a(y,z)$ is chosen in each set $\mathbf{S}(y,z)$ for all $y\in\mathbf{Y}, z\in\mathbf{Z}$. Indeed, in Ex.2.4 we find the fuzzy relation $S\in\mathbf{F}(\mathbf{Y}\times\mathbf{Z})$ given by

$$S = \begin{Vmatrix} 3 & 5 \\ 4 & 2 \\ 6 & 4 \end{Vmatrix},$$

which belongs to $\mathbf{S}$ and it is $S\neq S^*$. However, in order to find elements of $\mathbf{S}$ distinct from $S^*$, we devote special attention to $S^*$ because it allows us immediately to define a fuzzy relation $S_z^*\in\mathbf{F}(\mathbf{X}\times\mathbf{Y})$, for any $z\in\mathbf{Z}$, as

$$S_z^*(x,y) = Q(x,y) \wedge S^*(y,z)$$

for all $x\in\mathbf{X}$, $y\in\mathbf{Y}$. We obviously have:

$$\bigvee_{y \in \mathbf{Y}} S_z^*(x,y) = T(x,z)$$

for all $x \in \mathbf{X}$, $z \in \mathbf{Z}$. Moreover, $S^*(y,z)$ belongs to the set:

$$\mathbf{S}^*(y,z) = \{a(y,z) = a \in \mathbf{L} : Q(x,y) \wedge a = S_z^*(x,y),\ x \in \mathbf{X}\}$$

and then $\mathbf{S}^*(y,z) \neq \emptyset$ for any $y \in \mathbf{Y}$.

If $\mathbf{S}^*(y',z')$ is not a singleton for some $y' \in \mathbf{Y}$, $z' \in \mathbf{Z}$, then we are able to build an element S of $\mathbf{S}^*$ distinct from $S^*$ taking an arbitrary element $a(y,z)$ in each set $\mathbf{S}^*(y,z)$ and by setting $S(y,z)=a(y,z)$ for all $y \in \mathbf{Y}-\{y'\}$, $z \in \mathbf{Z}-\{z'\}$ and $S(y',z')=a(y',z') \neq S^*(y',z')$, as is shown in the following example:

EXAMPLE 2.5. Let L,**X**,**Y** and **Z** be as in Ex.2.1 and $Q \in \mathbf{F}(\mathbf{X}\times\mathbf{Y})$, $T \in \mathbf{F}(\mathbf{X}\times\mathbf{Z})$ be defined as

$$Q = \begin{Vmatrix} 4 & 2 & 8 \\ 6 & 3 & 2 \\ 18 & 5 & 4 \end{Vmatrix}, \qquad T = \begin{Vmatrix} 6 \\ 6 \\ 6 \end{Vmatrix}.$$

We have:

$$S^* = \begin{Vmatrix} 3 \\ 6 \\ 3 \end{Vmatrix} \quad \text{and} \quad S_z^* = \begin{Vmatrix} 12 & 6 & 24 \\ 6 & 6 & 6 \\ 18 & 30 & 12 \end{Vmatrix}.$$

Thus $\mathbf{S}^*(z)=\{S_z^*\}$ and $\mathbf{S}^*(y_1,z)= \mathbf{S}^*(y_3,z)=\{3,6\}$, $\mathbf{S}^*(y_2,z)=\{6\}$. Then we deduce the following other elements of $\mathbf{S}$ distinct from $S^*$:

$$S_1 = \begin{Vmatrix} 3 \\ 6 \\ 6 \end{Vmatrix}, \quad S_2 = \begin{Vmatrix} 6 \\ 6 \\ 3 \end{Vmatrix}, \quad S_3 = \begin{Vmatrix} 6 \\ 6 \\ 6 \end{Vmatrix}.$$

## 2.4. Some Theoretical Results

The Thms.2.3 and 2.4 ensure the existence of the greatest elements of $\mathbf{S}$ and $\mathbf{Q}$, respectively. We shall prove in Ch. 3 that if **L** is a linear lattice, $\mathbf{S}$ and $\mathbf{Q}$ have also minimal elements if Eq.(2.6) is assigned on finite spaces. Assuming also **L** to be a complete Brouwerian lattice (except in Corol.2.11), we give several results.

THEOREM 2.8. *If* $\mathbf{S}\neq\emptyset$, *then we have for all* $x\in\mathbf{X}$, $z\in\mathbf{Z}$, $S\in\mathbf{S}$:

$$T(x,z)\leq \bigvee_{y\in\mathbf{Y}} Q(x,y) \quad \text{and} \quad T(x,z)\leq \bigvee_{y\in\mathbf{Y}} S(y,z).$$

PROOF. The thesis follows immediately from Def.2.5 since $Q(x,y)\wedge S(y,z)\leq Q(x,y)$ and $Q(x,y)\wedge S(y,z)\leq S(y,z)$ for all $x\in\mathbf{X}$, $y\in\mathbf{Y}$, $z\in\mathbf{Z}$ and $S\in\mathbf{S}$. ■

Thm.2.8 corresponds to Thms.7 and 8 of [9].

COROLLARY 2.9. *If there exists some* $x'\in\mathbf{X}$ *such that* $Q(x',y)<T(x',z)$ *for all* $y\in\mathbf{Y}$, $z\in\mathbf{Z}$ *and* $T(x',z')\neq T(x',z'')$ *for some* $z',z''\in\mathbf{Z}$, *then* $\mathbf{S}=\emptyset$.

PROOF. If $\mathbf{S}\neq\emptyset$, we should deduce by Thm.2.8,

$$T(x',z')\leq \bigvee_{y\in\mathbf{Y}} Q(x',y) \leq T(x',z'') \leq \bigvee_{y\in\mathbf{Y}} Q(x',y) \leq T(x',z')$$

and this would imply $T(x',z')=T(x',z'')$, which contradicts the hypothesis. ■

Note that the assumption $T(x',z')\neq T(x',z'')$ for some $x'\in\mathbf{X}$, $z',z''\in\mathbf{Z}$ is necessary in Corol.2.9 otherwise $\mathbf{S}\neq\emptyset$ as is shown in the following example:

EXAMPLE 2.6. Let $\mathbf{X}=\{x_1,x_2,x_3\}$, $\mathbf{Y}=[0,1)$, $\mathbf{Z}=\{z_1,z_2\}$, $\mathbf{L}=[0,1]$ and $Q\in\mathbf{F}(\mathbf{X}\times\mathbf{Y})$, $T\in\mathbf{F}(\mathbf{X}\times\mathbf{Z})$ be given by $Q(x_i,y)=y/i$, $T(x_i,z_1)=T(x_i,z_2)=1/i$ for $i=1,2,3$ and for any $y\in\mathbf{Y}$. We have:

$$\bigvee_{0\leq y<1} \{Q(x_i,y)\} = \bigvee_{0\leq y<1} \{y/i\} = 1/i = T(x_i,z_k) > Q(x_i,y)$$

for $i=1,2,3$ and $k=1,2$. Thus $S^*(y,z_k)=1$ for any $y\in\mathbf{Y}$ and $k=1,2$ and

$$T(x_i,z_k) = \bigvee_{y\in\mathbf{Y}} Q(x_i,y) = \bigvee_{y\in\mathbf{Y}} [Q(x_i,y) \wedge S^*(y,z_k)]$$

for $i=1,2,3$ and $k=1,2$. This means that $\mathbf{S}\neq\emptyset$ by the fundamental Thm.2.3.

COROLLARY 2.10. *If there exists some* $x'\in\mathbf{X}$ *such that* $Q(x',y)<T(x',z)$ *for all* $y\in\mathbf{Y}$, $z\in\mathbf{Z}$ *and* $\bigvee_{y\in\mathbf{Y}} Q(x',y)=Q(x',y')$ *for some* $y'\in\mathbf{Y}$, *then* $\mathbf{S}=\emptyset$.

PROOF. By hypothesis, we have:

$$\bigvee_{y\in \mathbf{Y}} Q(x',y)=Q(x',y')<T(x',z)$$

and hence $\mathbb{S}=\emptyset$ by Thm.2.8. ■

The assumption $\bigvee_{y\in \mathbf{Y}} Q(x',y)=Q(x',y')$ for some $x'\in \mathbf{X}$, $y'\in \mathbf{Y}$ in the above Corollary arises, for instance, if we suppose **Y** to be a finite set and **L** to be a linear lattice, as is seen in the following example:

EXAMPLE 2.7. Let $\mathbf{L}=[0,1]$, $\mathbf{X}=\{x_1,x_2\}$, $\mathbf{Y}=\{y_1,y_2\}$, $\mathbf{Z}=\{z_1,z_2\}$ and $Q\in \mathbf{F}(\mathbf{X}\times\mathbf{Y})$, $T\in \mathbf{F}(\mathbf{X}\times\mathbf{Z})$ be given by

$$Q=\begin{Vmatrix} 0.4 & 0.6 \\ 0.9 & 0.8 \end{Vmatrix}, \qquad T=\begin{Vmatrix} 0.7 & 0.9 \\ 0.7 & 0.7 \end{Vmatrix}.$$

We have $Q(x_1,y_j)<T(x_1,z_k)$ for any $j,k\in \{1,2\}$ and

$$S^*=\begin{Vmatrix} 0.7 & 0.7 \\ 0.7 & 0.7 \end{Vmatrix} \quad \text{and} \quad S^*\odot Q=\begin{Vmatrix} 0.6 & 0.6 \\ 0.7 & 0.7 \end{Vmatrix} < T.$$

Therefore $\mathbb{S}=\emptyset$.

COROLLARY 2.11. *If* **L** *is a complete linear lattice and* $Q(x,y)>T(x,z)$ *for all* $x\in \mathbf{X}$, $y\in \mathbf{Y}$, $z\in \mathbf{Z}$, *then* $\mathbb{S}=\emptyset$ *if* $T(x',z')\neq T(x'',z')$ *for some* $z'\in \mathbf{Z}$ *and* $x', x''\in \mathbf{X}$.

PROOF. By Def.2.7 and since **L** is a complete linear lattice, we have that

$$S^*(y,z') = \bigwedge_{x\in \mathbf{X}} [Q(x,y)\ \alpha\ T(x,z')] = \bigwedge_{x\in \mathbf{X}} T(x,z')\leq T(x',z') < Q(x',y)$$

for any $y\in Y$. If $\mathbb{S}\neq\emptyset$, then we have by the basic Thm.2.3:

$$T(x',z') = \bigvee_{y\in \mathbf{Y}} [Q(x',y) \wedge S^*(y,z')] = \bigvee_{y\in \mathbf{Y}} \{Q(x',y) \wedge [\bigwedge_{x\in \mathbf{X}} T(x,z')]\}$$

$$= \bigvee_{y\in \mathbf{Y}} [\bigwedge_{x\in \mathbf{X}} T(x,z')] = \bigwedge_{x\in \mathbf{X}} T(x,z').$$

Similarly, we prove that

$$T(x'',z') = \bigwedge_{x\in \mathbf{X}} T(x,z')$$

and hence T(x',z)=T(x'',z), which contradicts the hypothesis. ■

If **L** is not a linear lattice, then Corol.2.11 does not hold as is proved in the following example:

EXAMPLE 2.8. Let **L** and **X** be as in the Ex.2.1, $\mathbf{Y}=\{y_1,y_2\}$, $\mathbf{Z}=\{z_1,z_2\}$ and $Q\in \mathbf{F}(\mathbf{X}\times\mathbf{Y})$, $T\in \mathbf{F}(\mathbf{X}\times\mathbf{Z})$ be given by

$$Q=\begin{Vmatrix} 6 & 4 \\ 3 & 9 \\ 2 & 1 \end{Vmatrix}, \qquad T=\begin{Vmatrix} 12 & 24 \\ 18 & 72 \\ 6 & 24 \end{Vmatrix}.$$

It is immediately seen that all the assumptions of Corol.2.11 are satisfied, but $\mathbf{S}\neq\emptyset$ since $S^*\odot Q=T$, where

$$S^*=\begin{Vmatrix} 36 & 72 \\ 6 & 24 \end{Vmatrix}.$$

Let $Q\in \mathbf{F}(\mathbf{X}\times\mathbf{Y})$, $S\in \mathbf{F}(\mathbf{Y}\times\mathbf{Z})$ and $T\in \mathbf{F}(\mathbf{X}\times\mathbf{Z})$ be fuzzy relations. Define two fuzzy relations $S_T$, $S_Q\in \mathbf{F}(\mathbf{Y}\times\mathbf{Z})$ as follows:

$$S_T(y,z) = S(y,z) \wedge [\bigvee_{x\in \mathbf{X}} T(x,z)],\ S_Q(y,z) = S(y,z) \wedge [\bigvee_{x\in \mathbf{X}} Q(x,y)]$$

for all $y\in \mathbf{Y}$, $z\in \mathbf{Z}$. Recalling that a complete lattice is Brouwerian iff the operation "$\vee$" is completely distributive on the operation "$\wedge$" [1, p.128] (cfr. also Prop.2.1), then the following result holds:

THEOREM 2.12. *If* $S\in \mathbf{S}$, *then* $S_T\in \mathbf{S}$ *and* $S_Q\in \mathbf{S}$.

PROOF. Let $S\in \mathbf{S}$ and then we have

$$(S_T\odot Q)\,(x,z) = \bigvee_{y\in \mathbf{Y}} [Q(x,y) \wedge S_T(y,z)] = \bigvee_{y\in \mathbf{Y}} \{Q(x,y) \wedge S(y,z) \wedge [\bigvee_{x\in \mathbf{X}} T(x,z)]\}$$

$$=\{\bigvee_{y\in \mathbf{Y}} [Q(x,y)\wedge S(y,z)]\}\wedge[\bigvee_{x\in \mathbf{X}} T(x,z)] = T(x,z)\wedge[\bigvee_{x\in \mathbf{X}} T(x,z)] = T(x,z)$$

for all $x\in \mathbf{X}$, $z\in \mathbf{Z}$ and this implies that $S_T\in \mathbf{S}$. Similarly, it is proved that $S_Q\in \mathbf{S}$. ■

## 2.5. Dual Composite Fuzzy Equations

In Secs.3 and 4, the fixed lattice **L** is complete Brouwerian in order to deal with max-min equations. To solve a dual composite fuzzy relation equation, we suppose **L** to be complete dually Brouwerian, i.e. for all $a,b\in \mathbf{L}$ the set of all $x\in \mathbf{L}$ such that $a\vee x\geq b$ contains the least element, denoted by $a\varepsilon b$, i.e. $a\varepsilon b=\inf\{x\in \mathbf{L} : a\vee x\geq b\}$.

Dualizing Defs.2.5 and 2.7, we have respectively:

DEFINITION 2.8. Let $Q\in \mathbf{F}(\mathbf{X}\times\mathbf{Y})$ and $S\in \mathbf{F}(\mathbf{Y}\times\mathbf{Z})$. We define the min-max composition of Q and S to be the fuzzy relation $T\in \mathbf{F}(\mathbf{X}\times\mathbf{Z})$, in symbols $T=S\Delta Q$, given by

$$T(x,z) = (S\Delta Q)(x,z) = \bigwedge_{y\in \mathbf{Y}} [Q(x,y)\vee S(y,z)] \tag{2.12}$$

for all $x\in \mathbf{X}$, $z\in \mathbf{Z}$.

DEFINITION 2.9. Let $Q\in \mathbf{F}(\mathbf{X}\times\mathbf{Y})$ and $T\in \mathbf{F}(\mathbf{X}\times\mathbf{Z})$. We define the ε-composition of $Q^{-1}$ and T to be the fuzzy relation $(Q^{-1}\varepsilon T)\in \mathbf{F}(\mathbf{Y}\times\mathbf{Z})$ defined as

$$(Q^{-1}\varepsilon T)(y,z) = \bigvee_{x\in \mathbf{X}} [Q^{-1}(y,x)\,\varepsilon\, T(x,z)]$$

for all $y\in \mathbf{Y}$, $z\in \mathbf{Z}$.

We observe that if **L** is a linear lattice, then $a\varepsilon b=b$ if $a<b$ and $a\varepsilon b=0$ if $a\geq b$.

We define the Δ-composite (or min-max) fuzzy relation equation Eq.(2.12) and with analogous proofs to those of Thms.2.3 and 2.4, one can prove the following fundamental theorems:

THEOREM 2.13. *Let* $\mathbf{S}'=\mathbf{S}'(Q,T)=\{S\in \mathbf{F}(\mathbf{Y}\times\mathbf{Z}) : S\Delta Q=T\}$. *Then* $\mathbf{S}'\neq\emptyset$ *iff* $(Q^{-1}\varepsilon T)\in \mathbf{S}'$. *Further,* $(Q^{-1}\varepsilon T)\leq S$ *for any* $S\in \mathbf{S}'$.

THEOREM 2.14. *Let* $\mathbf{Q}'=\mathbf{Q}'(S,T)=\{Q\in \mathbf{F}(\mathbf{X}\times\mathbf{Y}) : S\Delta Q=T\}$. *Then* $\mathbf{Q}'\neq\emptyset$ *iff* $(S\varepsilon T^{-1})^{-1}\in \mathbf{Q}'$. *Further,* $(S\varepsilon T^{-1})^{-1}\leq Q$ *for any* $Q\in \mathbf{Q}'$.

These simple results can be found in [8]. More generally, it is obvious that any result concerning a fuzzy equation can be *dualized*, i.e. formulated suitably for the dual fuzzy equation.

## References


[1] G. Birkhoff, *Lattice Theory*, 3rd Ed., Vol.XXV, AMS Colloquium Publications, Providence, Rhode Islands, 1967.

[2] T.S. Blyth, Matrices over ordered algebraic structures, *J.London Math. Soc.* 39 (1964), 427-432.

[3] A. Di Nola, On solving relational equations in Brouwerian lattices, *Fuzzy Sets and Systems*, to appear.

[4] A. Di Nola and A. Lettieri, Relation equations in residuated lattices, *BUSEFAL* 34 (1988), 95-106; final version in *Rend. Circ. Mat. Palermo*, to appear.

[5] J.A. Goguen, L-fuzzy sets, *J. Math. Anal. Appl.* 18 (1967), 145-174.

[6] R.D. Luce, A note on Boolean matrix theory, *Proc. Amer. Math. Soc.* 3 (1952), 382-388.

[7] S. Rudeanu, *Boolean Functions and Equations*, North-Holland, Amsterdam, 1974.

[8] E. Sanchez, Resolution of composite fuzzy relation equations, *Inform. and Control* 30 (1976), 38-48.

[9] E. Sanchez, Solutions in composite fuzzy relation equations: Application to medical diagnosis in Brouwerian logic, in: *Fuzzy Automata and Decision Processes* (M.M. Gupta, G.N. Saridis and B.R. Gaines, Eds.), North-Holland, New York (1977), pp.221-234.

[10] E. Sanchez, Compositions of fuzzy relations, in: *Advances in Fuzzy Set Theory and Applications* (M.M. Gupta, R.K. Ragade and R.R. Yager, Eds.), North-Holland, Amsterdam (1979), pp.421-433.

[11] E. Sanchez, Eigen fuzzy sets and fuzzy relations, *J. Math. Anal. Appl.* 81 (1981), 399-421.

[12] G. Szàsz, *Introduction to Lattice Theory*, 3rd Ed., Academic Press, New York, 1963.

[13] E. Turunen, On generalized fuzzy relation equations: necessary and sufficient conditions for the existence of solutions, *Acta Univ. Carolinae – Mathematica et Physica* 28 (1987), 33-37.

[14] L.A. Zadeh and C.A. Desoer, *Linear System Theory*, Mc.Graw-Hill, New York, 1963.

# CHAPTER 3

## LOWER SOLUTIONS OF MAX-MIN FUZZY EQUATIONS



*The whole of this Ch. is devoted to the study of the lower solutions of max-min fuzzy equations. In all the Secs., except Secs. 5 and 6, we assume* **L** *to be a linear lattice with universal bounds 0 and 1 and the domains, on which fuzzy sets and fuzzy relations are defined, to be finite sets, we denote by |**X**| the cardinality of a finite set* **X**. *In Secs. 1 and 2, we deal with max-min fuzzy equations of type (2.5) and (2.6), respectively. Further lattice results in the set* **S** *are given in Sec.3 and interesting properties of a particular fuzzy relation of* **S** *are pointed out in Sec.4. Secs.5 and 6 are devoted to the study of lower solutions of max-min fuzzy equations defined on complete Brouwerian lattices and on complete completely distributive lattices, respectively.*

### 3.1. Lower Solutions of Eq.(2.5) in Linear Lattices.

The results of this Sec. are due to Sanchez [9, Ch.2]. We now consider Eq.(2.5) defined on the finite sets **X** and **Y** and we assume that **L** is a linear lattice (we do not need the completeness of the lattice **L**, dealing here with equations on finite sets), so that Eq.(2.5) becomes a max-min fuzzy equation, i.e.

$$B(y) = (R \odot A)(y) = \bigvee_{x \in \mathbf{X}} [A(x) \wedge R(x, y)] \tag{3.1}$$

for any $y \in \mathbf{Y}$. Supposing that A and B are given, we denote by $\mathfrak{R}=\mathfrak{R}(A,B)$ the set of all the solutions $R \in \mathbf{F}(\mathbf{X} \times \mathbf{Y})$ of Eq.(3.1).

We define the following set:

$$G(y) = \{x \in \mathbf{X} : A(x) \geq B(y)\}$$

for any $y \in \mathbf{Y}$. Then the following simple result holds:

THEOREM 3.1. *If* $\mathfrak{R} \neq \emptyset$, *we have* $\mathbf{G}(y) \neq \emptyset$ *for any* $y \in \mathbf{Y}$.

PROOF. Arguing as in Thm.2.8, we deduce for any $y \in \mathbf{Y}$:

$$B(y) \leq \bigvee_{x \in \mathbf{X}} A(x) = A(x_y)$$

for some $x_y \in \mathbf{X}$ since $\mathbf{X}$ is finite. This means that $x_y \in \mathbf{G}(y)$. ■

We now introduce the operator $\sigma: \mathbf{L}^2 \rightarrow \mathbf{L}$ defined as $a\sigma b=0$ if $a<b$ and $a\sigma b=b$ if $a \geq b$, where $a,b \in \mathbf{L}$. This operator is different from the operator "ε", defined in Sec.2.5. Note that $a \wedge (a\sigma b) = a\sigma b$ and $a\sigma b \leq a \wedge b$ for all $a,b \in \mathbf{L}$.

THEOREM 3.2. *If* $\mathbf{G}(y) \neq \emptyset$ *for any* $y \in \mathbf{Y}$, *the fuzzy relation* $(A\sigma B) \in \mathbf{F}(\mathbf{X} \times \mathbf{Y})$, *pointwise defined by* $(A\sigma B)(x,y) = A(x)\sigma B(y)$ *for all* $x \in \mathbf{X}$, $y \in \mathbf{Y}$, *belongs to the set* $\mathfrak{R}$.

PROOF. We have:

$$[(A\sigma B) \odot A](y) = \bigvee_{x \in \mathbf{X}} \{A(x) \wedge [A(x)\,\sigma\, B(y)]\} =$$

$$\bigvee_{x \in \mathbf{X}} [A(x)\sigma B(y)] = \{\bigvee_{x \in \mathbf{G}(y)} [A(x)\,\sigma\, B(y)]\} \vee \{\bigvee_{x \notin \mathbf{G}(y)} [A(x)\,\sigma\, B(y)]\} =$$

$$\{\bigvee_{x \in \mathbf{G}(y)} B(y)\} \vee \{0\} = B(y)$$

for any $y \in \mathbf{Y}$.

THEOREM 3.3. $(A\alpha B) \in \mathfrak{R}$ *iff* $\mathfrak{R} \neq \emptyset$ *iff* $\bigcap_{y \in \mathbf{Y}} \mathbf{G}(y) \neq \emptyset$,

*where* $(A\alpha B)$ *is defined from formula* (2.7). *Further,* $(A\alpha B) \geq R$ *for any* $R \in \mathfrak{R}$.

PROOF. The thesis follows directly from Thms.2.3, 3.1 and 3.2 and observing that, since $\mathbf{X}$ is finite, there certainly exists an element $x' \in \mathbf{X}$ such that $A(x') = \sup\{A(x) : x \in \mathbf{X}\}$, i.e. $x' \in \mathbf{G}(y)$ for any $y \in \mathbf{Y}$. ■

We now define the following set:

$$\mathbf{Y}_1 = \{y \in \mathbf{Y} : B(y) > 0\}.$$

Obviously, $\mathbf{Y}_1 = \mathbf{Y} - \mathbf{B}^{-1}(0)$, where $\mathbf{B}^{-1}(0)=\{y\in \mathbf{Y}: B(y)=0\}$. Then

THEOREM 3.4. *If* $\mathfrak{R}\neq\emptyset$, *then* $\mathbf{Y}_1=\emptyset$ *iff* $(A\sigma B)(x, y)=0$ *for all* $x\in \mathbf{X}$, $y\in \mathbf{Y}$.

PROOF. If $\mathbf{Y}_1=\emptyset$, then $B(y)=0$ for any $y\in \mathbf{Y}$ and hence $A(x)\sigma B(y)=0$ for all $x\in \mathbf{X}$, $y\in \mathbf{Y}$. Vice versa, assume that $\mathbf{Y}_1\neq\emptyset$ and since $(A\sigma B)\in\mathfrak{R}$, we should have for some $y\in \mathbf{Y}_1$:

$$0<B(y) = \bigvee_{x\in \mathbf{X}} \{A(x) \wedge [A(x)\sigma B(y)]\} = \bigvee_{x\in \mathbf{X}} [A(x)\wedge 0] = 0,$$

a contradiction. ■

THEOREM 3.5. *Let* $\mathfrak{R}\neq\emptyset$ and $\mathbf{Y}_1\neq\emptyset$. *Then* $(A\sigma B)\leq R$ *for any* $R\in\mathfrak{R}$ *iff* $|\mathbf{G}(y)|=1$ *for any* $y\in \mathbf{Y}_1$.

PROOF. We first show that if $(A\sigma B)$ is the minimum of $\mathfrak{R}$, i.e. $(A\sigma B)\leq R$ for any $R\in\mathfrak{R}$, then $|\mathbf{G}(y)|=1$ for any $y\in \mathbf{Y}_1$.

Let us suppose $|\mathbf{G}(y')|\neq 1$ for some $y'\in \mathbf{Y}_1$ and then $|\mathbf{G}(y')|>1$ by Thm.3.1. Let $x',x''\in \mathbf{G}(y')$ with $x'\neq x''$ and we can define a fuzzy relation $R\in \mathbf{F}(\mathbf{X}\times\mathbf{Y})$ as follows:

$$R(x,y) = \begin{cases} B(y') & \text{if } x=x' \text{ and } y=y', \\ 0 & \text{if } x\neq x' \text{ and } y=y', \\ A(x)\sigma B(y) & \text{otherwise.} \end{cases}$$

We claim that $R\in\mathfrak{R}$. Indeed, we have for $y\neq y'$:

$$\bigvee_{x\in \mathbf{X}} [A(x) \wedge R(x,y)] = \bigvee_{x\in \mathbf{X}} \{A(x) \wedge [A(x)\sigma B(y)]\} = B(y)$$

since $(A\sigma B)\in\mathfrak{R}$. We have for $y=y'$:

$$\bigvee_{x\in \mathbf{X}} [A(x) \wedge R(x,y')] = [A(x') \wedge B(y')] \vee \{ \bigvee_{x\neq x'} [A(x) \wedge 0]\} =$$

$$A(x') \wedge B(y') = B(y').$$

Since $A\sigma B$ is the minimum of $\mathfrak{R}$, we have, in particular, for $x''\neq x'$:

$$B(y') = A(x'') \,\sigma\, B(y') \leq R(x'', y') = 0,$$

which is in opposition to the fact that $y \in \mathbf{Y}_1$. We now prove the converse implication.

Since $\mathbf{G}(y) = \mathbf{X}$ for any $y \in \mathbf{B}^{-1}(0)$, we have by Thm.3.3:

$$\bigcap_{y \in \mathbf{Y}} \mathbf{G}(y) = \bigcap_{y \in \mathbf{Y}_1} \mathbf{G}(y) \neq \emptyset.$$

Since $|\mathbf{G}(y)|=1$ for any $y \in \mathbf{Y}$, we have $\mathbf{G}(y)=\{x'\}$ for any $y \in \mathbf{Y}_1$, where $x' \in \mathbf{X}$ is such that $A(x')=\sup\{A(x): x \in \mathbf{X}\}$. We have $A(x')>A(x)$ for any $x \in \mathbf{X}-\{x'\}$ and

$$\sup\{A(x): x \in \mathbf{X}-\{x'\}\} < \inf\{B(y): y \in \mathbf{Y}_1\},$$

otherwise we should obtain for some $x'' \in \mathbf{X}-\{x'\}$ and $y' \in \mathbf{Y}_1$:

$$A(x'') = \sup\{A(x): x \in \mathbf{X}-\{x'\}\} \geq \inf\{B(y): y \in \mathbf{Y}_1\} = B(y') > 0.$$

Then $\mathbf{G}(y')=\{x'\}=\{x''\}$, a contradiction since $x' \neq x''$. Hence we have $A(x)<B(y)$ for all $x \in \mathbf{X}-\{x'\}$, $y \in \mathbf{Y}_1$ and since $A(x)\, \sigma B(y)=0$ for all $x \in \mathbf{X}$, $y \in \mathbf{B}^{-1}(0)$, we can say that, if R is an arbitrary element of $\mathcal{R}$, $0=(A\sigma B)(x,y) \leq R(x,y)$ for all $x \in \mathbf{X}-\{x'\}$, $y \in \mathbf{Y}$. Suppose that $R(x',y)<(A\sigma B)(x',y)=B(y) \leq A(x')$ for some $y \in \mathbf{Y}_1$. Then

$$B(y) = \bigvee_{x \in \mathbf{X}} [A(x) \wedge R(x,y)] = [A(x') \wedge R(x',y)] \vee \{ \bigvee_{x \neq x'} [A(x) \wedge R(x,y)]\}$$

$$\leq R(x',y) \vee [ \bigvee_{x \neq x'} A(x)] < B(y),$$

a contradiction. Thus $(A\sigma B)(x',y)=B(y) \leq R(x',y)$ and the proof is complete. ■

In all the examples of this Ch., unless otherwise specified, we assume $\mathbf{L}=[0,1]$ with the usual lattice operations. The following example illustrates Thm.3.5:

EXAMPLE 3.1. Let $\mathbf{X}=\{x_1,x_2,x_3\}$, $\mathbf{Y}=\{y_1,y_2,y_3\}$ and $A \in \mathbf{F}(\mathbf{X})$, $B \in \mathbf{F}(\mathbf{Y})$ be defined by $A(x_1)=0.9$, $A(x_2)=0.5$, $A(x_3)=0.4$, $B(y_1)=0.0$, $B(y_2)=0.6$, $B(y_3)=0.8$. We have $\mathbf{Y}_1=\{y_2,y_3\}$ and $\mathbf{G}(y_2)=\mathbf{G}(y_3)=\{x_1\}$. Then

$$(A\sigma B) = \begin{Vmatrix} 0.0 & 0.6 & 0.8 \\ 0.0 & 0.0 & 0.0 \\ 0.0 & 0.0 & 0.0 \end{Vmatrix} \leq (A\alpha B) = \begin{Vmatrix} 0.0 & 0.6 & 0.8 \\ 0.0 & 1.0 & 1.0 \\ 0.0 & 1.0 & 1.0 \end{Vmatrix}.$$

DEFINITION 3.1. *By minimal (or lower) solution, we mean a minimal element of the set of solutions of the fuzzy equation under discussion.*

We give a fundamental Theorem for the determination of the minimal solutions of Eq.(3.1).

THEOREM 3.6. *If* $\mathfrak{R}\neq\emptyset$, $\mathfrak{R}$ *has minimal elements.*

PROOF. If $\mathbf{Y}_1=\emptyset$, the thesis is straightforward by Thm.3.4. Therefore we may suppose $\mathbf{Y}_1\neq\emptyset$ and then $\mathbf{G}(y)\neq\emptyset$ for any $y\in\mathbf{Y}_1$ by Thm.3.1. For any $y\in\mathbf{Y}_1$, let $x_y$ be an arbitrary element of $\mathbf{G}(y)$; we define the following fuzzy relation $M\in\mathbf{F}(\mathbf{X}\times\mathbf{Y})$ with membership function given by

$$\begin{aligned} M(x,y) &= B(y) && \text{if } x=x_y \text{ and } y\in\mathbf{Y}_1, \\ &= 0 && \text{if } x\neq x_y \text{ and } y\in\mathbf{Y}_1, \\ &= 0 && \text{otherwise.} \end{aligned} \tag{3.2}$$

We claim that $M\in\mathfrak{R}$. Indeed, we get for any $y\in\mathbf{B}^{-1}(0)$:

$$\bigvee_{x\in\mathbf{X}} [A(x)\wedge M(x,y)] = 0 = B(y)$$

and for any $y\in\mathbf{Y}_1$:

$$\bigvee_{x\in\mathbf{X}} [A(x)\wedge M(x,y)] = [A(x_y)\wedge M(x_y,y)] \vee \{\bigvee_{x\neq x_y} [A(x)\wedge M(x,y)]\} =$$

$$= [A(x_y)\wedge B(y)] \vee \{0\} = B(y).$$

Now let $R\in\mathfrak{R}$ be such that $R\leq M$. We must show that $R=M$. It suffices to prove only that $M(x_y,y)=R(x_y,y)$ for any $y\in\mathbf{Y}_1$ since $M(x,y)=0$ otherwise. Indeed, we have for any $y\in\mathbf{Y}_1$:

$$B(y) = \bigvee_{x\in\mathbf{X}} [A(x)\wedge R(x,y)] = [A(x_y)\wedge R(x_y,y)] \vee \{\bigvee_{x\neq x_y} [A(x)\wedge R(x,y)]\}$$

$$= [A(x_y)\wedge R(x_y,y)] \vee \{0\} \leq R(x_y,y) \leq M(x_y,y) = B(y)$$

and hence $R(x_y,y)=B(y)=M(x_y,y)$, i.e. $R=M$.

Let R be a minimal element of $\mathfrak{R}$. We shall prove that R is a fuzzy relation of type (3.2). We have for any $y\in\mathbf{Y}_1$:

$$B(y) = \bigvee_{x\in\mathbf{X}} [A(x)\wedge R(x,y)] = A(x')\wedge R(x',y)$$

for some $x' \in \mathbf{X}$ (since $\mathbf{X}$ is finite). This implies that $x' \in \mathbf{G}(y)$ and, supposing that x' plays the same role as $x_y$, we are able to define, as in (3.2), a fuzzy relation $M' \in \mathcal{R}$ such that $M'(x',y)=B(y) \leq R(x',y)$ for any $y \in \mathbf{Y}_1$ and $M'(x,y)=0$, otherwise. So $M' \leq R$ and then $M'=R$ since R is minimal in $\mathcal{R}$. ■

The following example drawn from [6] clarifies Thm.3.6:

EXAMPLE 3.2. Let $\mathbf{X}=\{x_1,x_2,x_3\}$, $\mathbf{Y}=\{y_1,y_2,y_3,y_4\}$ and $A \in \mathbf{F(X)}$, $B \in \mathbf{F(Y)}$ be defined by $A(x_1)=0.4$, $A(x_2)=0.5$, $A(x_3)=1.0$, $B(y_1)=0.3$, $B(y_2)=0.5$, $B(y_3)=0.0$, $B(y_4)=1.0$. Then we have:

$$(A\sigma B) = \begin{Vmatrix} 0.3 & 0.0 & 0.0 & 0.0 \\ 0.3 & 0.5 & 0.0 & 0.0 \\ 0.3 & 0.5 & 0.0 & 1.0 \end{Vmatrix}.$$

Consequently we deduce the following minimal elements of $\mathcal{R}$:

$$M_1 = \begin{Vmatrix} 0.3 & 0.0 & 0.0 & 0.0 \\ 0.0 & 0.5 & 0.0 & 0.0 \\ 0.0 & 0.0 & 0.0 & 1.0 \end{Vmatrix}, \qquad M_2 = \begin{Vmatrix} 0.0 & 0.0 & 0.0 & 0.0 \\ 0.3 & 0.5 & 0.0 & 0.0 \\ 0.0 & 0.0 & 0.0 & 1.0 \end{Vmatrix},$$

$$M_3 = \begin{Vmatrix} 0.0 & 0.0 & 0.0 & 0.0 \\ 0.0 & 0.5 & 0.0 & 0.0 \\ 0.3 & 0.0 & 0.0 & 1.0 \end{Vmatrix}, \qquad M_4 = \begin{Vmatrix} 0.3 & 0.0 & 0.0 & 0.0 \\ 0.0 & 0.0 & 0.0 & 0.0 \\ 0.0 & 0.5 & 0.0 & 1.0 \end{Vmatrix},$$

$$M_5 = \begin{Vmatrix} 0.0 & 0.0 & 0.0 & 0.0 \\ 0.3 & 0.0 & 0.0 & 0.0 \\ 0.0 & 0.5 & 0.0 & 1.0 \end{Vmatrix}, \qquad M_6 = \begin{Vmatrix} 0.0 & 0.0 & 0.0 & 0.0 \\ 0.0 & 0.0 & 0.0 & 0.0 \\ 0.3 & 0.5 & 0.0 & 1.0 \end{Vmatrix}.$$

It is easily seen that the number $\nu$ of the minimal elements of $\mathcal{R}$ is obtained by the formula [6]:

$$\nu = \prod_{y \in \mathbf{Y}_1} |\mathbf{G}(y)|,$$

resulting $\nu=1$ iff $(A\sigma B)$ is the minimum of $\mathcal{R}$ (cfr. Thm.3.5).

See also the papers [2], [9] and [28] where other results can be found.

It is obvious that the following theorems hold:

THEOREM 3.7. *The fuzzy union of the minimal elements of* $\mathcal{R}$ *is equal to* $(A\sigma B)$.

THEOREM 3.8. *For any* $R \in \mathcal{R}$ (if $\mathcal{R} \neq \emptyset$), *there exists a minimal element* $M \in \mathcal{R}$

*such that* $M \leq \mathfrak{R}$.

The problem of the determination of the lower solutions of a fuzzy equation has been studied by several authors: here we cite the papers of Czogala, Drewniak and Pedrycz [2], Drewniak [10], Higashi and Klir [14], Luo [17], Miyakoshi and Shimbo [18], Pappis and Sugeno [20], Tsukamoto and Terano [25] and the book of Dubois and Prade [11].

We conclude this Section by pointing out a simple result of [22].

THEOREM 3.9. *If* $\mathfrak{R} \neq \emptyset$, *then* $(A\sigma B)=(A\alpha B)$ *iff either* $A(x)>B(y)$ *for all* $x \in \mathbf{X}$, $y \in \mathbf{Y}$ *if* $B(y)<1$ *for any* $y \in \mathbf{Y}$ *or* $A(x)=1$ *for any* $x \in \mathbf{X}$ *if* $B(y')=1$ *for some* $y' \in \mathbf{Y}$.

PROOF. The necessity follows from the fact that if $B(y')=1$ for some $y' \in \mathbf{Y}$, then $A(x)=1$ for any $x \in \mathbf{X}$ otherwise if $A(x')<1$ for some $x' \in \mathbf{X}$, we would have $0=A(x')\sigma B(y')=A(x')\alpha B(y')=1$, a contradiction. If $B(y)<1$ for any $y \in \mathbf{Y}$, then $A(x)>B(y)$ for all $x \in \mathbf{X}$, $y \in \mathbf{Y}$, otherwise we would have $1=A(x')\alpha B(y')=A(x')\sigma B(y')=B(y')$ (resp.0) if $A(x')=B(y')$ (resp. $A(x')<B(y')$) for some $x' \in \mathbf{X}$, $y' \in \mathbf{Y}$, also a contradiction. The sufficiency is trivial. ■

### 3.2. Lower Solutions of Eq.(2.6) in Linear Lattices

Let **X**, **Y** and **Z** be finite sets of cardinality n, m and p, respectively. By continuing to suppose **L** as a linear lattice (the completeness of **L** is not necessary here either), Eq.(2.6) is converted into the max-min fuzzy equation:

$$T(x, z) = (S \odot Q)(x, z) = \bigvee_{y \in \mathbf{Y}} [Q(x, y) \wedge S(y, z)] \tag{3.3}$$

for all $x \in \mathbf{X}$, $z \in \mathbf{Z}$.

Thm.2.3 ensures the existence of $S^*$, the greatest element of the set $\mathbf{S}$. Our aim is to find the minimal element of $\mathbf{S}$ based on the knowledge of $S^*$ and on the lower solutions of n equations of type (3.1), which constitute a suitable system associated to the assigned Eq.(3.3).

As underlined in [4], [5] and [7], Eq.(3.3) is equivalent to the following system of n equations of type (3.1):

$$S \odot Q_x = T_x, \qquad x \in \mathbf{X}, \tag{3.4}$$

where $Q_x \in \mathbf{F}(\{x\} \times \mathbf{Y})$ and $T_x \in \mathbf{F}(\{x\} \times \mathbf{Z})$ are fuzzy sets whose membership functions are, of course, the restrictions of Q and T to the specified Cartesian products.

We have:

$$\mathbf{S} = \bigcap_{x \in \mathbf{X}} \mathcal{R}_x, \tag{3.5}$$

where $\mathcal{R}_x=\mathcal{R}_x(Q_x,T_x)$ is, for any $x\in\mathbf{X}$, the set of all the solutions of each Eq.(3.4). Then we can show the following result ([9], [12], [22], [13, Ch.2]):

THEOREM 3.10. *If* $\mathbf{S}\neq\emptyset$, *then*

$$S^* = \bigwedge_{x\in\mathbf{X}} (Q_x \alpha T_x).$$

PROOF. If $\mathbf{S}\neq\emptyset$, then $\mathcal{R}_x\neq\emptyset$ for any $x\in\mathbf{X}$ by (3.5). Consequently, $S^*\in\mathcal{R}_x$ and $S^*\leq(Q_x\alpha T_x)$ for any $x\in\mathbf{X}$ by Thm.3.3, i.e.

$$S^* \leq \bigwedge_{x\in\mathbf{X}} (Q_x\alpha T_x) \leq (Q_x\alpha T_x).$$

By Lemma 2.5 (applied to the set $\mathcal{R}_x$), we see that the fuzzy relation $\wedge_{x\in\mathbf{X}}(Q_x\alpha T_x)$ belongs to $\mathcal{R}_x$ for any $x\in\mathbf{X}$, i.e. belongs to $\mathbf{S}$ and hence by Thm.2.3:

$$\bigwedge_{x\in\mathbf{X}} (Q_x\alpha T_x) \leq S^*,$$

which, together with the previous opposite inequality, implies the thesis. ■

As already noted in the proof of Thm.3.10, then $S^*\in\mathcal{R}_x$ for any $x\in\mathbf{X}$ and by Thm.3.8, there exists certainly a minimal element $M_x\in\mathcal{R}_x$ such that $M_x\leq S^*$ for any $x\in\mathbf{X}$, i.e. the following set:

$$\mathcal{M}_x = \mathcal{M}_x(Q_x, T_x) = \{M_x\in\mathcal{R}_x: M_x\leq S^*, x\in\mathbf{X}\}$$

is nonempty for any $x\in\mathbf{X}$. Then the following basic theorem of [4] holds:

THEOREM 3.11. *If* $\mathbf{S}\neq\emptyset$, *then the set*

$$\mathcal{M} = \mathcal{M}(Q, T) = \{M\in\mathbf{F}(\mathbf{Y}\times\mathbf{Z}): M = \bigvee_{x\in\mathbf{X}} M_x, M_x\in\mathcal{M}_x\}$$

*is a finite subset of* $\mathbf{S}$. *An element* M *is minimal in* $\mathcal{M}$ *iff it is minimal in* $\mathbf{S}$.

PROOF. We first prove that the set $\mathcal{M}$ is a subset of $\mathbf{S}$ and this follows immediately by the fact that $M_x\leq M\leq S^*$ for any $x\in\mathbf{X}$ and $M\in\mathcal{M}$. Since $M_x$ and $S^*$ belong to $\mathcal{R}_x$, M lies in $\mathcal{R}_x$ for any $x \in \mathbf{X}$ by Lemma 2.5 (applied to the set $\mathcal{R}_x$) and therefore $M \in \mathcal{S}$ by (3.5). $\mathcal{M}$ is a finite poset (ordered by fuzzy inclusion) and hence possesses minimal elements.

Let M be minimal in $\mathcal{M}$. We claim that M is minimal in $\mathcal{S}$ and indeed, let $S \in \mathcal{S}$ such that $S \leq M$. Since $S \in \mathcal{R}_x$ for any $x \in \mathbf{X}$, there exists a minimal element $M_x' \in \mathcal{R}_x$ such that $M_x' \leq S$ for any $x \in \mathbf{X}$ by Thm.3.8. Hence the fuzzy relation $M' = \vee_{x \in \mathbf{X}} M_x'$ is an element of $\mathcal{M}$ such that $M' \leq S \leq M$, then $M' = M$ since M is minimal in $\mathcal{M}$ and, a fortiori, M=S.

Vice versa, let S be minimal in $\mathcal{S}$ and arguing as above, we can find an element $M \in \mathcal{M}$ such that $M \leq S$, which implies M=S since S is minimal in $\mathcal{S}$. Thus S is an element of $\mathcal{M}$ and of course is also minimal in $\mathcal{M}$. ■

A direct verification of this theorem is shown in the following example of [4]:

EXAMPLE 3.3. Let $\mathbf{X}=\{x_1,x_2,x_3\}$, $\mathbf{Y}=\{y_1,y_2,y_3\}$, $\mathbf{Z}=\{z_1,z_2,z_3\}$, i.e. n=m=p=3 and $Q \in \mathbf{F}(\mathbf{X}\times\mathbf{Y})$, $T \in \mathbf{F}(\mathbf{X}\times\mathbf{Z})$ be defined as

$$Q = \begin{Vmatrix} 0.1 & 0.0 & 0.5 \\ 0.2 & 0.7 & 0.3 \\ 0.4 & 0.9 & 1.0 \end{Vmatrix}, \qquad T = \begin{Vmatrix} 0.5 & 0.1 & 0.5 \\ 0.6 & 0.5 & 0.6 \\ 0.6 & 0.5 & 0.9 \end{Vmatrix}.$$

For brevity, we put:

$$Q_{x_i} = Q_i,\ T_{x_i} = T_i,\ M_{x_i} = M_i,\ \mathcal{M}_{x_i} = \mathcal{M}_i,\ \mathcal{R}_{x_i} = \mathcal{R}_i \qquad \text{for } i=1,2,3.$$

(These notations will appear also in some examples). We have S* and $(Q_i \sigma T_i) \in \mathcal{R}_i$, for i=1,2,3, given by

$$S^* = \begin{Vmatrix} 1.0 & 1.0 & 1.0 \\ 0.6 & 0.5 & 0.6 \\ 0.6 & 0.1 & 0.9 \end{Vmatrix}, \qquad (Q_1 \sigma T_1) = \begin{Vmatrix} 0.0 & 0.1 & 0.0 \\ 0.0 & 0.0 & 0.0 \\ 0.5 & 0.1 & 0.5 \end{Vmatrix},$$

$$(Q_2 \sigma T_2) = \begin{Vmatrix} 0.0 & 0.0 & 0.0 \\ 0.6 & 0.5 & 0.6 \\ 0.0 & 0.0 & 0.0 \end{Vmatrix}, \qquad (Q_3 \sigma T_3) = \begin{Vmatrix} 0.0 & 0.0 & 0.0 \\ 0.6 & 0.5 & 0.9 \\ 0.6 & 0.5 & 0.9 \end{Vmatrix}.$$

Consequently each set $\mathcal{R}_i$ has the following minimal elements in accordance with Thm.3.6:

$$M_1^{I} = \begin{Vmatrix} 0.0 & 0.1 & 0.0 \\ 0.0 & 0.0 & 0.0 \\ 0.5 & 0.0 & 0.5 \end{Vmatrix}, \quad M_1^{II} = \begin{Vmatrix} 0.0 & 0.0 & 0.0 \\ 0.0 & 0.0 & 0.0 \\ 0.5 & 0.1 & 0.5 \end{Vmatrix}, \quad M_2 = (Q_2 \sigma T_2),$$

$$M_3^{I} = \begin{Vmatrix} 0.0 & 0.0 & 0.0 \\ 0.6 & 0.5 & 0.9 \\ 0.0 & 0.0 & 0.0 \end{Vmatrix}, \quad M_3^{II} = \begin{Vmatrix} 0.0 & 0.0 & 0.0 \\ 0.0 & 0.5 & 0.9 \\ 0.6 & 0.0 & 0.0 \end{Vmatrix}, \quad M_3^{III} = \begin{Vmatrix} 0.0 & 0.0 & 0.0 \\ 0.6 & 0.0 & 0.9 \\ 0.0 & 0.5 & 0.0 \end{Vmatrix},$$

$$M_3^{IV} = \begin{Vmatrix} 0.0 & 0.0 & 0.0 \\ 0.6 & 0.5 & 0.0 \\ 0.0 & 0.0 & 0.9 \end{Vmatrix}, \quad M_3^{V} = \begin{Vmatrix} 0.0 & 0.0 & 0.0 \\ 0.6 & 0.0 & 0.0 \\ 0.0 & 0.5 & 0.9 \end{Vmatrix}, \quad M_3^{VI} = \begin{Vmatrix} 0.0 & 0.0 & 0.0 \\ 0.0 & 0.0 & 0.9 \\ 0.6 & 0.5 & 0.0 \end{Vmatrix},$$

$$M_3^{VII} = \begin{Vmatrix} 0.0 & 0.0 & 0.0 \\ 0.0 & 0.0 & 0.0 \\ 0.6 & 0.5 & 0.9 \end{Vmatrix}, \quad M_3^{VIII} = \begin{Vmatrix} 0.0 & 0.0 & 0.0 \\ 0.0 & 0.5 & 0.0 \\ 0.6 & 0.0 & 0.9 \end{Vmatrix}.$$

A fast examination excludes $M_3^I$, $M_3^{II}$, $M_3^{III}$, $M_3^V$, $M_3^{VI}$, $M_3^{VII}$ since these matrices are greater, in some entry, than the corresponding entry of $S^*$. Then we have:

$$\mathcal{M}_1 = \{M_1^I, M_1^{II}\}, \qquad \mathcal{M}_2 = \{M_2\}, \qquad \mathcal{M}_3 = \{M_3^{IV}, M_3^{VIII}\}.$$

Hence $\mathcal{M}$ has the following fuzzy relations:

$$M = \begin{Vmatrix} 0.0 & 0.1 & 0.0 \\ 0.6 & 0.5 & 0.6 \\ 0.5 & 0.0 & 0.9 \end{Vmatrix} = M_1^I \vee M_2 \vee M_3^{IV}, \quad P = \begin{Vmatrix} 0.0 & 0.0 & 0.0 \\ 0.6 & 0.5 & 0.6 \\ 0.5 & 0.1 & 0.9 \end{Vmatrix} = M_1^{II} \vee M_2 \vee M_3^{IV},$$

$$N = \begin{Vmatrix} 0.0 & 0.1 & 0.0 \\ 0.6 & 0.5 & 0.6 \\ 0.6 & 0.0 & 0.9 \end{Vmatrix} = M_1^I \vee M_2 \vee M_3^{VIII}, U = \begin{Vmatrix} 0.0 & 0.0 & 0.0 \\ 0.6 & 0.5 & 0.6 \\ 0.6 & 0.1 & 0.9 \end{Vmatrix} = M_1^{II} \vee M_2 \vee M_3^{VIII}.$$

The minimal elements of $\mathcal{M}$ are M and P since $M \le N$ and $P \le U$.

In [4], the following fuzzy relation is defined:

$$\Sigma = [\bigvee_{x \in \mathbf{X}} (Q_x \,\sigma\, T_x)] \wedge S^*. \tag{3.6}$$

Of course, if $\mathbf{S} \neq \emptyset$, we have $(Q_x \,\sigma\, T_x) \le \Sigma \le S^*$ for any $x \in \mathbf{X}$, i.e. $\Sigma \in \mathbf{R}_x$ for any $x \in \mathbf{X}$ by Lemma 2.5. Therefore $\Sigma$ belongs to the set $\mathbf{S}$, which has analogous properties to the ones enjoyed by $(A\sigma B)$ in the set $\mathbf{R}$. These algebraic properties of $\Sigma$ are investigated in the next Secs. We conclude this Sec. by recalling the papers of Di Nola [3] and Xu, Wu and Cheng [29], where algorithms of calculation of the minimal elements of $\mathbf{S}$ are proposed. Prévot [21] gives an algorithm which determines the membership functions of some elements of $\mathbb{S}$; in [1], [8], [9], [13], [17], [19], [23], [24] and [26] the study of the minimal elements is made by different approaches. A good presentation of the present literature is to be found in [15]. Later we shall study conditions of existence of lower solutions of fuzzy equations defined on complete and completely distributive lattices (cfr. Sec.3.6).

### 3.3. Further Lattice Results in $\mathbb{S}$

Parallel to the results established in Sec.3.1 for Eq.(3.1), following Sessa [22], we define the following sets:

$$\mathbf{G}(x,z) = \{y \in \mathbf{Y}: Q(x,y) \geq T(x,z)\}$$

for all $x \in \mathbf{X}$, $z \in \mathbf{Z}$. Interpreting Eq.(3.3) as the system of Eqs. (3.4), certain results of Sec.3.1 can be given also for Eq.(3.3) but we are able to give only sufficient conditions, because the converse results are not generally true, as is shown with the help of suitable examples. Some results and examples of this Sec. can be found in [22].

THEOREM 3.12. *If* $\mathbb{S} \neq \emptyset$, *then* $\mathbf{G}(x,z) \neq \emptyset$ *for all* $x \in \mathbf{X}$, $z \in \mathbf{Z}$.

PROOF. Since $\mathbf{Y}$ is a finite set, for any $x \in \mathbf{X}$ we clearly have $\vee_{y \in \mathbf{Y}} Q(x,y) = Q(x,y')$ for some $y' \in \mathbf{Y}$. By Thm.2.8, we have $Q(x,y') \geq T(x,z)$ for any $z \in \mathbf{Z}$ and hence $y' \in \mathbf{G}(x,z)$. ■

This theorem parallels Thm.3.1, but it is not invertible, as is proved in the next example:

EXAMPLE 3.4. Let n=m=p=3 and $Q \in \mathbf{F}(\mathbf{X} \times \mathbf{Y})$, $T \in \mathbf{F}(\mathbf{X} \times \mathbf{Z})$ be given by

$$Q = \begin{Vmatrix} 0.0 & 0.9 & 1.0 \\ 0.2 & 0.0 & 0.8 \\ 0.5 & 0.0 & 0.9 \end{Vmatrix}, \qquad T = \begin{Vmatrix} 0.8 & 0.3 & 0.6 \\ 0.6 & 0.5 & 0.8 \\ 0.6 & 0.5 & 0.9 \end{Vmatrix}.$$

Then we deduce $\mathbf{G}(x_1,z_k)=\{y_2,y_3\}$ for k=1,2,3, $\mathbf{G}(x_2,z_k)=\mathbf{G}(x_3,z_1)=\mathbf{G}(x_3,z_3)=\{y_3\}$ for k=1,2,3 and $\mathbf{G}(x_3,z_2)=\{y_1,y_3\}$, i.e. $\mathbf{G}(x,z) \neq \emptyset$ for all $x \in \mathbf{X}$, $z \in \mathbf{Z}$. Since

$$S^* = \begin{Vmatrix} 1.0 & 1.0 & 1.0 \\ 0.8 & 0.3 & 0.6 \\ 0.6 & 0.3 & 0.6 \end{Vmatrix},$$

we have:

$$0.5 = T(x_2, z_2) > \bigvee_{j=1}^{3} [Q(x_2, y_j) \wedge S^*(y_j, z_2)] = 0.3,$$

i.e. $S^* \notin \mathbf{S}$ and then $\mathbf{S}=\emptyset$ by the basic Thm.2.3.

Similarly to Thm.3.3, one could guess that

$$\mathbf{S}\neq\emptyset \quad \text{iff} \quad \bigcap_{x\in \mathbf{X}} \bigcap_{z\in \mathbf{Z}} \mathbf{G}(x,z)\neq\emptyset,$$

but this condition is false in both directions, as is shown in the following examples:

EXAMPLE 3.5. Let n=m=2, p=4 and $Q\in \mathbf{F}(\mathbf{X}\times\mathbf{Y})$, $T\in \mathbf{F}(\mathbf{X}\times\mathbf{Z})$ be given by

$$Q = \begin{Vmatrix} 1.0 & 0.3 \\ 0.4 & 0.8 \end{Vmatrix}, \qquad T = \begin{Vmatrix} 0.8 & 0.9 & 0.4 & 0.5 \\ 0.6 & 0.7 & 0.5 & 0.8 \end{Vmatrix}.$$

Then we have:

$$S^* = \begin{Vmatrix} 0.8 & 0.9 & 0.4 & 0.5 \\ 0.6 & 0.7 & 0.5 & 1.0 \end{Vmatrix},$$

and $S^*\circledcirc Q=T$, i.e. $\mathbf{S}\neq\emptyset$ but $\mathbf{G}(x_1,z_k)=\{y_1\}$, $\mathbf{G}(x_2,z_k)=\{y_2\}$ for k=1,2,3,4 and hence

$$\bigcap_{i=1}^{2} \bigcap_{k=1}^{4} \mathbf{G}(x_i,z_k) = \emptyset.$$

EXAMPLE 3.6. By recalling Ex.3.4, we have $\mathbf{S}=\emptyset$ but

$$\bigcap_{i=1}^{3} \bigcap_{k=1}^{3} \mathbf{G}(x_i,z_k) = \{y_3\}.$$

THEOREM 3.13. *If either* $|\mathbf{G}(x,z)|=1$ *or* $T(x,z)=0$ *for all* $x\in \mathbf{X}$, $z\in \mathbf{Z}$, *then the fuzzy relation* $\Sigma$ *defined by* (3.6) *is the minimum of* $\mathbf{S}$, *i.e.* $\Sigma\leq S$ *for any* $S\in \mathbf{S}$ *if* $\mathbf{S}\neq\emptyset$.

PROOF. By Thms.3.4. and 3.5, $(Q_x \sigma T_x)$ is the minimum of each set $\mathfrak{R}_x$ for any $x \in \mathbf{X}$. Since $S^* \in \mathfrak{R}_x$ (by (3.5)) for any $x \in \mathbf{X}$, it follows that $(Q_x \sigma T_x) \leq S^*$ for any $x \in \mathbf{X}$ and then

$$\Sigma = \bigvee_{x \in \mathbf{X}} (Q_x \, \sigma \, T_x).$$

Let $S \in \mathbf{S}$, i.e. $S \in \mathfrak{R}_x$ for any $x \in \mathbf{X}$. Thus $(Q_x \sigma T_x) \leq S$ for any $x \in \mathbf{X}$ and then we have $\Sigma \leq S$. ■

Thm.3.13 parallels Thms.3.4 and 3.5, but it is not invertible, as is proved in the following example:

EXAMPLE 3.7. Let n=m=p=3 and $Q \in \mathbf{F}(\mathbf{X} \times \mathbf{Y})$, $T \in \mathbf{F}(\mathbf{X} \times \mathbf{Z})$ be given by

$$Q = \begin{Vmatrix} 0.0 & 0.2 & 1.0 \\ 0.2 & 0.5 & 0.8 \\ 0.5 & 0.0 & 0.9 \end{Vmatrix}, \qquad T = \begin{Vmatrix} 0.6 & 0.3 & 1.0 \\ 0.6 & 0.5 & 0.8 \\ 0.6 & 0.5 & 0.9 \end{Vmatrix}.$$

It is easily seen, by using Thm.3.11, that each set $\mathfrak{M}_i$, i=1,2,3, is singleton, hence the set $\mathfrak{M}$ has a unique element, necessarily equal to $\Sigma$, i.e.

$$\Sigma = \begin{Vmatrix} 0.0 & 0.5 & 0.0 \\ 0.0 & 0.5 & 0.0 \\ 0.6 & 0.3 & 1.0 \end{Vmatrix}.$$

Hence $\Sigma$ is the minimum of $\mathbf{S}$, but $\mathbf{G}(x_2,z_2)=\{y_2,y_3\}$ and $\mathbf{G}(x_3,z_2)=\{y_1,y_3\}$.

A similar result to Thm.3.7 in the set $\mathbf{S}$ will be established later (cfr. Thm.3.19). By arguing as in the proof of Thm.3.11, it is easily seen that the following result, parallel to Thm.3.8, holds.

THEOREM 3.14. *For any* $S \in \mathbf{S}$ (if $\mathbf{S} \neq \emptyset$), *there exists a minimal element* $M \in \mathfrak{M}$ *such that* $M \leq S$.

### 3.4. Further Properties of the Fuzzy Relation $\Sigma$

The results of this Sec. are due to Sessa [22], and here we always suppose $\mathbf{S} \neq \emptyset$.

THEOREM 3.15. *If* $\mathbf{X}=\mathbf{Y}$ *and* $\mathbf{G}(x,z)=\{x\}$ *for all* $x \in \mathbf{X}$, $z \in \mathbf{Z}$, *we have* $\Sigma = T$. *Further,* $T \odot Q = T$.

PROOF. We have:

$$\bigvee_{x\in \mathbf{X}} [(Q_x \sigma T_x)(x',z)] = [(Q_{x'} \sigma T_{x'})(x',z)] \vee \{ \bigvee_{x\neq x'} [(Q_x \sigma T_x)(x',z)]\}$$

$$= [Q(x',x') \sigma T(x',z)] \vee \{ \bigvee_{x\neq x'} [(Q(x,x') \sigma T(x,z)]\}$$

$$= T(x',z) \vee \{0\} = T(x',z)$$

for all $x'\in \mathbf{X}$, $z\in \mathbf{Z}$ since $Q(x',x')\geq T(x',z)$ and $x'\notin \mathbf{G}(x,z)$.

We also have by Def. 2.7,

$$S^*(x',z) = [Q(x',x') \alpha T(x',z)] \wedge \{ \bigwedge_{x\neq x'} [Q(x,x') \alpha T(x,z)]\}$$

$$= [Q(x',x') \alpha T(x',z)] \wedge 1$$

$$= [Q(x',x') \alpha T(x',z)] \geq T(x',z)$$

for all $x'\in \mathbf{X}$, $z\in \mathbf{Z}$ since $x'\notin \mathbf{G}(x,z)$.

Thus

$$S^* \geq T = \bigvee_{x\in \mathbf{X}} (Q_x \sigma T_x)$$

and this implies $\Sigma=T\wedge S^*=T$ by (3.6). ■

In Sec.2.2, we have seen that $S_1\wedge S_2$ does not generally belong to $\mathbf{S}$ if $S_1,S_2\in \mathbf{S}$, however the following result holds:

THEOREM 3.16. $(S\wedge\Sigma)\in \mathbf{S}$ *for any* $S\in \mathbf{S}$.

PROOF. Let $S\in \mathbf{S}$. By Thm.3.14, there exists a minimal element M of $\mathbf{S}$ such that $M\leq S$. Since $M\leq S^*$ and

$$M \leq \bigvee_{x\in \mathbf{X}} (Q_x \sigma T_x),$$

we have $M\leq S\wedge\Sigma\leq S$ and hence $(S\wedge\Sigma)\in \mathbf{S}$ by Lemma 2.5. ■

We now put for any $x\in \mathbf{X}$:

$$\bigvee_{M_x \in \mathcal{M}_x} M_x = W_x,$$

where $\mathcal{M}_x$ is the set defined before Thm.3.11. One can easily verify that

$$W_x(y,z) = 0 \qquad \text{if } [Q(x,y)\ \sigma\ T(x,z)] > S^*(y,z),$$
$$= Q(x,y)\ \sigma\ T(x,z) \qquad \text{otherwise,}$$

for all $y \in \mathbf{Y}$, $z \in \mathbf{Z}$. The following Lemmas hold:

LEMMA 3.17. *Let* $S^*(y',z') < 1$ *for some* $y' \in \mathbf{Y}$, $z' \in \mathbf{Z}$. *Then there exists an element* $x' \in \mathbf{X}$ *such that* $S^*(y',z') = (Q_{x'} \sigma T_{x'})(y',z')$.

PROOF. By Def. 2.7, there exists an element $x' \in \mathbf{X}$ such that $S^*(y',z') = [Q(x',y') \alpha T(x',z')] < 1$ and hence $T(x',z') < Q(x',y')$. This implies that

$$(Q_{x'}\ \sigma\ T_{x'})\ (y',z') = Q(x',y')\ \sigma\ T(x',z') = T(x',z') = S^*(y',z'),$$

i.e. the thesis. ■

LEMMA 3.18. *We have:* $\bigvee_{x \in \mathbf{X}} W_x = \Sigma$.

PROOF. We define the following sets:

$$\mathbf{X}(y,z) = \{x \in \mathbf{X}: (Q_x\ \sigma\ T_x)\ (y,z) \leq S^*(y,z)\}$$

for all $y \in \mathbf{Y}$, $z \in \mathbf{Z}$. We necessarily deduce $\mathbf{X}(y,z) \neq \emptyset$, otherwise, if there exist two elements $y' \in \mathbf{Y}$, $z' \in \mathbf{Z}$, such that

$$1 \geq (Q_x\ \sigma\ T_x)\ (y',z') > S^*(y',z')$$

for any $x \in \mathbf{X}$, we would obtain a contradiction to Lemma 3.17.

For any $y \in \mathbf{Y}$ and $z \in \mathbf{Z}$, distinguishing two situations, we can have either $\mathbf{X}(y,z) = \mathbf{X}$ or $\mathbf{X}(y,z)$ is a nonempty proper subset of $\mathbf{X}$. In the first case, we deduce that

$$\bigvee_{x \in \mathbf{X}} W_x(y,z) = \bigvee_{x \in \mathbf{X}} [Q(x,y)\ \sigma\ T(x,z)]$$

$$= \bigvee_{x \in \mathbf{X}} \{[(Q_x\ \sigma\ T_x)\ (y,z)] \wedge S^*(y,z)\}$$

$$= \{ \bigvee_{x \in \mathbf{X}} [(Q_x \,\sigma\, T_x)\,(y,z)] \} \wedge S^*(y,z) = \Sigma(y,z).$$

In the second case, since $\mathbf{X}\text{-}\mathbf{X}(y,z) \neq \emptyset$, there exists at least one element $x \in \mathbf{X}$ such that

$$1 \geq (Q_x \,\sigma\, T_x)\,(y,z) > S^*(y,z)$$

and hence, by Lemma 3.17, in $\mathbf{X}(y,z)$ lies an element $x' \in \mathbf{X}$ such that

$$[(Q_{x'} \,\sigma\, T_{x'})\,(y,z)] = S^*(y,z) = W_{x'}(y,z).$$

Since $W_x(y,z)=0$ for any $x \in \mathbf{X}\text{-}\mathbf{X}(y,z)$, then we have:

$$\bigvee_{x \in \mathbf{X}} W_x(y,z) = \bigvee_{x \in \mathbf{X}(y,z)} W_x(y,z) = \{ \bigvee_{x \neq x'} W_x(y,z)\} \vee W_{x'}(y,z)\} = S^*(y,z)$$

since

$$W_x(y,z) = (Q_x \,\sigma\, T_x)\,(y,z) \leq S^*(y,z)$$

for any $x \in \mathbf{X}(y,z)\text{-}\{x'\}$. On the other hand,

$$\Sigma(y,z) = \bigvee_{x \in \mathbf{X}(y,z)} \{[(Q_x \,\sigma\, T_x)\,(y,z)] \wedge S^*(y,z)\} \vee$$

$$\bigvee_{x \notin \mathbf{X}(y,z)} \{[(Q_x \,\sigma\, T_x)\,(y,z)] \wedge S^*(y,z)\}$$

$$= \{ \bigvee_{x \in \mathbf{X}(y,z)} [(Q_x \,\sigma\, T_x)\,(y,z)]\} \vee S^*(y,z) = S^*(y,z).$$

This and the previous equality imply that

$$\bigvee_{x \in \mathbf{X}} W_x(y,z) = \Sigma(y,z)$$

and hence, in both situations, we get the thesis. ■

Lemma 3.18 implies the following theorem, parallel to Thm.3.7.

THEOREM 3.19. *The fuzzy union of the elements of the set* $\mathcal{M}$ *is equal to* $\Sigma$, *i.e.*

$$\bigvee_{M \in \mathcal{M}} M = \Sigma.$$

PROOF. Since each element $M \in \mathcal{M}$ is (cfr. Thm.3.11) a fuzzy union of elements $M_x \in \mathcal{M}_x$, we clearly have:

$$M \le [\bigvee_{x \in \mathbf{X}} (Q_x \sigma T_x)] \wedge S^* = \Sigma.$$

We must prove that $\Sigma$ is the l.u.b. of $\mathcal{M}$. Indeed, let $S \in \mathbf{S}$ be such that $M \le S$ for any $M \in \mathcal{M}$. Hence $M_x \le S$ for any $M_x \in \mathcal{M}_x$ and $x \in \mathbf{X}$, i.e.

$$\bigvee_{M_x \in \mathcal{M}_x} M_x = W_x \le S$$

for any $x \in \mathbf{X}$. By using Lemma 3.18, we deduce that

$$\Sigma = \bigvee_{x \in \mathbf{X}} W_x \le S$$

and this completes the proof. ■

We explicitly observe that $\Sigma$ is not generally the fuzzy union of the minimal elements of $\mathbf{S}$, i.e. of $\mathcal{M}$ (by Thm.3.11), as is proved in the next example:

EXAMPLE 3.8. By returning to Ex.3.3, we have $W_1=(Q_1\sigma T_1)$, $W_2=(Q_2\sigma T_2)$ and

$$W_3 = \begin{Vmatrix} 0.0 & 0.0 & 0.0 \\ 0.6 & 0.5 & 0.0 \\ 0.6 & 0.0 & 0.9 \end{Vmatrix}.$$

We have:

$$W_1 \vee W_2 \vee W_2 = \Sigma = \begin{Vmatrix} 0.0 & 0.1 & 0.0 \\ 0.6 & 0.5 & 0.6 \\ 0.6 & 0.1 & 0.9 \end{Vmatrix} > \begin{Vmatrix} 0.0 & 0.1 & 0.0 \\ 0.6 & 0.5 & 0.6 \\ 0.5 & 0.1 & 0.9 \end{Vmatrix} = M \vee P,$$

where M and P are the minimal elements of $\mathbf{S}$.

On the other hand, it is possible to exhibit other examples in which $\Sigma$ is simply equal to the fuzzy union of the minimal elements of $\mathbf{S}$ (i.e. of $\mathcal{M}$). For instance, this happens if all the elements of $\mathcal{M}$ are distinct from each other and each of them is minimal in $\mathbb{S}$.

Further properties of $\Sigma$ will be illustrated in Secs.4.4 and 5.2.

## 3.5. On Lower Solutions of Eq.(3.3) in Complete Brouwerian Lattices

Thm.3.11 characterizes the minimal elements of $\mathbb{S}$ if Eq.(3.3) is assigned on linear lattices and on finite sets. Nowadays, to the best of our knowledge, it seems that the study of the minimal elements of $\mathbb{S}$, when Eq.(3.3) is given on a complete Brouwerian lattice, has not yet been attached by any authors. Here we give some results of Di Nola [3, Ch.2] who has determined a necessary and sufficient condition for the minimality of a solution.

Let $\mathbb{S}\neq\emptyset$. By Thm.2.7, then $\mathcal{P}(z)\neq\emptyset$ for any $z\in \mathbf{Z}$ and let $P_Z$ be an arbitrary element of $\mathcal{P}(z)$. Also $\mathcal{P}(y,z)\neq\emptyset$ for all $y\in\mathbf{Y}$, $z\in\mathbf{Z}$ and we observe that

$$\mathcal{P}(y,z) = [\mu(y,z), \rho(y,z)] \tag{3.7}$$

for all $y\in\mathbf{Y}$, $z\in\mathbf{Z}$, where

$$\mu(y,z) = \bigvee_{x\in\mathbf{X}} P_Z(x,y) ,$$

$$\rho(y,z) = \bigwedge_{x\in\mathbf{X}} [Q(x,y)\ \alpha\ P_Z(x,y)] .$$

Indeed, we have $a\geq P_Z(x,y)$ for all $a\in\mathcal{P}(y,z)$, $x\in\mathbf{X}$ and hence

$$a \geq \bigvee_{x\in\mathbf{X}} P_Z(x,y) = \mu(y,z) ,$$

which implies that

$$P_Z(x,y) = Q(x,y) \wedge a \geq Q(x,y) \wedge \mu(y,z) \geq Q(x,y) \wedge P_Z(x,y) = P_Z(x,y) \tag{3.8}$$

for all $a\in\mathcal{P}(y,z)$, $x\in\mathbf{X}$.

Further, we have $a\leq[Q(x,y)\alpha P_Z(x,y)]$ for all $a\in\mathcal{P}(y,z)$, $x\in\mathbf{X}$ and hence

$$a \leq \bigwedge_{x\in\mathbf{X}} [Q(x,y)\ \alpha\ P_Z(x,y)] = \rho(y,z),$$

which implies that

$$P_Z(x,y) = Q(x,y) \wedge a \leq Q(x,y) \wedge \rho(y,z) \leq Q(x,y) \wedge [Q(x,y)\ \alpha\ P_Z(x,y)]$$

$$\leq P_Z(x,y) \tag{3.9}$$

for all $a \in \mathbf{P}(y,z)$, $x \in \mathbf{X}$. The inequalities (3.8) and (3.9) imply that (3.7) holds.

It is easily seen that every fuzzy relation $S \in \mathbf{F}(\mathbf{Y}\times\mathbf{Z})$ such that $\mu(y,z) \leq S(y,z) \leq \rho(y,z)$ for all $y \in \mathbf{Y}$, $z \in \mathbf{Z}$, is an element of $\mathbf{S}$. In particular, a fuzzy relation M such that $M(y,z)=\mu(y,z)$ for all $y \in \mathbf{Y}$, $z \in \mathbf{Z}$ belongs to $\mathbf{S}$. If we denote by $\mathcal{M}$ the set of all such M fuzzy relations, we are in a position to state the following basic theorem.

THEOREM 3.20. *If* $\mathcal{M}$ *has minimal elements, then these elements are minimal in* $\mathbf{S}$. *If* $\mathbf{S}$ *has minimal elements, then these elements are minimal in* $\mathcal{M}$.

PROOF. Let M be minimal in $\mathcal{M}$ and $S \in \mathbf{S}$ such that $S \leq M$. From the proof of Thm.2.7 (cfr.(2.11)), we can find a fuzzy relation $P_Z \in \mathbf{P}(z)$ for any $z \in \mathbf{Z}$ and hence a fuzzy relation $M' \in \mathcal{M}$ such that

$$M'(y,z) = \bigvee_{x \in \mathbf{X}} P_Z(x,y) \leq S(y,z)$$

for all $y \in \mathbf{Y}$, $z \in \mathbf{Z}$, i.e. $M' \leq S$ and hence $M' \leq S \leq M$. This implies that $M'=M$ since M is minimal in $\mathcal{M}$ and then $M=S$. Thus M is minimal in $\mathbf{S}$. Vice versa is straightforward. ■

As a comment to the above theorem, we must add that, generally, we do not know if $\mathbf{S}$ or $\mathcal{M}$ has or does not have minimal elements. However, if one supposes a priori the existence of minimal elements in $\mathbf{S}$, these elements must be sought among the minimal elements of $\mathcal{M}$, which is a subset of $\mathbf{S}$.

Further, we give a criterion in order to state when a given solution is a minimal element in $\mathbf{S}$, but we first need to define certain subsets of $\mathbf{L}$. Precisely, we put:

$$\mathbf{H}(S,y',z') = \{a \in \mathbf{L}: [Q(x,y') \wedge a] \vee \{ \bigvee_{y \neq y'} [Q(x,y) \wedge S(y,z')]\} = T(x,z'),\quad x \in \mathbf{X}\}$$

for all $y' \in \mathbf{Y}$, $z' \in \mathbf{Z}$, $S \in \mathbf{S}$. Since $\mathbf{L}$ is complete, we put:

$$\gamma(S,y',z') = \min\ \{a : a \in \mathbf{H}(S,y',z')\}.$$

THEOREM 3.21. *Let* $S \in \mathbf{S}$. S *is minimal in* $\mathbf{S}$ *iff* $S(y',z')=\gamma(S,y',z')$ *for all* $y' \in \mathbf{Y}$, $z' \in \mathbf{Z}$.

PROOF. Let y' and z' be arbitrary elements of $\mathbf{Y}$ and $\mathbf{Z}$, respectively and S be minimal in $\mathbf{S}$. We first observe that $S(y',z') \in \mathbf{H}(S,y',z')$ and further we have:

$$[S(y',z') \wedge a] \in \mathbf{H}(S,y',z')$$

for any $a \in \mathbf{H}(S,y',z')$. Indeed, we obtain:

$$[Q(x,y') \wedge S(y',z') \wedge a] \vee \{ \bigvee_{y \neq y'} [Q(x,y) \wedge S(y,z')]\} =$$

$$\{[Q(x,y') \wedge a] \vee \{ \bigvee_{y \neq y'} [Q(x,y) \wedge S(y,z')]\}\} \wedge \{[Q(x,y') \wedge S(y',z')]$$

$$\vee \{ \bigvee_{y \neq y'} [Q(x,y) \wedge S(y,z')]\}\} =$$

$$= T(x,z') \wedge T(x,z') = T(x,z')$$

for all $x \in \mathbf{X}$, $a \in \mathbf{H}(S,y',z')$.

Then the fuzzy relation $S' \in \mathbf{F}(\mathbf{Y} \times \mathbf{Z})$ defined by

$$\begin{aligned} S'(y,z) &= S(y',z') \wedge a && \text{if } y=y',\ z=z', \\ &= S(y',z') && \text{otherwise,} \end{aligned}$$

is an element of $\mathbf{S}$ such that $S' \leq S$. Since S is minimal in $\mathbf{S}$, it follows that $S'(y',z')= S(y',z')$, i.e. $S(y',z') \leq a$ for any $a \in \mathbf{H}(S,y',z')$ and hence $S(y',z')=\gamma(S,y',z')$.

Vice versa, assume that $S(y',z')=\gamma(S,y',z')$ for all $y' \in \mathbf{Y}$, $z' \in \mathbf{Z}$. Let $S' \in \mathbf{S}$ be such that $S' \leq S$ and define a fuzzy relation $S'' \in \mathbf{F}(\mathbf{X} \times \mathbf{Y})$ such that

$$\begin{aligned} S''(y,z) &= S'(y',z') && \text{if } y=y',\ z=z', \\ &= S(y,z) && \text{otherwise.} \end{aligned}$$

Thus $S' \leq S'' \leq S$ and then $S'' \in \mathbf{S}$ by Lemma 2.5. This implies that

$$[Q(x,y') \wedge S''(y',z')] \vee \{ \bigvee_{y \neq y'} [Q(x,y) \wedge S(y,z')]\} = T(x,z')$$

for any $x \in \mathbf{X}$, i.e. $S'(y',z')=S''(y',z') \in \mathbf{H}(S,y',z')$ and then $S'(y',z') \geq S(y',z')$ for all $y' \in \mathbf{Y}$, $z' \in \mathbf{Z}$, i.e. $S' \geq S$. Since $S' \leq S$, we have $S'=S$ and then S is minimal in $\mathbf{S}$. ■

## 3.6. On Lower Solutions of Eq.(3.3) in Complete Completely Distributive Lattices

We recall that a lattice L is *completely* (or *infinitely*) *distributive* if for any $a \in \mathbf{L}$ and any set of indices **I**, we have

$$a \wedge \left( \bigvee_{i \in \mathbf{I}} x_i \right) = \bigvee_{i \in \mathbf{I}} (a \wedge x_i), \tag{3.10}$$

$$a \vee \left( \bigwedge_{i \in \mathbf{I}} x_i \right) = \bigwedge_{i \in \mathbf{I}} (a \vee x_i). \tag{3.11}$$

We have already used property (3.10) in Thm.2.12 and recalled that ([1, p.128], Ch.2) (3.10) characterizes the class of the complete Brouwerian lattices. It is also known that the property (3.11) characterizes the class of the complete dually Brouwerian lattices (cfr. Sec.2.5). Therefore a complete lattice **L** is completely distributive iff it is Brouwerian and dually Brouwerian. If Eq.(3.3) is defined on such types of lattices, a further result can be established.

THEOREM 3.22. *Let* $S \in \mathbf{S}$, $y' \in \mathbf{Y}$, $z' \in \mathbf{Z}$ *and* $M \in \mathbf{F}(\mathbf{X} \times \mathbf{Z})$ *be a fuzzy relation defined as*

$$\begin{aligned} M(y,z) &= \gamma(S,y',z') && \text{if } y=y',\ z=z', \\ &= S(y,z) && \text{otherwise,} \end{aligned}$$

*for all* $y \in \mathbf{Y}$, $z \in \mathbf{Z}$. *Then we have* $M \in \mathbf{S}$.

PROOF. Since $S(y',z') \in \mathbf{H}(S,y',z')$, we have $\mathbf{H}(S,y',z') \neq \emptyset$. Then we deduce:

$$\bigvee_{y \in \mathbf{Y}} [Q(x,y) \wedge M(y,z')] = \{ \bigvee_{y \neq y'} [Q(x,y) \wedge M(y,z')]\} \vee [Q(x,y') \wedge M(y',z')]$$

$$= \{ \bigvee_{y \neq y'} [Q(x,y) \wedge M(y,z')]\} \vee [Q(x,y') \wedge \gamma(S,y',z')]$$

$$= \{ \bigvee_{y \neq y'} [Q(x,y) \wedge M(y,z')]\} \vee \{ \bigwedge_{a \in \mathbf{H}(S,y',z')} [Q(x,y') \wedge a]\}$$

$$= \bigwedge_{a \in \mathbf{H}(S,y',z')} \{\{ \bigvee_{y \neq y'} [Q(x,y) \wedge S(y,z')]\} \vee [Q(x,y') \wedge a]\}$$

$$= \bigwedge_{a \in \mathbf{H}(S,y',z')} T(x,z') = T(x,z')$$

for all $x \in \mathbf{X}$, $z' \in \mathbf{Z}$ ($z'$ is chosen arbitrarily in **Z**). This means that $M \in \mathbf{S}$. ■

The determination of the minimal elements of **S** becomes much more effective if Eq.(3.3) is assigned on finite sets. As usually, we assume that $\mathbf{X}=\{x_1,x_2,\ldots,x_n\}$, $\mathbf{Y}=\{y_1,y_2,\ldots,y_m\}$ and $\mathbf{Z}=\{z_1,z_2,\ldots,z_p\}$. We first need to make a simple remark.

As stressed by Czogala, Drewniak and Pedrycz [2], Higashi and Klir [14] and Lettieri and Liguori [16], it is possible to translate the problem (q1) of Sec.2.2 in p problems of the same type, if we define the following fuzzy relations:

$$S_z(y,z) = S(y,z) \quad \text{and} \quad T_z(x,z) = T(x,z)$$

for all $x\in\mathbf{X}$, $y\in\mathbf{Y}$, $z\in\mathbf{Z}$ where $S_z\in\mathbf{F}(\mathbf{Y}\times\{z\})$, $T_z\in\mathbf{F}(\mathbf{X}\times\{z\})$ are the restrictions of S and T to the sets $\mathbf{Y}\times\{z\}$ and $\mathbf{X}\times\{z\}$, respectively. Eq.(3.3) is equivalent to solving p fuzzy equations of the same type:

$$S_z \circ Q_z = T_z \tag{3.12}$$

for any $z\in\mathbf{Z}$. Indeed, if $S\in\mathbf{F}(\mathbf{X}\times\mathbf{Y})$ solves Eq.(3.3), it is obvious that its p restrictions solve p Eqs.(3.12). Vice versa, if $S_z\in\mathbf{F}(\mathbf{Y}\times\{z\})$ solves Eq.(3.12) for any $z\in\mathbf{Z}$, then we put :

$$\begin{aligned} M_z(y,z') &= 0 && \text{if } z\neq z', \\ &= S_z(y,z) && \text{if } z=z', \end{aligned}$$

we obtain p fuzzy relations $M_z\in\mathbf{F}(\mathbf{X}\times\mathbf{Y})$ and then by putting:

$$M = \bigvee_{z\in\mathbf{Z}} M_z,$$

we have a fuzzy relation $M\in\mathbf{F}(\mathbf{Y}\times\mathbf{Z})$ which belongs to **S** since, by using property (3.10), we deduce:

$$\begin{aligned} \bigvee_{y\in\mathbf{Y}} [Q(x,y)\wedge M(y,z')] &= \bigvee_{y\in\mathbf{Y}} \{Q(x,y)\wedge [\bigvee_{z\in\mathbf{Z}} M_z(y,z')]\} \\ &= \bigvee_{y\in\mathbf{Y}} \{\bigvee_{z\in\mathbf{Z}} [Q(x,y)\wedge M_z(y,z')]\} \\ &= \bigvee_{y\in\mathbf{Y}} \{\{\bigvee_{z\neq z'} [Q(x,y)\wedge M_z(y,z')]\} \vee [Q(x,y)\wedge M_{z'}(y,z')]\} \\ &= \bigvee_{y\in\mathbf{Y}} \{0 \vee [Q(x,y)\wedge M_{z'}(y,z')]\} = \bigvee_{y\in\mathbf{Y}} [Q(x,y)\wedge M_{z'}(y,z')] \end{aligned}$$

$$= T_{z'}(x,z') = T(x,z').$$

Thus one can study Eq.(3.12) instead of Eq.(3.3) and this motivates the choice of p=1, i.e. **Z** is singleton as is assumed in Exs.2.1 and 2.5, without loss of generality. However, in the sequel and as has been our practice hitherto, we do not always make this choice because other questions and applications can be treated by handling Eq.(3.3) directly.

Returning to the main topic of this Sec., we present a procedure to build a minimal element of $\mathbf{S}$ starting from a fixed fuzzy relation $S \in \mathbf{S}$. For simplicity, we illustrate this procedure directly on an example in which we assume m=3 and p=1. For notational convenience, we put $Q(x_i,y_j)=Q_{ij}$, $S(y_j,z)=S_j$, $T(x_i,z)=T_i$, $\mathbf{H}(S,y_j,z)=\mathbf{H}_j(S)$, $\gamma_j(S)=\min\{a: a\in \mathbf{H}_j(S)\}$ for any $i=1,2,\ldots,n$, $j=1,2,3$ and $S\in\mathbf{S}$. Further, the elements of $\mathbf{S}$ are denoted by means of 3-tuples.

Since

$$\mathbf{H}_1(S) = \{a\in L: (Q_{i1} \wedge a) \vee (Q_{i2} \wedge S_2) \vee (Q_{i3} \wedge S_3) = T_i, \quad i=1,\ldots,n\},$$

we have $\mathbf{H}_1(S)\neq\emptyset$ because $S_1\in \mathbf{H}_1(S)$. By Thm.3.22, the fuzzy relation:

$$S^{(1)} = (\gamma_1(S), S_2, S_3)$$

is an element of $\mathbf{S}$. Similarly, we can consider:

$$\gamma_2(S^{(1)}) = \min\{a: a\in \mathbf{H}_2(S^{(1)})\}$$

and by Thm.3.22 again, the fuzzy relation:

$$S^{(12)} = (\gamma_1(S), \gamma_2(S^{(1)}), S_3)$$

is an element of $\mathbf{S}$. Finally, we can consider:

$$\gamma_3(S^{(12)}) = \min\{a: a\in \mathbf{H}_3(S^{(12)})\}$$

and also by Thm.3.22, the fuzzy relation:

$$S^{(123)} = (\gamma_1(S), \gamma_2(S^{(1)}), \gamma_3(S^{(12)}))$$

is an element of $\mathbf{S}$ and clearly $S^{(123)}\leq S$. We now claim that $S^{(123)}$ is minimal in $\mathbf{S}$. Indeed, let $M\in\mathbf{S}$ such that $M\leq S^{(123)}$ and consider the fuzzy relation:

$$M^{(1)} = (M_1, S_2, S_3).$$

Thus $M^{(1)} \in \mathbf{S}$ by Lemma 2.5 since $M \leq M^{(1)} \leq S$. This implies that $M_1 \geq \gamma_1(S)$ and hence $M_1 = \gamma_1(S)$ since the opposite inequality holds by hypothesis. Similarly, considering the fuzzy relations:

$$M^{(2)} = (\gamma_1(S), M_2, S_3) \quad \text{and} \quad M^{(3)} = (\gamma_1(S), \gamma_2(S^{(1)}), M_3),$$

one proves that $M_2 = \gamma_2(S^{(1)})$ and $M_3 = \gamma_3(S^{(12)})$. This means that $M = S^{(123)}$.

We put the following question: is $S^{(123)}$ the unique lower solution reachable from any fixed element $S \in \mathbf{S}$ using this procedure?

It is almost obvious that we can consider all the permutations of the set {1,2,3}, instead of the 3-tuple (1,2,3), and another minimal element of **S** can be built by repeating the above procedure. Further details about this method can be found in [3, Ch.2]. We conclude this Sec. by pointing out that if **L** has the additional property that any element has irredundant finite decomposition into join and meet-irreducible elements, Zhao [30] determines entirely the set **S** if Eq.(3.3) is assigned on finite sets.

## 3.7. Concluding Comments

Although the basic Thm.2.3 guarantees the existence of the greatest element of **S** (if $\mathbf{S} \neq \emptyset$), the determination of the minimal elements of **S** when Eq.(3.3) is assigned on complete Brouwerian lattices remains open in the finite case as well as in the infinite case. Thm.2.3 holds in both cases. As already seen in Thm.3.20, we do not know generally if **S** has minimal elements but these elements (if they exist) are minimal in $\mathcal{M}$.

If Eq.(3.3) is assigned on infinite referential sets and on complete linear lattices, it is not yet known if **S** possesses minimal elements. This question has been studied partially by Wang and Yuan [27], but these authors suspect only a class of elements of **S** *which probably possesses some minimal characterization* (cfr. also [17]). The same question is studied by several authors (cfr., e.g., [18]), who, studying systems of fuzzy equations, assume the existence of minimal elements of each equation in order to characterize the minimal elements of the system.

Of course, if one solves the problem of the determination of the minimal elements of **S** in the case of complete linear lattices and for infinite sets, one does not solve the main problem identified in the case of complete Brouwerian lattices, because the class of these lattices includes properly the class of complete linear lattices.

## References

[1] L.C. Cheng and B. Peng, The fuzzy relation equation with union and intersection preserving operator, *Fuzzy Sets and Systems* 25 (1988), 191-204.

[2] E. Czogala, J. Drewniak and W. Pedrycz, Fuzzy relation equations on a finite set, *Fuzzy Sets and Systems* 7(1982), 89-101.

[3] A. Di Nola, An algorithm of calculation of lower solutions of fuzzy relation equation, *Stochastica* 3 (1984), 33-40.

[4] A. Di Nola, Relational equations in totally ordered lattices and their complete resolution, *J. Math. Anal. Appl.* 107 (1985), 148-155.

[5] A. Di Nola and S. Sessa, On measures of fuzziness of solutions of composite fuzzy relation equations, in: *Fuzzy Information, Knowledge Representation and Decision Analysis* (M.M. Gupta and E.Sanchez, Eds.), IX IFAC Symposium, Marseille, 19-21 July 1983, Pergamon Press, pp.275-279.

[6] A. Di Nola and S. Sessa, On the fuzziness of solutions of σ-fuzzy relation equations on finite spaces, *Fuzzy Sets and Systems* 11 (1983), 65-77.

[7] A. Di Nola, W. Pedrycz and S. Sessa, Some theoretical results of fuzzy relation equations describing fuzzy systems, *Inform. Sciences* 34 (1984), 241-264.

[8] A. Di Nola, W. Pedrycz and S. Sessa, On measures of fuzziness of solutions of fuzzy relation equations with generalized connectives, *J. Math. Anal. Appl.* 2 (1985), 443-453.

[9] A. Di Nola, W. Pedrycz, S. Sessa and P.Z. Wang, Fuzzy relation equations under a class of triangular norms: a survey and new results, *Stochastica* 8 (1984), 99-145.

[10] J. Drewniak, Systems of equations in linear lattices, *BUSEFAL* 15 (1983), 88-96.

[11] D. Dubois and H. Prade, *Fuzzy Sets and Systems: Theory and Applications*, Academic Press, New York, 1980.

[12] S. Gottwald, On the existence of solutions of systems of fuzzy equations, *Fuzzy Sets and Systems* 12 (1984), 301-302.

[13] S.Z. Guo, P.Z. Wang, A Di Nola and S. Sessa, Further contributions to the study of finite fuzzy relations equations, *Fuzzy Sets and Systems* 26 (1988), 93-104.

[14] M. Higashi and G.J. Klir, Resolution of finite fuzzy relation equations, *Fuzzy Sets and Systems* 13 (1984), 65-82.

[15] G. Hirsch, *Equations de Relation Floue et Measures d'Incertain en Reconnaissance de Formes*, Thèse de Doctorat d'Etat, Univ. de Nancy I, 1987.

[16] A Lettieri and F. Liguori, Characterization of some fuzzy relation equations provided with one solution on a finite set, *Fuzzy Sets and Systems* 13 (1984), 83-94.

[17] C.Z. Luo, Reachable solution set of a fuzzy relation equation, *J. Math. Anal. Appl.* 103 (1984), 524-532.

[18] M. Miyakoshi and M. Shimbo, Lower solutions of systems of fuzzy equations, *Fuzzy Sets and Systems* 19 (1986), 37-46.

[19] M. Miyakoshi and M. Shimbo, Sets of solution-set-invariant coefficient matrices of simple fuzzy relation equations, *Fuzzy Sets and Systems* 21 (1987), 59-83.

[20] C.P. Pappis and M. Sugeno, Fuzzy relation equations and the inverse problem, *Fuzzy Sets and Systems* 15 (1985), 79-90.

[21] M. Prévot, Algorithm for the solution of fuzzy relations, *Fuzzy Sets and Systems* 5 (1981), 319-322.

[22] S. Sessa, Some results in the setting of fuzzy relation equations theory, *Fuzzy Sets and Systems* 14 (1984), 281-297.

[23] T. Tashiro, Method of solution to inverse problem of fuzzy correspondence model, in: *Summary of Papers on General Fuzzy Problems*, Working Group Fuzzy Syst., Tokyo, no.3 (1977), 70-79.

[24] Y. Tsukamoto and T. Tashiro, A method of solution to fuzzy inverse problems, *Trans. Soc. Instrum. Control. Eng.* 15 (1) (1979).

[25] Y. Tsukamoto and T. Terano, Failure diagnosis by using fuzzy logic, *Proc. IEEE Conf. Decision Control*, New Orleans 2 (1977), 1390-1395.

[26] M. Wagenknecht and K. Hartmann, On direct and inverse problems for fuzzy equation systems with tolerances, *Fuzzy Sets and Systems* 24 (1987), 93-102.

[27] P.Z. Wang and M. Yuan, Relation equations and relation inequalities, *Selected Papers on Fuzzy Subsets*, Beijing Normal University (1980), 20-31.

[28] P.Z. Wang, A. Di Nola, W. Pedrycz and S. Sessa, How many lower solutions does a fuzzy relation equation have? *BUSEFAL* 18 (1984), 67-74.

[29] W.L. Xu, C.F. Wu and W.M. Cheng, An algorithm to solve the max-min composite fuzzy relational equations, in: *Approximate Reasoning in Decision Analysis* (M.M. Gupta and E. Sanchez, Eds.), North-Holland, Amsterdam (1982), pp.47-49.

[30] C.K. Zhao, On matrix equations in a class of complete and completely distributive lattices, *Fuzzy Sets and Systems* 22 (1987), 303-320.

# CHAPTER 4

## MEASURES OF FUZZINESS OF SOLUTIONS OF MAX-MIN FUZZY RELATION EQUATIONS ON LINEAR LATTICES



*In this Ch. we continue to study the max-min fuzzy relation equation (3.3) defined on finite sets and assuming L to be a linear lattice with universal bounds 0 and 1 but endowed with an additional structure L' of a complete linear lattice with universal bounds 0' and 1', tied to the foregoing one by accurate and reasonable requirements. These are basic preliminaries in order to solve some important optimization problems in the set of solutions of a fuzzy equation. Strictly speaking, introducing a suitable functional which measures the "fuzziness content" of a fuzzy relation, we characterize all the solutions of a max-min composite fuzzy equation possessing the smallest and the greatest value of such fuzziness.*

### 4.1. Basic Preliminaries

The results of the Secs. 4.1, 4.2 and 4.3 of this Ch. are due to Di Nola [9].

Let $\mathbf{L}\equiv(L,\wedge,\vee,\leq)$ be a linear lattice with universal bounds 0 and 1 and $a,b\in\mathbf{L}$ such that $a\leq b$. As usual, we denote by $[a,b]=\{x\in\mathbf{L}: a\leq x\leq b\}$ the related closed interval having a and b as extremes. Assume that on the same support of the lattice **L**, there is another structure $\mathbf{L}'\equiv(L,\sqcap,\sqcup,\sqsubseteq)$ which is a complete linear lattice (whose universal bounds are denoted by 0' and 1') such that the following properties hold:

(i) $\sqcap\{x: x\in[a,b]\} = a\sqcap b$,

(ii) $\sqcup\{x: x\in[a,b]\}\in[a,b]$,

where $\sqcap\{x: x\in[a,b]\}$ and $\sqcup\{x: x\in[a,b]\}$ denote, of course, the infimum and the supremum of the set [a,b] in the lattice **L'**, respectively.

Further, we postulate the existence of a real isotone functional $\Phi: \mathbf{L}'\rightarrow[0,1]$, i.e. $a\sqsubseteq b$ implies $\Phi(a)\leq\Phi(b)$, $a,b\in\mathbf{L}'$. We now give some examples of lattices **L** and **L'** satisfying the above properties and related functionals $\Phi$.

EXAMPLE 4.1. Let $\mathbf{L}\equiv([0.1], \wedge, \vee, \leq)$ be the usual lattice on the unit interval and $\mathbf{L}'\equiv\mathbf{L}$ (i.e. $\wedge\equiv\sqcap$, $\vee\equiv\sqcup$, $\leq\equiv\sqsubseteq$, $0\equiv0'$, $1\equiv1'$). The above properties are trivially verified and as isotone functional it suffices to consider a nondecreasing function $\Phi$: $[0,1]\rightarrow[0,1]$.

EXAMPLE 4.2. Let $a'=1-a$ for any $a\in[0,1]$. Define a binary (symmetric) operation "$\sqcap$" in $[0,1]$ as $a\sqcap b=a$ if $a\wedge a'<b\wedge b'$ and $a\sqcap b=a\wedge b$ if $a\wedge a'=b\wedge b'$. We can dually define the operation "$\sqcup$" and by putting $a\sqsubseteq b$ iff $a\sqcap b=a$ with $a,b\in[0,1]$ (i.e. the total ordering is the one induced by the above operations), it is seen [12] that $\mathbf{L}'\equiv([0,1],\sqcap,\sqcup,\sqsubseteq)$ is a complete linear lattice with universal bounds $0'\equiv0$ and $1'\equiv1/2$ (cfr. also [13]). The properties (i) and (ii) are satisfied if we assume $\mathbf{L}=([0,1],\wedge,\vee,\leq)$ with the usual lattice operations (cfr. [9], [7, Ch.3]).

Following Knopfmacher [23], let $\Phi:[0,1]\rightarrow[0,1]$ be a real function such that $\Phi(0)=\Phi(1)=0$, $\Phi(a)=\Phi(a')$ for any $a\in[0,1]$ and strictly increasing in $[0,1/2]$. By thinking of $\Phi$ as a functional from $\mathbf{L}'$ into $[0,1]$, it turns out to be isotone since $a\sqsubseteq b$ implies $a\wedge a'\leq b\wedge b'\leq1/2$ and hence $\Phi(a)=\Phi(a\wedge a')\leq\Phi(b\wedge b')=\Phi(b)$. Here $\Phi$ represents the classical functional used by De Luca and Termini ([4], [5]) for measuring the entropy of a fuzzy set. For further details, we refer the reader to recent literature, e.g., [1]÷[3], [6]÷[8], [14]÷[19], [21], [22], [24]÷[33].

EXAMPLE 4.3. Let $\gamma\in(0,1)$ and let us define two binary operations "$\sqcap$" and "$\sqcup$" in $[0,1]$ as follows:

$$a\sqcap\gamma=\gamma\sqcap a=a,\quad a\sqcup\gamma=\gamma\sqcup a=\gamma \qquad \text{if } a\in[0,1],$$

$$a\sqcap1=1\sqcap a=1,\quad a\sqcup1=1\sqcup a=a \qquad \text{if } a\in[0,1],$$

$$a\sqcap0=0\sqcap a=0,\quad a\sqcup0=0\sqcup a=a \qquad \text{if } a\in[0,1),$$

and for any $a,b\in(0,1)-\{\gamma\}$,

$$\begin{aligned} a\sqcap b &= a\vee b, & a\sqcup b &= a\wedge b, &&\text{if } a\wedge b>\gamma,\\ &= a\wedge b, & &= a\vee b, &&\text{if } a\wedge b<\gamma, \end{aligned}$$

where $\vee\equiv\max$ and $\wedge\equiv\min$.

It is easy to show that $\mathbf{L}'\equiv([0,1],\sqcap,\sqcup,\sqsubseteq)$ is a complete linear lattice with universal bounds $0'\equiv1$ and $1'\equiv\gamma$ if we define the total ordering $a\sqsubseteq b$ iff $a\sqcap b=a$ for any $a,b\in[0,1]$. By taking the usual lattice $\mathbf{L}\equiv([0,1],\wedge,\vee,\leq)$, we now prove that the properties (i) and (ii) hold. We first show that $(a\sqcap b)\sqcap x=a\sqcap b$ for any $x\in[a,b]$. This is certainly true for $x=a$ and $x=b$. Hence we may suppose $a<x<b$, distinguishing two situations: either $b<1$ or $b=1$. In the first case, the following equalities hold:

$$a \sqcap b = a \wedge b = a = a \wedge x = a \sqcap x = (a \sqcap b) \sqcap x \qquad \text{if } a<\gamma,$$

$$a \sqcap b = \gamma \sqcap b = b = b \vee x = b \sqcap x = (a \sqcap b) \sqcap x \qquad \text{if } a=\gamma,$$

$$a \sqcap b = a \vee b = b = b \vee x = b \sqcap x = (a \sqcap b) \sqcap x \qquad \text{if } a>\gamma.$$

In the second case, we have $a \sqcap b = a \sqcap 1 = 1 = 1 \sqcap x = (a \sqcap b) \sqcap x$. Thus $(a \sqcap b) \sqsubseteq x$ for any $x \in [a,b]$ in all cases and we also observe that if $y \sqsubseteq x$ in **L**' for any $x \in [a,b]$, then $y \sqsubseteq (a \sqcap b)$ since $(a \sqcap b) \in \{a,b\} \subseteq [a,b]$. This implies the validity of the property (i). The property (ii) follows from the fact that

$$\begin{aligned} \sqcup\{x\colon x \in [a,b]\} &= b && \text{if } b<\gamma, \\ &= \gamma && \text{if } \gamma \in [a,b], \\ &= a && \text{if } \gamma<a. \end{aligned}$$

By taking a real function $\Phi\colon [0,1] \to [0,1]$ such that $\Phi(0) = \Phi(1) = 0$ and $\Phi(x)=1$ for any $x \in (0,1)$, we have clearly an isotone functional from **L**' into $[0,1]$.

### 4.2. Two Optimization Problems

Let **X**, **Y**, **Z** be nonempty finite sets, S: **Y**×**Z**→**L** where **L** is the same support of the above lattices and let $\Phi\colon$ **L**'→[0,1] be an isotone real functional. We define the Φ-measure of fuzziness of the fuzzy relation S by the number

$$\Phi(S) = \frac{1}{|Y| \cdot |Z|} \cdot \sum_{z \in Z} \sum_{y \in Y} \Phi(S(y,z)). \tag{4.1}$$

By taking into account the max-min composite fuzzy equation (3.3) defined in the sets **X**, **Y**, **Z** and on the linear lattice $(\mathbf{L}, \wedge, \vee, \leq)$ and assuming that the properties (i) and (ii) of Sec.4.1 hold, in this Sec. we solve the following problems:

(Φ1) *characterize all the elements of* **S** *having the smallest Φ-measure of fuzziness,*

(Φ2) *characterize all the elements of* **S** *having the greatest Φ-measure of fuzziness.*

If $\mathbf{S} \neq \emptyset$, then $M \leq S^*$ (by Thm.2.3) for any $M \in \mathcal{M}$, where $\mathcal{M}$ is the finite subset of **S** defined in the statement of Thm.3.11. Thus $M(y,z) \leq S^*(y,z)$ in **L** for all $y \in \mathbf{Y}$, $z \in \mathbf{Z}$, i.e. we can consider the closed interval $[M(y,z), S^*(y,z)]$ in **L** for all $y \in \mathbf{Y}$, $z \in \mathbf{Z}$. By setting in **L**',

$$M'(y,z) = M(y,z) \sqcap S^*(y,z), \tag{4.2}$$

$$M''(y,z) = \sqcup \{x: x \in [M(y,z), S^*(y,z)]\} \tag{4.3}$$

for all $y \in \mathbf{Y}$, $z \in \mathbf{Z}$, we have $M'(y,z)$, $M''(y,z) \in [M(y,z), S^*(y,z)]$ for all $y \in \mathbf{Y}$, $z \in \mathbf{Z}$, since the properties (i) and (ii) hold, i.e. the fuzzy relations $M', M'' \in \mathbf{F}(\mathbf{Y} \times \mathbf{Z})$ defined pointwise by (4.2) and (4.3), are elements of $\mathbf{S}$ for any $M \in \mathcal{M}$. We now define the following finite subsets of $\mathcal{M}$:

$$\mathcal{M}' = \{M \in \mathcal{M}: \Phi(M') = \min \{\Phi(N'): N \in \mathcal{M}\}\},$$

$$\mathcal{M}'' = \{M \in \mathcal{M}: \Phi(M'') = \max \{\Phi(N''): N \in \mathcal{M}\}\}.$$

Then the following result, here enunciated in a slightly different form from Thm.1 of [9], holds:

THEOREM 4.1. *Let* $\mathbf{S} \neq \emptyset$. *Then* $\Phi(M') = \min\{\Phi(S): S \in \mathbf{S}\}$ *for any* $M \in \mathcal{M}'$.

PROOF. Of course, $\min\{\Phi(S): S \in \mathbf{S}\} \leq \Phi(M')$ for any $M \in \mathcal{M}'$. Let S be arbitrary in $\mathbf{S}$. By Thm.3.14, there exists an element $P \in \mathcal{M}$ such that $P \leq S \leq S^*$. By property (i), $P'(y,z) \sqsubseteq S(y,z)$ for all $y \in \mathbf{Y}$, $z \in \mathbf{Z}$, in L' and this implies $\Phi(P'(y,z)) \leq \Phi(S(y,z))$ for all $y \in \mathbf{Y}$, $z \in \mathbf{Z}$, since $\Phi$ is isotone. Then $\Phi(P') \leq \Phi(S)$ and this implies that $\Phi(M') = \min\{\Phi(N'): N \in \mathcal{M}\} \leq \Phi(P') \leq \Phi(S)$ for any $S \in \mathbf{S}$, i.e. $\Phi(M') \leq \min\{\Phi(S): S \in \mathbf{S}\}$ for any $M \in \mathcal{M}'$. This gives the thesis. ■

Using property (ii), it is analogously proved that

THEOREM 4.2. *Let* $\mathbf{S} \neq \emptyset$. *Then* $\Phi(M'') = \max\{\Phi(S): S \in \mathbf{S}\}$ *for any* $M \in \mathcal{M}''$.

Thm.4.1 (resp. 4.2) solves the problem ($\Phi$1) (resp. ($\Phi$2)) since an element $S \in \mathbf{S}$ has the smallest (resp. greatest) $\Phi$-measure of fuzziness iff $\Phi(S) = \min\{\Phi(N'): N \in \mathcal{M}\}$ (resp. $\Phi(S) = \max\{\Phi(N''): N \in \mathcal{M}\}$).

### 4.3. Some Measures of Fuzziness

Here we apply the results of Sec.4.2 to the Exs. of Sec. 4.1.

In Ex.4.1, by imposing some additional properties on the functional $\Phi$ such as $\Phi(0)=0$ and $\Phi(1)=1$, it seems natural, since $\Phi$ is nondecreasing, to define for any $S \in \mathbf{S}$ the number $\Phi(S)$ as the *$\Phi$-energy measure of fuzziness* of the fuzzy relation S in accordance with the axioms of De Luca and Termini [7] postulated for an energy measure of a fuzzy set. Since we have supposed $\wedge \equiv \sqcap$ and $\vee \equiv \sqcup$, it is easily seen that $M \in \mathcal{M}'$ iff M is minimal in $\mathcal{M}$ (or in $\mathbf{S}$) and $\mathcal{M}=\mathcal{M}''$ because $M'=M \wedge S^*=M$ and $M''=M \vee S^*=S^*$ for any $M \in \mathcal{M}$. Then $\Phi(M')=\min\{\Phi(S): S \in \mathbf{S}\}$ iff M is minimal in $\mathcal{M}$ and $\Phi(M'')=\max\{\Phi(S): S \in \mathbf{S}\}=\Phi(s^*)$ iff $M \in \mathcal{M}$.

In Ex.4.2, the number $\Phi(S)$ represents the *Φ-entropy measure of fuzziness* of the fuzzy relation $S \in \mathbf{S}$.

By recalling Ex.3.3, we have the following fuzzy relations:

$$M' = \begin{Vmatrix} 0.0 & 1.0 & 0.0 \\ 0.6 & 0.5 & 0.6 \\ 0.6 & 0.0 & 0.9 \end{Vmatrix}, \qquad P' = \begin{Vmatrix} 0.0 & 0.0 & 0.0 \\ 0.6 & 0.5 & 0.6 \\ 0.6 & 0.1 & 0.9 \end{Vmatrix},$$

$$M'' = \begin{Vmatrix} 1.0 & 0.1 & 1.0 \\ 0.6 & 0.5 & 0.6 \\ 0.5 & 0.1 & 0.9 \end{Vmatrix}, \qquad P'' = \begin{Vmatrix} 1.0 & 1.0 & 1.0 \\ 0.6 & 0.5 & 0.6 \\ 0.5 & 0.1 & 0.9 \end{Vmatrix},$$

which have entropy measure $\Phi(M')=0.2$, $\Phi(P')=0.211\ldots$, $\Phi(M'')=0.233\ldots$, $\Phi(P'')=0.22\ldots$, respectively, if we define $\Phi(a)=a \wedge a'$ for any $a \in [0,1]$. Thus we have $\mathcal{M}'=\mathcal{M}''=\{M\}$.

In Ex.4.3, we define the *Boolean degree* of the fuzzy relation $S \in \mathbf{S}$ as the number $\beta(S)=1-\Phi(S)$.

By recalling again Ex.3.3, we first exhibit all the elements of $\mathbf{S}$ which have minimum Boolean degree, i.e. maximum $\Phi$-measure. By assuming $\gamma=0.9$, we have:

$$M''=P''= \begin{Vmatrix} 0.9 & 0.9 & 0.9 \\ 0.6 & 0.5 & 0.6 \\ 0.6 & 0.1 & 0.9 \end{Vmatrix}$$

and $\beta(M'')=1-\Phi(M'')=1-1=0$. Note that the fuzzy relations $S \in \mathbf{S}$ for which $\Phi(S)=1$ have membership functions such that $S_{11}, S_{12}, S_{13} \in (0,1)$, $S_{21}=0.6$, $S_{22}=0.5$, $S_{23}=0.6$, $S_{31} \in [0.5,0.6]$, $S_{32} \in (0,0.1]$, $S_{33}=0.9$, where $S_{jk}=S(y_j,z_k)$ for j,k=1,2,3. Further we have:

$$M' = \begin{Vmatrix} 1.0 & 1.0 & 1.0 \\ 0.6 & 0.5 & 0.6 \\ 0.5 & 0.0 & 0.9 \end{Vmatrix}, \qquad P' = \begin{Vmatrix} 1.0 & 1.0 & 1.0 \\ 0.6 & 0.5 & 0.6 \\ 0.5 & 0.1 & 0.9 \end{Vmatrix},$$

and hence $\beta(M')=4>3=\beta(P')$, i.e. $\mathcal{M}'=\mathcal{M}''=\{M\}$. On the other hand, one can easily show that the elements of $\mathbf{S}$ having the smallest fuzzy entropy are also elements of $\mathbf{S}$ which possess maximum Boolean degree.

## 4.4. Further Lattice Results

In this Sec. we establish other properties of the fuzzy relation $\Sigma$ defined and widely studied in Ch.3. Considering the linear lattices **L** and **L**' of Sec.4.1 and supposing that the properties (i) and (ii) hold, we point out the following Lemma (cfr. [6, Ch.3], [22, Ch.3]):

LEMMA 4.3. *Let* $a,b,c\in \mathbf{L}$ *such that* $a\leq c$, $b\leq c$. *Then*

$$(a \sqcap c) \vee (b \sqcap c) = (a\vee b) \sqcap c.$$

PROOF. If $a \sqcap c=a$ and $b \sqcap c=c$, then we have $(a \sqcap c)\vee(b \sqcap c)=a\vee c=c=b \sqcap c=(a\vee b) \sqcap c$ since $b\leq a\vee b\leq c$ and the property (i) implies that $x \sqcap (b \sqcap c)=b \sqcap c$ for any $x\in [b,c]$. Similarly, the thesis is proved if $a \sqcap c=c$ and $b \sqcap c=b$. If $a \sqcap c=b \sqcap c=c$ and since $a\leq a\vee b\leq c$, we have $(a\vee b) \sqcap c=(a\vee b) \sqcap (a \sqcap c)=a \sqcap c=c=c\vee c=(a \sqcap c)\vee(b \sqcap c)$.

If $a \sqcap c=a$ and $b \sqcap c=b$, then we have

$$\begin{aligned}(a \sqcap c) \vee (b \sqcap c) = a\vee b &= a = a \sqcap c = (a\vee b) \sqcap c \qquad \text{if } b\leq a,\\ &= b = b \sqcap c = (a\vee b) \sqcap c \qquad \text{if } a<b.\end{aligned}$$

Therefore the thesis is proved in any case. ■

Then, if we define the fuzzy relation $\Sigma \sqcap S^*$ in the same way as we defined the fuzzy relations M' by using (4.2), we obtain the following result of [22, Ch.3]:

THEOREM 4.4. *Let* $\mathbf{S}\neq\emptyset$. *Then*

$$\bigvee_{M\in\mathcal{M}} M' = \Sigma \sqcap S^*.$$

PROOF. By Thm.3.19 and Lemma 4.3, since $\mathcal{M}$ is finite, we have

$$\bigvee_{M\in\mathcal{M}} M' = \bigvee_{M\in\mathcal{M}} (M \sqcap S^*) = (\bigvee_{M\in\mathcal{M}} M) \sqcap S^* = \Sigma \sqcap S^*.$$ ■

Note that if **L**=**L**', which is the lattice of Ex.4.1, then Thm.4.4 is Thm.3.19. Thm.4.4 was proved for max-min fuzzy equations of type (3.1) in [6, Ch.3].

## 4.5. Concluding Remarks

The problem of the reduction of the entropy of the solutions of a max-min fuzzy equation of type (3.3) was begun in [11]. This problem was completely solved for fuzzy equations of type (3.1) in the finite case and on linear lattices in [10] and [6, Ch.3]. In [20], it was resolved under a different approach and in [8, Ch.3] and [9, Ch.3] it was solved for fuzzy equations under a particular class of triangular norms (cfr. Sec.8.5). Summarizing, we can say that the problem of the determination of the elements having the smallest measure of fuzzy entropy is completely solvable if Eq.(3.3) (or Eq.(3.1)) is assigned on finite sets and on linear lattices. Unfortunately, to the best of our knowledge, there remains the open question of max-min fuzzy equations of type (3.3) defined on complete Brouwerian lattices in the finite case as well as in the infinite case.

However, we shall illustrate in Sec.5.3 an algorithm which reduces the fuzzy entropy of the solutions of a max-min fuzzy equation (3.3) assigned on finite sets and on complete Brouwerian lattices determining elements of **S** with maximum Boolean degree.

## References

[1] N. Batle and E. Trillas, Entropy and fuzzy integral, *J. Math. Anal. Appl.* 69 (1979), 469-474.

[2] R.M. Capocelli and A. De Luca, Fuzzy sets and decision theory, *Inform. and Control* 23 (1973), 446-473.

[3] A. De Luca, Dispersion measures of fuzzy sets, in *Approximate Reasoning in Expert Systems* (M.M. Gupta, A. Kandel, W. Bandler and J.B. Kiszka, Eds.), Elsevier Science Publishers B.V. (North-Holland), Amsterdam (1985), pp.199-216.

[4] A. De Luca and S. Termini, A definition of a non-probabilistic entropy in the setting of fuzzy sets theory, *Inform. and Control* 20 (1972), 301-312.

[5] A. De Luca and S. Termini, Entropy of L-fuzzy sets, *Inform. and Control* 24 (1974), 55-73.

[6] A. De Luca and S. Termini, On the convergence of entropy measures of a fuzzy set, *Kybernetes* 6 (1977), 219-227.

[7] A. De Luca and S. Termini, Entropy and energy measures of fuzzy sets, in: *Advances in fuzzy Sets Theory and Applications* (M.M. Gupta, R.K. Ragade and R.R. Yager, Eds.), North-Holland, Amsterdam (1979), pp.321-328.

[8] A. De Luca and S. Termini, On some algebraic aspects of the measure of fuzziness, in: *Fuzzy Information and Decision Processes* (M.M. Gupta and E. Sanchez, Eds.), North-Holland, Amsterdam (1982), pp.17-24.

[9] A. Di Nola, On functionals measuring the fuzziness of solutions in relational equations, *Fuzzy Sets and Systems* 14 (1984), 249-258.

[10] A. Di Nola and W. Pedrycz, Entropy and energy measures characterization of resolution of some fuzzy relational equations, *BUSEFAL* 10 (1982), 44-53.

[11] A. Di Nola and S. Sessa, On the set of solutions of composite fuzzy relation equations, *Fuzzy Sets and Systems* 9 (1983), 275-286.

[12] A. Di Nola and S. Sessa, On the fuzziness measure and negation in totally ordered lattices, *J. Math. Anal. Appl.* 144 (1986), 156-170.

[13] A. Di Nola and A. Ventre, On some chains of fuzzy sets, *Fuzzy Sets and Systems* 4 (1980), 185-191.

[14] A. Di Nola and A. Ventre, Ordering via fuzzy entropy, in: *Fuzzy Information and Decision Processes* (M.M. Gupta and E. Sanchez, Eds.), North-Holland, Amsterdam (1982), pp.25-28.

[15] A. Di Nola and A. Ventre, Towards an algebraic setting of measures of fuzziness, in: *The Mathematics of Fuzzy Systems* (A. Di Nola and A. Ventre, Eds.), Verlag TÜV, Köln (1986), pp.87-102.

[16] J. Dombi, A general class of fuzzy operators, the De Morgan class of fuzzy operators and fuzziness measures induced by fuzzy operators, *Fuzzy Sets and Systems* 8 (1982), 149-163.

[17] B.R. Ebanks, On measures of fuzziness and their representations, *J. Math. Anal. Appl.* 94 (1983), 24-37.

[18] H. Emptoz, Nonprobabilistic entropies and indetermination measures in the setting of fuzzy set theory, *Fuzzy Sets and Systems* 5 (1981), 307-317.

[19] G. Gerla, Nonstandard entropy and energy measures of fuzzy-sets, *J. Math. Anal. Appl.* 104 (1984), 107-116.

[20] M. Higashi, A. Di Nola, W. Pedrycz and S. Sessa, Ordering fuzzy sets by consensus concepts and fuzzy relation equations, *Internat. J. Gen. Systems* 10 (1984), 47-56.

[21] M. Higashi and G.J Klir, On measures of fuzziness and fuzzy complements, *Internat. J. Gen. Systems* 8 (1982), 169-180.

[22] G.J. Klir, Ed., Special Issue on Measures of Uncertainty, *Fuzzy Sets and Systems* 24 (1987).

[23] J. Knopfmacher, On measures of fuzziness, *J. Math. Anal. Appl.* 49 (1975), 529-534.

[24] K. Kuriyama, Entropy of a finite partition of fuzzy sets, *J. Math. Anal. Appl.* 94 (1983), 38-43.

[25] W. Pedrycz, S. Gottwald and E. Czogala, Contribution to applications of energy measure of fuzziness, *Fuzzy Sets and Systems* 8 (1982), 205-214.

[26] W. Pedrycz, S. Gottwald and E. Czogala, Measures of fuzziness and operations with fuzzy sets, *Stochastica* 3(1982), 187-205.

[27] T. Riera and E. Trillas, From measures of fuzziness to Booleanity control, in *Fuzzy Information and Decision Processes* (M.M. Gupta and E. Sanchez, Eds.) North-Holland, Amsterdam (1982), pp.3-16.

[28] T. Riera and E. Trillas, Entropies in finite fuzzy sets, *Inform. Sciences* 15(1978), 159-168.

[29] F. Suarez Garcia and P. Gil Alvarez, Measures of fuzziness of fuzzy events, *Fuzzy Sets and Systems* 21 (1987), 147-157.

[30] Z.X. Wang, Fuzzy measures and measures of fuzziness, *J. Math. Anal. Appl.* 104 (1984), 589-601.

[31] S. Weber, Measures of fuzzy sets and measures of fuzziness, *Fuzzy Sets and*

*Systems* 13 (1984), 247-271.

[32] R.R. Yager, On the fuzziness measure and negation, Part I: membership in the unit interval, *Internat. J. Gen. Systems* 5 (1979), 221-229.

[33] R.R. Yager, On the fuzziness measure and negation, Part II: Lattices, *Inform. and Control* 44 (1980), 236-260.

# CHAPTER 5

## BOOLEAN SOLUTIONS OF MAX-MIN FUZZY EQUATIONS



*This Ch. is entirely dedicated to the study of the Boolean (i.e. non-fuzzy) solutions of a max-min fuzzy equation defined on a complete Brouwerian lattice* **L** *with universal bounds 0 and 1. In Sec.1, we give a simple necessary and sufficient condition for the existence of the greatest Boolean solution of* **S** *and in Sec.2, using some results of Ch.3, we determine the minimal Boolean elements of* **S** *if* **L** *is a linear lattice. In Sec.3, if Boolean solutions do not exist, an algorithm is presented to find the elements of* **S** *with maximum Boolean degree, i.e. with the maximum number of 0 and 1. The referential sets are assumed to be necessarily finite in Secs.2 and 3.*

### 5.1. A Fundamental Theorem.

In Sec.4.2 we have already determined and characterized all the solutions of a max-min fuzzy equation assigned on finite sets and on linear lattices, which have the smallest and greatest fuzzy entropy measure $\Phi$. Here we characterize the subset $\mathbf{B}$ of **S** constituted by the fuzzy relations $S \in \mathbf{S}$ such that $\Phi(S)=0$, i.e. $\Phi(S(y,z))=0$ for all $y \in \mathbf{Y}$, $z \in \mathbf{Z}$.

De Luca and Termini [4, Ch.4] define a functional $\Phi: \mathbf{F(X)} \rightarrow [0,1]$, where $\mathbf{F(X)}=\{A: \mathbf{X} \rightarrow [0,1]\}$, to be a fuzzy entropy measure if it satisfies the following conditions:

(i) $\Phi(A)=0$ *iff* $A \in \mathbf{F}(X)$ *is a sharp fuzzy set, i.e.* $A(x) \in \{0,1\}$ *for any* $x \in X$,

(ii) *if* $A \in \mathbf{F}(X)$ *is such that* $A(x)=1/2$ *for all* $x \in X$, *then* $\Phi(A) \geq \Phi(B)$ *for all* $B \in \mathbf{F}(X)$,

(iii) *if B is sharper than A, i.e.* $A(x) \leq B(x)$ *if* $A(x) \geq 1/2$ *and* $A(x) \geq B(x)$ *if* $A(x) \leq 1/2$, *then* $\Phi(A) \geq \Phi(B)$.

By setting:

$$\Phi(A) = \frac{1}{|X|} \cdot \sum_{x \in X} \Phi(A(x)), \tag{5.1}$$

where $\Phi$: $[0,1] \rightarrow [0,1]$ is a real function having the properties given in the Ex.4.2, it is clear that the functional (5.1) satisfies the above requirements and of course (4.1) is an extension of (5.1) from fuzzy sets upon fuzzy relations.

In accordance with axiom (i), we have:

$$S \in \mathbf{B} \text{ iff } S(y,z) \in \{0,1\} \text{ for all } y \in \mathbf{Y},\ z \in \mathbf{Z}.$$

The elements of $\mathbf{B}$ are defined as Boolean solutions of Eq.(3.3), considered here to be assigned on a complete Brouwerian lattice with universal bounds 0 and 1 and over sets not necessarily finite.

The following fundamental result holds.

THEOREM 5.1. *Let* $\mathbf{S} \neq \emptyset$. *Then* $\mathbf{B} \neq \emptyset$ *iff the fuzzy relation* $B^* \in \mathbf{F}(\mathbf{Y} \times \mathbf{Z})$ *defined as* $B^*(y,z)=1$ *if* $S^*(y,z)=1$ *and* $B^*(y,z)=0$ *otherwise, for all* $y \in \mathbf{Y}$, $z \in \mathbf{Z}$, *belongs to* $\mathbf{S}$. *Further,* $S \leq B^*$ *for any* $S \in \mathbf{B}$.

PROOF. It suffices to prove that if $\mathbf{B} \neq \emptyset$, then $B^*$ exists and lies in $\mathbf{S}$. Indeed, let $S \in \mathbf{B}$ and then $S \leq B^* \leq S^*$ since if $S(y,z)=0$ for some $y \in \mathbf{Y}$ and $z \in \mathbf{Z}$, then clearly $0=S(y,z) \leq B^*(y,z)$ and if $S(y,z)=1$ for some $y \in \mathbf{Y}$ and $z \in \mathbf{Z}$, then we have necessarily $S^*(y,z)=1$ and this implies that $B^*(y,z)=1=S^*(y,z)$. We deduce the thesis using Lemma 2.5. ■

This result can be found in [2] and we verify Thm.5.1 in the following example [2]:

EXAMPLE 5.1. Let $\mathbf{L},\mathbf{X},\mathbf{Y}$ be as in Ex.2.1, $\mathbf{Z}=\{z_1,z_2,z_3\}$ and $Q \in \mathbf{F}(\mathbf{X} \times \mathbf{Y})$, $T \in \mathbf{F}(\mathbf{X} \times \mathbf{Z})$ be defined as

$$Q = \begin{Vmatrix} 4 & 5 & 2 \\ 7 & 1 & 3 \\ 3 & 0 & 12 \end{Vmatrix}, \qquad T = \begin{Vmatrix} 2 & 1 & 5 \\ 1 & 1 & 1 \\ 3 & 12 & 0 \end{Vmatrix},$$

$$S^* = \begin{Vmatrix} 1 & 4 & 0 \\ 2 & 1 & 1 \\ 1 & 1 & 0 \end{Vmatrix}, \qquad B^* = \begin{Vmatrix} 1 & 0 & 0 \\ 0 & 1 & 1 \\ 1 & 1 & 0 \end{Vmatrix}$$

and it is easily seen that $B^* \odot Q=T$.

## 5.2. Minimal Boolean Solutions

In this Sec. we use the results of Sec.3.2 to characterize the whole set of the Boolean solutions of the max-min fuzzy equation (3.3) assigned on finite referential sets and on a linear lattice **L** (we recall that the completeness of **L** is not necessary here) with universal bounds 0 and 1.

From Sec.5.1, we already know the greatest Boolean solution and we now characterize the minimal elements of $\mathfrak{B}$.

Following Di Nola and Ventre [1], we give the following binary operation "ω" in $\mathfrak{S}$, pointwise defined as

$$(S\omega S')\,(y,z) = \begin{cases} S'(y,z) & \text{if } S(y,z)>0,\\ 0 & \text{otherwise,}\end{cases}$$

for all $y\in \mathbf{Y}$, $z\in \mathbf{Z}$, where $S,S'\in \mathfrak{S}$.

Assume $\mathfrak{B}\neq\emptyset$ and consider the following set:

$$\mathfrak{M}^* = \{M\in\mathfrak{M}: M \le B^*\}.$$

By Thm.3.14, $\mathfrak{M}^*$ is a nonempty set and then the following Lemma holds:

LEMMA 5.2. *For any* $M\in\mathfrak{M}^*$, *the fuzzy relation* $(M\omega B^*)\in \mathbf{F}(\mathbf{Y}\times\mathbf{Z})$ *belongs to the set* $\mathfrak{B}^*$ *and further,*

$$(M_S\omega B^*) = (M_S\omega S) \tag{5.2}$$

*for any* $S\in\mathfrak{B}$ *and for any* $M_S\in\mathfrak{M}^*$ *such that* $M_S\le S$.

PROOF. It is easily seen that $M\le(M\omega B^*)\le B^*$ for any $M\in\mathfrak{M}^*$. By Lemma 2.5, then $(M\omega B^*)$ is in $\mathfrak{S}$ and it is clearly an element of $\mathfrak{B}$. Let now S be an arbitrary element of $\mathfrak{B}$.

By Thm.3.14, there exists a fuzzy relation $M_S\in\mathfrak{M}$ such that $M_S\le S\le B^*$ and thus $M_S\in\mathfrak{M}^*$. Since $M_S\le M_S\omega S\le S$, then $(M_S\omega S)$ is in $\mathfrak{S}$ by Lemma 2.5 and of course $(M_S\omega S)$ is in $\mathfrak{B}$. Hence both the fuzzy relations $(M_S\omega B^*)$ and $(M_S\omega S)$, where $M_S\le S$, are elements of $\mathfrak{B}$ and it is easy to verify the equality (5.2). ■

Then Lemma 5.2 ensures that the following set:

$$\mathfrak{B}^* = \{(M\omega B^*): M\in\mathfrak{M}^*\}$$

is a finite subset of $\mathfrak{B}$. The following theorem characterizes the minimal Boolean solutions of $\mathfrak{S}$:

THEOREM 5.3. *If* $\mathfrak{B}\neq\emptyset$, *an element* $S\in\mathfrak{B}$ *is minimal in* $\mathfrak{B}$ *iff it is minimal in* $\mathfrak{B}^*$.

PROOF. Let $(M\omega B^*)$ be a minimal element in $\mathfrak{B}^*$ with $M\in\mathcal{M}^*$. We have to prove that $M\omega B^*$ is minimal in $\mathfrak{B}$, i.e. if $S\leq(M\omega B^*)$ for some $S\in\mathfrak{B}$, then

$$S = (M\omega B^*) . \tag{5.3}$$

By Thm.3.14, there exists a fuzzy relation $M_S\in\mathcal{M}$ such that $M_S\leq S\leq B^*$ and using Lemma 5.2, we deduce:

$$(M_S\omega B^*) = (M_S\omega S) \leq S \leq (M\omega B^*).$$

Since $(M_S\omega B^*)$ belongs to $\mathfrak{B}^*$ (because $M_S\in\mathcal{M}^*$) and $(M\omega B^*)$ is minimal in $\mathfrak{B}^*$, we have that

$$(M\omega B^*) = (M_S\omega B^*) = S,$$

i.e. the equality (5.3). Vice versa, let S be minimal in $\mathfrak{B}$. It suffices to show that S belongs to the set $\mathfrak{B}^*$. Indeed, by Thm.3.14, there exists a fuzzy relation $M_S\in\mathcal{M}$ such that $M_S\leq S\leq B^*$.

By Lemma 5.2, we have:

$$(M_S\omega B^*) = (M_S\omega S) \leq S$$

and this implies that

$$(M_S\omega B^*) = S \tag{5.4}$$

since $(M_S\omega B^*)\in\mathfrak{B}$ and S is minimal in $\mathfrak{B}$. The equality (5.4) proves that S is in the set $\mathfrak{B}^*$ since $M_S\in\mathcal{M}^*$. ■

In other words, Thm.5.3 ensures that the determination of the minimal elements of $\mathfrak{B}$ leads to the determination of the minimal elements of $\mathfrak{B}^*$, as is shown in the following example:

EXAMPLE 5.2. Let $\mathbf{L}=[0,1]$ with the usual operations, $\mathbf{X}=\{x_1,x_2\}$, $\mathbf{Y}=\{y_1,y_2,y_3\}$, $\mathbf{Z}=\{z_1,z_2\}$ and $Q\in\mathbf{F}(\mathbf{X}\times\mathbf{Y})$, $T\in\mathbf{F}(\mathbf{X}\times\mathbf{Z})$ be given by

$$Q = \begin{Vmatrix} 0.8 & 0.8 & 0.8 \\ 0.9 & 0.9 & 1.0 \end{Vmatrix}, \qquad T = \begin{Vmatrix} 0.8 & 0.8 \\ 0.9 & 1.0 \end{Vmatrix},$$

$$S^* = \begin{Vmatrix} 1.0 & 1.0 \\ 1.0 & 1.0 \\ 0.9 & 1.0 \end{Vmatrix}, \qquad B^* = \begin{Vmatrix} 1.0 & 1.0 \\ 1.0 & 1.0 \\ 0.0 & 1.0 \end{Vmatrix}.$$

It is easily seen that $S^* \odot Q = B^* \odot Q = T$, hence $\mathfrak{B} \neq \emptyset$ by Thm.5.1. Using the results of Sec.3.2, it is immediately seen that the minimal elements of $\mathcal{M}$ are the following fuzzy relations:

$$M_1 = \begin{Vmatrix} 0.9 & 0.0 \\ 0.0 & 0.0 \\ 0.0 & 1.0 \end{Vmatrix}, \qquad M_2 = \begin{Vmatrix} 0.0 & 0.0 \\ 0.9 & 0.0 \\ 0.0 & 1.0 \end{Vmatrix}, \qquad M_3 = \begin{Vmatrix} 0.0 & 0.0 \\ 0.0 & 0.0 \\ 0.9 & 1.0 \end{Vmatrix}.$$

Then $\mathcal{M}^* = \{M_1, M_2\}$ and consequently $\mathfrak{B}^*$ has only two elements given by

$$S_1 = \begin{Vmatrix} 1.0 & 0.0 \\ 0.0 & 0.0 \\ 0.0 & 1.0 \end{Vmatrix}, \qquad S_2 = \begin{Vmatrix} 0.0 & 0.0 \\ 1.0 & 0.0 \\ 0.0 & 1.0 \end{Vmatrix}.$$

The results of this Sec. can be found in [2]. We conclude this Sec. by establishing a further property of the fuzzy relation $\Sigma$ defined by formula (3.6) but we first point out an obvious Lemma.

LEMMA 5.4. *For all* $S, S_1, S_2 \in \mathbf{S}$, *we have*

$$(S_1 \,\omega\, S) \vee (S_2 \,\omega\, S) = (S_1 \vee S_2) \,\omega\, S.$$

PROOF. It is immediate by the same definition of the operation "ω". ■

THEOREM 5.5. *If* $\mathfrak{B} \neq \emptyset$, *then*

$$\bigvee_{M \in \mathcal{M}} (M \,\omega\, B^*) = \Sigma \vee B^*.$$

PROOF. By Thm.3.19, $\Sigma$ is the fuzzy union of the elements $M \in \mathcal{M}$ and using Lemma 5.4, we have

$$\bigvee_{M \in \mathcal{M}} (M \,\omega\, B^*) = \Big(\bigvee_{M \in \mathcal{M}} M\Big) \,\omega\, B^* = \Sigma \,\omega\, B^*.$$ ■

## 5.3. Solutions with Maximum Boolean Degree

In this Sec. we consider Eq.(3.3) given on finite referential sets. In Sec.4.3, if **L** is a linear lattice, we already proved that the elements of $\mathbb{S}$ having maximum Boolean degree $\beta$, i.e. with the greatest number of 0 and 1, are the elements of $\mathbb{S}$ with the smallest fuzzy entropy. If $\mathbb{B}\neq\emptyset$ and **L** is a complete Brouwerian lattice, the results of Ch.4, there proved under the assumption that **L** is a linear lattice, are not applicable in order to find the elements of $\mathbb{S}$ with maximum Boolean degree. However, we solve this question by illustrating an algorithm of Di Nola and Ventre [1], but we also need the following Lemma.

LEMMA 5.6. *Let* $\mathbb{S}\neq\emptyset$. *For any* $S\in\mathbb{S}$, *we have* $(S\omega S^*)\in\mathbb{S}$ *and* $\beta(S)\leq\beta(S\omega S^*)$.

PROOF. By definition of the operator "$\omega$", we have $S\leq(S\omega S^*)\leq S^*$ for any $S\in\mathbb{S}$ and hence $(S\omega S^*)\in\mathbb{S}$ by Lemma 2.5. The second part of the thesis follows immediately observing that if $S(y,z)=1$, then $(S\omega S^*)(y,z)=S^*(y,z)=1$ and if $S(y,z)=0$, then $(S\omega S^*)(y,z)=0$. ■

Without loss of generality, we may assume $\mathbf{Z}=\{z\}$ to be singleton (cfr. Sec.3.6). We now define the following subset of $\mathbf{Y}$:

$$\mathbf{Y}^*=\{y\in\mathbf{Y}: S^*(y,z)<1\}.$$

Of course $\mathbf{Y}^*\neq\emptyset$ otherwise we should obtain $B^*=S^*$ and hence $\mathbb{B}\neq\emptyset$ by Thm.5.1, a contradiction to the hypothesis $\mathbb{B}=\emptyset$.

Let $\mathbf{K}^*$ be a subset of $\mathbf{Y}^*$ such that the fuzzy relation $S^{\mathbf{K}^*}\in\mathbf{F}(\mathbf{Y}\times\{z\})$ defined by

$$\begin{aligned} S^{\mathbf{K}^*}(y,z) &= 0 && \text{if } y\in\mathbf{K}^*,\\ &= S^*(y,z) && \text{if } y\notin\mathbf{K}^*, \end{aligned}$$

is an element of $\mathbb{S}$. Further, assume that the following property holds:

(i) for any subset $\mathbf{K}\subset\mathbf{Y}^*$ such that $|\mathbf{K}|>|\mathbf{K}^*|$, then the fuzzy relation $S^{\mathbf{K}}$, defined similarly as $S^{\mathbf{K}^*}$, does not belong to $\mathbb{S}$.

Then the following theorem holds:

THEOREM 5.7. *Let* $\mathbb{S}\neq\emptyset$. *Then* $\beta(S^{\mathbf{K}^*})\geq\beta(S)$ *for any* $S\in\mathbb{S}$.

PROOF. Assume the existence of an element $S\in\mathbb{S}$ such that $\beta(S^{\mathbf{K}^*})<\beta(S)$ and define the fuzzy relation $S^{**}\in\mathbf{F}(\mathbf{Y}\times\{z\})$ as

$$\begin{aligned} S^{**}(y,z) &= (S\omega S^*)(y,z) && \text{if } y\in\mathbf{Y}^*,\\ &= 1 && \text{if } y\notin\mathbf{Y}^*. \end{aligned}$$

Thus $(S\omega S^*) \leq S^{**} \leq S^*$ and $S^{**} \in \mathbf{S}$ by Lemmas 2.5 and 5.6. Of course, $\beta(S\omega S^*) \leq \beta(S^{**})$ and then we have by Lemma 5.6,

$$\beta(S^{\mathbf{K}^*}) < \beta(S) \leq \beta(S\omega S^*) \leq \beta(S^{**}),$$

which implies that

$$|\mathbf{K}^*| + |\mathbf{Y} - \mathbf{Y}^*| = \beta(S^{\mathbf{K}^*}) < \beta(S^{**}) = |\mathbf{K}| + |\mathbf{Y} - \mathbf{Y}^*|,$$

where $\mathbf{K}$ is a subset of $\mathbf{Y}$ such that

$$\mathbf{K} = \{y \in \mathbf{Y}: S^{**}(y,z) = 0\}.$$

Since $\mathbf{K} \subset \mathbf{Y}^*$, let $\mathbf{K}^{**} = \mathbf{Y}^* - \mathbf{K}$. For any $y \in \mathbf{K}^{**}$, we have

$$S^{**}(y,z) = (S\omega S^*)(y,z) > 0$$

and then $S^{**}(y,z) = S^*(y,z)$. Further, we have $S^{**}(y,z) = 1 = S^*(y,z)$ for any $y \in \mathbf{Y} - \mathbf{Y}^*$. Since $(\mathbf{Y} - \mathbf{Y}^*) \cup \mathbf{K}^{**} = \mathbf{Y} - \mathbf{K}$, we deduce that $S^{**} = S^{\mathbf{K}}$ is a fuzzy relation, solution of Eq.(3.3), with $|\mathbf{K}| > |\mathbf{K}^*|$, which contradicts property (i).

The proof is complete. ■

The following algorithm shows how to build such a subset $\mathbf{K}^*$; of course $\mathbf{K}^* = \emptyset$ is possible but the proposed algorithm tests this situation too.

1. Definition of the set $\mathbf{Y}^* = \{y_1', y_2', \ldots, y_r'\}$, $0 \leq r \leq m$
2. $h := r$.
3. $j(a) := a$, $a = 1, 2, \ldots, h$.
4. Definition of the set $\mathbf{K} = \{y_{j(a)}', a = 1, 2, \ldots, h\}$.
5. Definition of $S^{\mathbf{K}}$.
6. If $S^{\mathbf{K}}$ is solution of Eq.(3.3), then $\mathbf{K}^* := \mathbf{K}$. Go to 22. Otherwise go to 7.
7. If $j(h) = r$, go to 9. Otherwise go to 8.
8. $j(h) := j(h) + 1$. Go to 4.
9. If $h = 1$, go to 17. Otherwise go to 10.

10. q: =h-1.

11. If j(q)=r+q-h, go to 14. Otherwise go to 12.

12. j(q): =j(q)+1.

13. j(a): =j(q)+a-q, a=q+1,...,h. Go to 4.

14. If q=1, go to 15. Otherwise go to 16.

15. h: =h-1. Go to 3.

16. q: =q-1. Go to 11.

17. j(1): =1

18. **K**$=\{y_{j(1)}'\}$.

19. If $S^{\mathbf{K}}$ is solution of Eq.(3.3), then **K***: =**K**. Go to 22. Otherwise go to 20.

20. If j(1)=2, then **K***=Ø and S* has maximum Boolean degree. Otherwise go to 21.

21. j(1): =j(1)+1. Go to 18.

22. END.

The following example illustrates the above algorithm.

EXAMPLE 5.3. Let **L**,**X** be as in Ex.2.1, **Z** be singleton, $\mathbf{Y}=\{y_1,y_2,y_3,y_4,y_5,y_6\}$ and Q∈**F**(**X**×**Y**), T∈**F**(**X**×**Z**) given by

$$Q=\begin{Vmatrix} 3 & 6 & 8 & 5 & 4 & 2 \\ 4 & 7 & 9 & 2 & 6 & 1 \\ 2 & 5 & 3 & 9 & 1 & 4 \end{Vmatrix}, \qquad T=\begin{Vmatrix} 3 \\ 2 \\ 1 \end{Vmatrix}.$$

We have $(S^*)^{-1}=[1\ \ 2\ \ 6\ \ 3\ \ 3\ \ 6]$ and the following steps:

$\mathbf{Y}^*=\{y_1'=2,\ y_2'=6,\ y_3'=3,\ y_4'=3,\ y_5'=6\}$.

r: =5.

h: =5.

j(1): =1, j(2): =2, j(3): =3, j(4): =4, j(5): =5.

**K**={$y_{j(1)}'=2$, $y_{j(2)}'=6$, $y_{j(3)}'=3$, $y_{j(4)}'=3$, $y_{j(5)}'=6$}.

$(S^K)^{-1}=[1\ \ 0\ \ 0\ \ 0\ \ 0\ \ 0]$.

*Comment*. This fuzzy relation $S^K$ is not solution of the given equation since

$(S^K \odot Q)(x_2,z) = Q(x_2,y_1) \wedge S^K(y_1,z) = 4 \wedge 1 = 4 < 2 = T(x_2,z)$.

j(5)=5? (Yes).

h=1? (No).

q: =h-1=5-1=4.

j(4): =r+q-h=5+4-5=4? (Yes).

q=1? (No).

q: =q-1=4-1=3.

j(3): =r+q-h=5+3-5=3? (Yes).

q=1? (No).

q: =q-1=3-1=2.

j(2): =2+q-h=5+2-5=2? (Yes).

q=1? (No).

q: =q-1=2-1=1.

j(1): =5+1-5=1? (Yes).

q=1? (Yes).

h: =5-1=4.

j(1): =1, j(2): =2, j(3): =3, j(4): =4.

$\mathbf{K}=\{y_{j(1)}'=2,\ y_{j(2)}'=6,\ y_{j(3)}'=3,\ y_{j(4)}'=3\}$.

$(S^{\mathbf{K}})^{-1}=[1\quad 0\quad 0\quad 0\quad 0\quad 6]$.

*Comment.* This fuzzy relation is not solution of the given equation since

$$(S^{\mathbf{K}}\odot Q)\,(x_3,z) = [Q(x_3,y_1)\wedge S^{\mathbf{K}}(y_1,z)] \vee [Q(x_3,y_6)\wedge S^{\mathbf{K}}(y_6,z)]$$

$$= (2\wedge 1)\vee(4\wedge 6) = 2\vee 12 = 2 < 1 = T(x_3,z).$$

$j(4)=5$? (NO).

$j(4):=j(4)+1=4+1=5$.

$\mathbf{K}=\{y_{j(1)}'=2,\ y_{j(2)}'=6,\ y_{j(3)}'=3,\ y_{j(4)}'=y_{j(5)}'=6\}$.

$(S^{\mathbf{K}})^{-1}=[1\quad 0\quad 0\quad 0\quad 3\quad 0]$.

*Comment.* This fuzzy relation is a solution of the given equation since

$(S^{\mathbf{K}}\odot Q)\,(x_1,z) = (3\wedge 1)\vee(4\wedge 3) = 3\vee 12 = 3 = T(x_1,z)$,

$(S^{\mathbf{K}}\odot Q)\,(x_2,z) = (4\wedge 1)\vee(6\wedge 3) = 4\vee 6 = 2 = T(x_2,z)$,

$(S^{\mathbf{K}}\odot Q)\,(x_3,z) = (2\wedge 1)\vee(1\wedge 3) = 2\vee 3 = 1 = T(x_3,z)$.

$\mathbf{K}^*:=\mathbf{K}$.

$S^{\mathbf{K}^*}:=S^{\mathbf{K}}$.

END.

As pointed out in [1], the problem of the minimization of the fuzzy entropy measure $\Phi$ of the solutions of Eq.(3.3) in the context of complete Brouwerian lattices (which is still an open question, cfr.Sec.4.5) does not coincide with the problem of the maximization of the Boolean degree of the solutions S of Eq.(3.3). Certainly, maximizing the "Booleanity" of these solutions on S, we reduce the amount of their entropy $\Phi$, but generally $\Phi$ does not achieve the smallest value since its amount depends also on the non-Boolean entries of S.

## References

[1] A. Di Nola and A. Ventre, On Booleanity of relational equation in Brouwerian lattices, *Boll. Un. Mat. Ital.* (6) **3** - B (1984), 871-882.

[2] S. Sessa, Characterizing the Boolean solutions of relational equations in Brouwerian lattices, *Boll. Un. Mat. Ital.* (6) **5** - B (1986), 39-49.

# CHAPTER 6

## α–FUZZY RELATION EQUATIONS AND DECOMPOSABLE FUZZY RELATIONS




*In this Ch., we investigate a new type of fuzzy relation equation defined on finite sets and with membership functions on a complete Brouwerian lattice. We characterize the entire set of the solutions when the fuzzy equation is assigned on a linear lattice. We also study when a fuzzy relation is pointwise decomposable in the intersection (resp. union) of two fuzzy sets and we show that such relations are max-min transitive (resp. compact). Finally, properties of convergence of powers of these fuzzy relations are established.*


### 6.1. α-Fuzzy Relation Equations on Complete Brouwerian Lattices

Let **X**, **Y** be finite sets, A∈**F**(**X**) and B∈**F**(**Y**) two fuzzy sets assuming values in a complete Brouwerian lattice (**L**,∧,∨, ≤) with universal bounds 0 and 1. In this Ch., we study the following equation:

$$B = R \,\alpha\, A, \qquad (6.1)$$

defined as

$$B(y) = \bigwedge_{x\in \mathbf{X}} [A(x)\,\alpha\, R(x,y)]$$

for any y∈**Y**, where R∈**F**(**X**×**Y**) and "α" is the operator defined in Sec.2.1. This equation was introduced in [4] and its study is motivated by applications which will be presented in Ch.14. Eq.(6.1) is defined as an α-fuzzy relation equation and assuming A and B known, we denote by 𝔑'=𝔑'(A,B) the set of the solutions R∈**F**(**X**×**Y**) which satisfy it. Here we enunciate only some basic results whose proofs are omitted since they are similar to those already established for max-min equations in Chs. 2 and 3.

It is not difficult, following the lines of the proof of the basic Thm.2.3 and using properties (i), (ii), (iii) of the "$\alpha$" operator of Sec.2.1, to prove that [4]:

THEOREM 6.1. $\mathfrak{R}'\neq\emptyset$ *iff the fuzzy relation* $(A\times B)\in \mathbf{F}(\mathbf{X}\times\mathbf{Y})$, *defined by* $(A\times B)(x,y)=A(x)\wedge B(y)$ *for any* $x\in \mathbf{X}$, $y\in \mathbf{Y}$, *is an element of* $\mathfrak{R}'$. *Further,* $(A\times B)\leq R$ *for any* $R\in\mathfrak{R}'$.

If $R\in \mathbf{F}(\mathbf{X}\times\mathbf{Y})$ is such that $R_1\leq R\leq R_2$ with $R_1,R_2\in\mathfrak{R}'$, then $R\in\mathfrak{R}'$ (cfr. Lemma 2.5) since the property (ii) of the "$\alpha$" operator gives

$$B = R_1 \,\alpha\, A \leq R \,\alpha\, A \leq R_2 \,\alpha\, A = B.$$

This remark implies that $\mathfrak{R}'$ is a meet semilattice with respect to the fuzzy minimum $\wedge$ (cfr. Thm.2.6) and on the other hand it is easily seen with examples that $R_1\vee R_2$ in general is not an element of $\mathfrak{R}'$ if $R_1,R_2\in\mathfrak{R}'$ (cfr. Ex.2.3).

## 6.2. Related Results on Linear Lattices

In this Sec. we assume $\mathbf{L}$ to be a linear lattice (not necessarily complete). Analogously to Sec.3.1, we define the following set:

$$\mathbf{G}'(y) = \{x\in \mathbf{X}: A(x) \,\alpha\, B(y) = B(y)\}$$

for any $y\in \mathbf{Y}$. It is easily seen (cfr. Thms.3.1 and 3.2) that

THEOREM 6.2. $\mathfrak{R}'\neq\emptyset$ *iff* $\mathbf{G}'(y)\neq\emptyset$ *for any* $y\in\mathbf{Y}$.

In virtue of this theorem, then, assuming $\mathfrak{R}'\neq\emptyset$, we can choose an element $x_y\in \mathbf{G}'(y)$, for any $y\in\mathbf{Y}$, such that

$$A(x_y) \,\alpha\, B(y) = B(y)$$

and hence it is possible to define the following fuzzy relation $M'\in \mathbf{F}(\mathbf{X}\times\mathbf{Y})$ as

$$\begin{aligned} M'(x,y) &= 1 && \text{if } y\in \mathbf{B}^{-1}(1),\\ &= 1 && \text{if } y\notin \mathbf{B}^{-1}(1) \text{ and } x\neq x_y,\\ &= B(y) && \text{if } y\notin \mathbf{B}^{-1}(1) \text{ and } x=x_y, \end{aligned} \tag{6.2}$$

where $\mathbf{B}^{-1}(1)=\{y\in \mathbf{Y}: B(y)=1\}$. Then we have (cfr. Thm.3.6 and 3.7, respectively):

THEOREM 6.3. *The fuzzy relation* M', *defined from* (6.2), *is a maximal element of the set* $\mathfrak{R}'$.

THEOREM 6.4. *If* $\mathfrak{R}'\neq\emptyset$, *the fuzzy intersection of the maximal elements of* $\mathfrak{R}'$ *is* $(A\alpha B)\in\mathfrak{R}'$ *(cfr. Def.2.6).*

In other terms, we can also say that, in order to write the fuzzy relations M', defined from formula (6.2), it suffices to keep an element B(y) (<1) in any column of (AαB) for which $y\notin \mathbf{B}^{-1}(1)$ and to put 1 everywhere in the same column. If $y\in\mathbf{B}^{-1}(1)$, then it suffices to assume 1 along the column under consideration.

Concluding this Sec., we point out that in general the results here established in the context of linear lattices are not extendable to the case in which Eq.(6.1) is defined on complete Brouwerian lattices. For instance, Thm.6.2 does not hold in this case as is proved in the following example:

EXAMPLE 6.1. Let **L** be the lattice of Ex.2.1, $\mathbf{X}=\{x_1,x_2\}$, $\mathbf{Y}=\{y_1,y_2\}$ and $A\in\mathbf{F}(\mathbf{X})$, $B\in\mathbf{F}(\mathbf{Y})$ be defined as $A(x_1)=6$, $A(x_2)=10$, $B(y_1)=15$, $B(y_2)=12$. We have:

$$(A\times B)=\begin{Vmatrix} 30 & 12 \\ 30 & 60 \end{Vmatrix}.$$

Since $(A\times B)\alpha A=B$, then $\mathfrak{R}'\neq\emptyset$ by Thm.6.1. Note that $\mathbf{G}'(y_1)=\mathbf{G}'(y_2)=\emptyset$ since

$$A(x_1)\ \alpha\ B(y_1) = 6\ \alpha\ 15 = 5 \neq 15 = B(y_1),$$

$$A(x_2)\ \alpha\ B(y_1) = 10\ \alpha\ 15 = 3 \neq 15 = B(y_1),$$

$$A(x_1)\ \alpha\ B(y_2) = 6\ \alpha\ 12 = 4 \neq 12 = B(y_2),$$

$$A(x_2)\ \alpha\ B(y_2) = 10\ \alpha\ 12 = 6 \neq 12 = B(y_2).$$

### 6.3. Composite α-Fuzzy Relation Equations

Let **X**, **Y**, **Z** be three nonempty finite sets and $Q\in\mathbf{F}(\mathbf{X}\times\mathbf{Y})$, $S\in\mathbf{F}(\mathbf{Y}\times\mathbf{Z})$, $T\in\mathbf{F}(\mathbf{X}\times\mathbf{Z})$ be three fuzzy relations such that the following extended form of Eq.(6.1) holds:

$$T = S\ \alpha\ Q, \tag{6.3}$$

i.e.

$$T(x,z) = \bigwedge_{y\in\mathbf{Y}} [Q(x,y)\ \alpha\ S(y,z)]$$

for all $x \in \mathbf{X}$, $y \in \mathbf{Y}$, $z \in \mathbf{Z}$. If Q and T are given and S is unknown, we denote by $\mathbf{S}'=\mathbf{S}'(Q,T)$ the set of all the solutions $S \in \mathbf{F}(\mathbf{Y}\times\mathbf{Z})$ satisfying Eq.(6.3). As in the basic Thm.2.3, it is easily seen that

THEOREM 6.5. $\mathbf{S}' \neq \emptyset$ *iff the fuzzy relation* $(T \odot Q^{-1}) \in \mathbf{S}'$. *Further,* $(T \odot Q^{-1}) \leq S$ *for any* $S \in \mathbf{S}'$.

Note that this theorem holds also in complete Brouwerian lattices.

If $\mathbf{L}$ is a linear lattice, as in the results of Sec.3.3, it is possible to give an analogous algorithm for the calculation of the maximal elements of the set $\mathbf{S}'$.

## 6.4. Fuzzy Relations Decomposable into Two Fuzzy Sets

Let $\mathbf{X}$ and $\mathbf{Y}$ be two nonempty sets (not necessarily finite) and $(\mathbf{L},\wedge,\vee,\leq)$ be a complete lattice.

Here we consider the following problems solved in [5]:

*Given a fuzzy relation $R \in F(X \times Y)$, determine two fuzzy sets $A \in F(X)$ and $B \in F(Y)$ such that for all $x \in X$, $y \in Y$:*

$$R(x,y) = A(x) \wedge B(y), \tag{6.4}$$

$$R(x,y) = A(x) \vee B(y). \tag{6.5}$$

Since $\mathbf{L}$ is complete, we can define the following fuzzy sets:

$$A^{\cup}(x) = \bigvee_{y \in \mathbf{Y}} R(x,y) \qquad A^{\cap}(x) = \bigwedge_{y \in \mathbf{Y}} R(x,y)$$

$$B^{\cup}(x) = \bigvee_{x \in \mathbf{X}} R(x,y) \qquad B^{\cap}(x) = \bigwedge_{x \in \mathbf{X}} R(x,y)$$

for all $x \in \mathbf{X}$, $y \in \mathbf{Y}$. The following theorem holds:

THEOREM 6.6. *A fuzzy relation* $R \in \mathbf{F}(\mathbf{X}\times\mathbf{Y})$ *is decomposable in accordance with* (6.4) *iff* $A^{\cup} \times B^{\cup} = R$. *Further,* $A^{\cup}$ *and* $B^{\cup}$ *are the smallest fuzzy sets satisfying* (6.4).

PROOF. We show only the non-trivial implication. We first observe that

$$A^{\cup}(x) \wedge B^{\cup}(y) = [\bigvee_{y \in \mathbf{Y}} R(x,y)] \wedge [\bigvee_{x \in \mathbf{X}} R(x,y)] \geq R(x,y),$$

i.e.

$$A^{\cup}(x) \wedge B^{\cup}(y) \geq R(x,y) \tag{6.6}$$

for all $x \in \mathbf{X}$, $y \in \mathbf{Y}$. If there exist two fuzzy sets $A \in \mathbf{F}(\mathbf{X})$ and $B \in \mathbf{F}(\mathbf{Y})$ satisfying (6.4), then we have $B(y) \geq A(x) \wedge B(y) = R(x,y)$ for any $x \in \mathbf{X}$ and $A(x) \geq A(x) \wedge B(y) = R(x,y)$ for any $y \in \mathbf{Y}$.

This implies that

$$B(y) \geq \bigvee_{x \in \mathbf{X}} R(x,y) = B^{\cup}(y) \quad \text{and} \quad A(x) \geq \bigvee_{y \in \mathbf{Y}} R(x,y) = A^{\cup}(x)$$

for all $x \in \mathbf{X}$, $y \in \mathbf{Y}$. From (6.6), it follows that

$$R(x,y) = A(x) \wedge B(y) \geq A^{\cup}(x) \wedge B^{\cup}(y) \geq R(x,y)$$

for all $x \in \mathbf{X}$, $y \in \mathbf{Y}$, i.e. $A^{\cup} \times B^{\cup} = R$. ■

We illustrate Thm.6.6 with the following example of [5]:

EXAMPLE 6.2. Let $\mathbf{L}$ be the complete lattice of Ex.2.1, $\mathbf{X} = \{x_1, x_2, x_3\}$, $\mathbf{Y} = \{y_1, y_2, y_3\}$ and $A \in \mathbf{F}(\mathbf{X})$, $B \in \mathbf{F}(\mathbf{Y})$ defined as $A(x_1)=3$, $A(x_2)=7$, $A(x_3)=8$, $B(y_1)=2$, $B(y_2)=14$, $B(y_3)=12$. We have:

$$R = A \times B = \begin{Vmatrix} 6 & 42 & 12 \\ 14 & 14 & 84 \\ 8 & 56 & 24 \end{Vmatrix}.$$

Since $A^{\cup}(x_1)=6$, $A^{\cup}(x_2)=14$, $A^{\cup}(x_3)=8$ and $B^{\cup}=B$, it is easily seen that $A^{\cup} \times B^{\cup} = R$ and $A^{\cup} \leq A$.

If $\mathbf{L}$ is a linear lattice and $\mathbf{X}$ and $\mathbf{Y}$ are finite, we can say that a fuzzy relation $R \in \mathbf{F}(\mathbf{X} \times \mathbf{Y})$ is decomposable in accordance with (6.4) iff $R(x,y)$ is the maximum either of the x-th row or of the y-th column of R, since $R(x,y)$ is an element of the set $\{A^{\cup}(x), B^{\cup}(y)\}$ for all $x \in \mathbf{X}$, $y \in \mathbf{Y}$.

EXAMPLE 6.3. Let $\mathbf{L} = ([0,1], \wedge, \vee, \leq)$, $\mathbf{X} = \{x_1, x_2, x_3\}$, $\mathbf{Y} = \{y_1, y_2, y_3\}$ and $R \in \mathbf{F}(\mathbf{X} \times \mathbf{Y})$ given by

$$R = \begin{Vmatrix} 0.4 & 0.3 & 0.1 \\ 0.8 & 0.2 & 0.7 \\ 0.6 & 0.5 & 0.4 \end{Vmatrix}.$$

Then R is not decomposable in accordance with (6.4) since

$$R(x_1,y_3) = 0.1 < 0.4 = 0.4 \vee 0.3 \vee 0.1 = \bigvee_{j=1}^{3} R(x_1,y_j) = A^{\cup}(x_1),$$

and

$$R(x_1,y_3) = 0.1 < 0.7 = 0.1 \vee 0.7 \vee 0.4 = \bigvee_{i=1}^{3} R(x_i,y_3) = B^{\cup}(y_3).$$

Now we would like to point out a simple result, due to Kolodziejczyk [17].

THEOREM 6.7. *Let* **L** *be a linear lattice and* **X**, **Y** *be finite sets. If* $R \in \mathbf{F}(\mathbf{X}\times\mathbf{Y})$ *is decomposable in accordance with* (6.4), *then there exists a line of* R *with all the entries equal to* $\wedge\{R(x,y): x\in\mathbf{X}, y\in\mathbf{Y}\}$.

PROOF. Since **L** is a linear lattice and **X**, **Y** are finite, we have:

$$A^{\cup}(x_0) = \bigwedge_{x\in\mathbf{X}} A^{\cup}(x) \qquad \text{and} \qquad B^{\cup}(y_0) = \bigwedge_{y\in\mathbf{Y}} B^{\cup}(y)$$

for some $x_0\in\mathbf{X}$, $y_0\in\mathbf{Y}$. Without loss of generality, we can assume that $A^{\cup}(x_0)\le B^{\cup}(y_0)$. Since R is decomposable in accordance with (6.4), we have, by Thm.6.6, that

$$R(x_0,y) = A^{\cup}(x_0) \wedge B^{\cup}(y) = A^{\cup}(x_0) = A^{\cup}(x_0) \wedge B^{\cup}(y_0) =$$

$$\Big(\bigwedge_{x\in\mathbf{X}} A^{\cup}(x)\Big) \wedge \Big(\bigwedge_{y\in\mathbf{Y}} B^{\cup}(y)\Big) = \bigwedge_{\substack{x\in\mathbf{X}\\ y\in\mathbf{Y}}} [A^{\cup}(x) \wedge B^{\cup}(y)] = \bigwedge_{\substack{x\in\mathbf{X}\\ y\in\mathbf{Y}}} R(x,y)$$

for any $y\in\mathbf{Y}$. ■

Dually one can prove the following theorem in a complete lattice.

THEOREM 6.8. *A fuzzy relation* $R\in\mathbf{F}(\mathbf{X}\times\mathbf{Y})$ *is decomposable in accordance with* (6.5) *iff* $A^{\cap}(x) \vee B^{\cap}(y) = R(x,y)$ *for all* $x\in\mathbf{X}$, $y\in\mathbf{Y}$. *Further,* $A^{\cap}$ *and* $B^{\cap}$ *are the greatest fuzzy sets satisfying* (6.5).

### 6.5. Decomposable Fuzzy Relations and the Transitivity Property

From now on, we assume **L**=[0,1] with the usual lattice operations "$\wedge$", "$\vee$" and we study fuzzy relations $R \in \mathbf{F}(\mathbf{X}\times\mathbf{X})$ decomposable in accordance with (6.4) and (6.5), **X** being a finite set and $A,B \in \mathbf{F}(\mathbf{X})$.

We first consider fuzzy relations $R \in \mathbf{F}(\mathbf{X}\times\mathbf{X})$ such that $R=A\times B$ since they are max-min transitive (cfr. Thm.6.9). The transitive relations are important in clustering, information retrieval, preference and fuzzy orderings ([25], [29], [31] ÷ [33], [35], [36]). Their properties and the correlated mentioned topics have been investigated mainly by Chakraborty and Das [2], [3], Hashimoto [6]÷[15], Kandel [cfr. book cited in Sec.1.2], Kolodziejczyk [16]÷[21], Ovchinnikov [26], [27], Ovchinnikov and Ozernoy [28], Roubens and Vincke [30].

We recall the following definition of Zadeh [36]:

DEFINITION 6.1. *A fuzzy relation* $R \in \mathbf{F}(\mathbf{X}\times\mathbf{X})$ *is max-*$\blacklozenge$ *transitive if* $R(x,y) \geq [R(x,z) \blacklozenge R(z,y)]$ *for any* $z \in \mathbf{X}$, *where* "$\blacklozenge$" *is a binary operation on* [0,1].

If "$\blacklozenge$" is a triangular norm (cfr. Ch.8), then Def.6.1 reduces to that of Ovchinnikov [27]. Note also that a conorm (cfr. Ch.8) can be taken for "$\blacklozenge$". However, in accordance with Kolodziejczyk [17] and using the symbology of Bezdek and Harris [1] (except for "$\Delta$"), one can use one of the following operations:

$$a \oplus b = a + b - ab,$$

$$a \mathbin{\triangle} b = \max\{0, a + b - 1\},$$

$$a \bullet b = ab,$$

$$a \,\square\, b = (a + b)/2,$$

where $a,b \in [0,1]$. We will use different notations for some of these operations in Ch.8.

Let $\mathfrak{R}_\blacklozenge$ be the family of the max-$\blacklozenge$ transitive fuzzy relations $R \in \mathbf{F}(\mathbf{X}\times\mathbf{X})$. It is easily seen that

$$a \mathbin{\triangle} b \leq ab \leq a \wedge b \leq (a+b)/2 \leq a \vee b \leq a + b - ab$$

for any $a,b \in [0,1]$, so that the following hierarchy of inclusions holds [1]:

$$\mathfrak{R}_\oplus \subseteq \mathfrak{R}_\vee \subseteq \mathfrak{R}_\square \subseteq \mathfrak{R}_\wedge \subseteq \mathfrak{R}_\bullet \subseteq \mathfrak{R}_\triangle.$$

We have the following result [17]:

LEMMA 6.8. $R \in \mathfrak{R}_{\square}$ *iff* $R(x,y)=$ *const. for all* $x,y \in \mathbf{X}$.

PROOF. Of course, we prove only the non-trivial implication. We must show that $R(x,y)=R(w,z)$ for all $x,y,w,z \in \mathbf{X}$ if $R \in \mathfrak{R}_{\square}$.

Since R is max-$\square$ transitive, we have $R(x,y) \geq [R(x,y)+R[y,y)]/2$ and hence $R(x,y) \geq R(y,y)$. Similarly, $R(y,x) \geq R(y,y)$ and then $R(y,y) \geq [R(y,x)+R(x,y)]/2 \geq [R(y,y)+R(x,y)]/2$, i.e. $R(y,y) \geq R(x,y)$.

Thus $R(x,y)=R(y,y)$ and one can show analogously that $R(x,y)=R(x,x)$ for all $x,y \in \mathbf{X}$. Therefore $R(x,y)=R(y,y)=R(w,y)=R(w,w)=R(w,z)$ for all $x,y,w,z \in \mathbf{X}$. ■

Note that $\mathfrak{R}_{\vee}=\mathfrak{R}_{\square}$ and $\mathfrak{R}_{\oplus}=\{R_1,R_2\}$ where $R_1(x,y)=0$ and $R_2(x,y)=1$ for all $x,y \in \mathbf{X}$. Further, $R \in \mathfrak{R}_{\wedge}$ iff $R^2 \leq R$.

We denote by $\mathfrak{R}_D$ the family of all the fuzzy relations $R \in \mathbf{F}(\mathbf{X} \times \mathbf{X})$ decomposable in accordance with (6.4).

Then we have [16]:

THEOREM 6.9. $\mathfrak{R}_{\square} \subset \mathfrak{R}_D \subset \mathfrak{R}_{\wedge}$.

PROOF. The first inclusion follows from Lemma 6.8 while the second inclusion follows from the simple fact that if $R \in \mathfrak{R}_D$, i.e. $R=A \times B$ with $A,B \in \mathbf{F}(\mathbf{X})$, we have

$$R(x,y) = A^{\cup}(x) \wedge B^{\cup}(y) \geq A^{\cup}(x) \wedge B^{\cup}(z) \wedge A^{\cup}(z) \wedge B^{\cup}(y)$$

$$= R(x,z) \wedge R(z,y)$$

for all $x,y,z \in \mathbf{X}$. Easy examples prove that the above inclusions are strict. ■

Now we take into account the set $\mathfrak{R}_d$ of the fuzzy relations $R \in \mathbf{F}(\mathbf{X} \times \mathbf{X})$ decomposable in accordance with (6.5). Of course, since $\mathbf{L}$ is a linear lattice, $R \in \mathfrak{R}_d$ iff $R(x,y)$ is the minimum either of the x-th row or of the y-th column of R.

We recall the following well-known definition [21]:

DEFINITION 6.2. *A fuzzy relation* $R \in \mathbf{F}(\mathbf{X} \times \mathbf{X})$ *is compact if for all* $x,y \in \mathbf{X}$, *there exists a* $z \in \mathbf{X}$ *such that* $R(x,y) \leq R(x,z) \wedge R(z,y)$.

Let $\mathfrak{R}_c$ be the set of such fuzzy relations. Then the following result holds [21]:

THEOREM 6.10. $\mathfrak{R}_{\square} \subset \mathfrak{R}_d \subset \mathfrak{R}_c$.

PROOF. Lemma 6.8 implies the first inclusion. Now let $R \in \mathfrak{R}_d$ and assume that $R \notin \mathfrak{R}_c$. This means that $R(x,y) > R(x,z) \wedge R(z,y)$ for any $z \in \mathbf{X}$. In particular, $R(x,y) > R(x,y) \wedge R(y,y)$ and $R(x,y) > R(x,x) \wedge R(x,y)$, i.e. $R(x,y) > R(y,y)$ and $R(x,y) >$

R(x,x). This is a contradiction since one of the following cases holds:

$$R(x,y) = \bigwedge_{z \in \mathbf{X}} R(z,y) \qquad \text{and} \qquad R(x,y) = \bigwedge_{z \in \mathbf{X}} R(x,z).$$

This proves the second inclusion and it is easily seen with examples that these inclusions are strict. ■

The set $\mathfrak{R}_{\wedge}$ is furtherly enlarged by Kolodziejczyk [16] who introduced the notion of strong transitivity as a property of the fuzzy preference relations used for solving some decision problems, thus extending the theory of Orlovsky ([23], [24]). We give the following definition drawn from [16]:

DEFINITION 6.3. *A fuzzy relation* $R \in \mathbf{F}(\mathbf{X} \times \mathbf{X})$ *is strongly transitive if for all* $x,y,z \in \mathbf{X}$ *with* $x \neq y$, $y \neq z$, $x \neq z$ *such that* $R(x,y) > R(y,x)$ *and* $R(y,z) > R(z,y)$, *we have that* $R(x,z) > R(z,x)$.

Let $\mathfrak{R}_{\Delta}$ be the set of the strongly transitive fuzzy relations $R \in \mathbf{F}(\mathbf{X} \times \mathbf{X})$. Then the following result holds [16]:

THEOREM 6.11. $\mathfrak{R}_{\wedge} \subset \mathfrak{R}_{\Delta}$.

PROOF. Let $R \in \mathfrak{R}_{\wedge}$ be such that $R(x,y) > R(y,x)$ and $R(y,z) > R(z,y)$ with $x \neq y$, $y \neq z$, $x \neq z$. Assume that $R(x,z) \leq R(z,x)$.

Then $R(x,z) \geq R(x,y) \wedge R(y,z) > R(y,x) \wedge R(z,y) \geq R(y,z) \wedge R(z,x) \wedge R(z,y) = R(z,x) \wedge R(z,y)$, which implies $R(z,x) > R(z,y)$ otherwise from $R(z,x) \leq R(z,y)$, we would have $R(x,z) > R(z,x)$, a contradiction.

On the other hand, $R(x,y) > R(y,x) \geq R(y,z) \wedge R(z,x) \geq R(z,y) \wedge R(z,x) = R(z,y)$. Thus $R(z,x) \wedge R(x,y) > R(z,y)$, a contradiction because $R \in \mathfrak{R}_{\wedge}$. Easy examples prove that the inclusion is strict. ■

Thus the class of strongly transitive fuzzy relations is essentially larger than the class of max-min transitive fuzzy relations, as is shown in Ex.3.1 of [18]. A more practical characterization of strong transitivity is given in [18], where a simple method is developed to determine whether a given fuzzy matrix is strongly transitive or not.

By Thm.6.9, we have: $\mathfrak{R}_{\square} \subset \mathfrak{R}_{D} \subset \mathfrak{R}_{\wedge} \subset \mathfrak{R}_{\Delta}$.

## 6.6. Convergence of Powers of Decomposable Fuzzy Relations

As pointed out by Kolodziejczyk [21], the convergence of powers of fuzzy relations was used in connection with graph-theoretic properties of fuzzy relations and for the convergence of systems represented by fuzzy matrices [34].

The powers of a fuzzy relation $R\in \mathbf{F}(\mathbf{X}\times\mathbf{X})$ are defined recursively as $R^1=R$, $R^{k+1}=R^k \odot R$ for $k=1,2,3,\ldots$, where "$\odot$" stands for the max-min composition.

DEFINITION 6.4. *A fuzzy relation* $R\in \mathbf{F}(\mathbf{X}\times\mathbf{X})$ *is called convergent if* $R^{k+1}=R^k$ *for some integer* k.

Thomason [34] pointed out that powers of fuzzy relations are convergent or oscillate with a finite period. In the literature, it is known that if $R\in \mathfrak{R}_c$, then $R^{n-1}=R^n$ [34], if $R\in \mathfrak{R}_\wedge$, then $R^n=R^{n+1}$ [8], if $R\in \mathfrak{R}_\Delta$, then $R^k$ oscillates with the smallest period equal to 2 [19]. See also [22] for further results.

We establish some results of [21] about the convergence of powers of fuzzy relations $R\in \mathfrak{R}_D$ and $R\in \mathfrak{R}_d$.

The following results hold:

THEOREM 6.12. *If* $R\in \mathfrak{R}_D$ *(resp.* $\mathfrak{R}_d$*), then* $R^2=R^3$.

PROOF. Let $R\in \mathfrak{R}_D$. Then

$$R^3(x,y) = (R^2 \odot R)(x,y) = \bigvee_{w\in \mathbf{X}} [R(x,w) \wedge R^2(w,y)]$$

$$= \bigvee_{w\in \mathbf{X}} \{R(x,w) \wedge \{ \bigvee_{z\in \mathbf{X}} [R(w,z) \wedge R(z,y)]\}\}$$

$$= \bigvee_{w\in \mathbf{X}} \{[A^\cup(x) \wedge B^\cup(w)] \wedge \{ \bigvee_{z\in \mathbf{X}} [A^\cup(w) \wedge B^\cup(z) \wedge A^\cup(z) \wedge B^\cup(y)]\}\}$$

$$= \bigvee_{w\in \mathbf{X}} \{[A^\cup(x) \wedge B^\cup(y)] \wedge [A^\cup(w) \wedge B^\cup(w)] \wedge \{ \bigvee_{z\in \mathbf{X}} [A^\cup(z) \wedge B^\cup(z)]\}\}$$

$$= \bigvee_{w\in \mathbf{X}} \{[A^\cup(x) \wedge B^\cup(y)] \wedge [A^\cup(w) \wedge B^\cup(w)] \}$$

$$= \bigvee_{w\in \mathbf{X}} \{[A^\cup(x) \wedge B^\cup(w)] \wedge [A^\cup(w) \wedge B^\cup(y)] \}$$

$$= \bigvee_{w\in \mathbf{X}} [R(x,w) \wedge R(w,y)] = R^2(x,y)$$

for all $x,y\in \mathbf{X}$, i.e. $R^2=R^3$.

Now let $R\in \mathfrak{R}_d$. Then

$$R^2(x,y) = \bigvee_{z\in \mathbf{X}} [R(x,z) \wedge R(z,y)] = \bigvee_{z\in \mathbf{X}} \{[A^\frown(x) \vee B^\frown(z)] \wedge [A^\frown(z) \vee B^\frown(y)]\}$$

$$= \{\bigvee_{z\in \mathbf{X}} [A^\frown(x) \wedge A^\frown(z)]\} \vee \{\bigvee_{z\in \mathbf{X}} [A^\frown(x) \wedge B^\frown(y)]\}$$

$$\vee \{\bigvee_{z\in \mathbf{X}} [A^\frown(z) \wedge B^\frown(z)]\} \vee \{\bigvee_{z\in \mathbf{X}} [B^\frown(z) \wedge B^\frown(y)]\}$$

$$= A^\frown(x) \vee B^\frown(y) \vee C$$

for all $x,y\in \mathbf{X}$, where

$$C = \bigvee_{z\in \mathbf{X}} [A^\frown(z) \wedge B^\frown(z)]. \tag{6.7}$$

On the other hand, we have:

$$R^3(x,y) = (R^2 \odot R)(x,y) = \bigvee_{z\in \mathbf{X}} [R(x,z) \wedge R^2(z,y)]$$

$$= \bigvee_{z\in \mathbf{X}} \{[A^\frown(x) \vee B^\frown(z)] \wedge [A^\frown(z) \vee B^\frown(y) \vee C]\}$$

$$= \{\bigvee_{z\in \mathbf{X}} [A^\frown(x) \wedge A^\frown(z)]\} \vee \{\bigvee_{z\in \mathbf{X}} [A^\frown(x) \wedge B^\frown(y)]\} \vee \{\bigvee_{z\in \mathbf{X}} [A^\frown(x) \wedge C]\}$$

$$\vee \{\bigvee_{z\in \mathbf{X}} [A^\frown(z) \wedge B^\frown(z)]\} \vee \{\bigvee_{z\in \mathbf{X}} [B^\frown(y) \wedge B^\frown(z)]\} \vee \{\bigvee_{z\in \mathbf{X}} [B^\frown(z) \wedge C]\}$$

$$= A^\frown(x) \vee B^\frown(y) \vee C$$

for all $x,y\in \mathbf{X}$, i.e. $R^2=R^3$. ■

THEOREM 6.13. *If* $R\in \mathfrak{R}_d$, *then* $R^2\in \mathfrak{R}_d$.

PROOF. From the proof of Thm.6.12, we have:

$$R^2(x,y) = A^\frown(x) \vee B^\frown(y) \vee C = [A^\frown(x) \vee C] \vee [B^\frown(y) \vee C],$$

where C is given by formula (6.7). The thesis is proved if we show that

$$A^{\frown}(x) \vee C = \bigwedge_{z \in \mathbf{X}} R^2(x,z) \qquad (6.8)$$

and

$$B^{\frown}(y) \vee C = \bigwedge_{z \in \mathbf{X}} R^2(y,z). \qquad (6.9)$$

We show the equality (6.8) since the proof of the equality (6.9) is analogous. We have:

$$C \geq \bigwedge_{y \in \mathbf{X}} B^{\frown}(y), \qquad (6.10)$$

otherwise we would have:

$$\bigwedge_{y \in \mathbf{X}} B^{\frown}(y) > A^{\frown}(z) \wedge B^{\frown}(z) = A^{\frown}(z)$$

for any $z \in \mathbf{X}$ and $R(x,y)=A^{\frown}(x) \vee B^{\frown}(y)=B^{\frown}(y)$ for all $x,y \in \mathbf{X}$.

Thus

$$A^{\frown}(x) = \bigwedge_{y \in \mathbf{X}} R(x,y) = \bigwedge_{y \in \mathbf{X}} B^{\frown}(y)$$

for any $x \in \mathbf{X}$ and this would imply:

$$C = \bigvee_{z \in \mathbf{X}} [A^{\frown}(z) \wedge B^{\frown}(z)] = \bigvee_{z \in \mathbf{X}} \{[\bigwedge_{y \in \mathbf{X}} B^{\frown}(y)] \wedge B^{\frown}(z)\}$$

$$= \bigvee_{z \in \mathbf{X}} [\bigwedge_{y \in \mathbf{X}} B^{\frown}(y)] = \bigwedge_{y \in \mathbf{X}} B^{\frown}(y) > C,$$

a contradiction. By (6.10), it follows that

$$\bigwedge_{z \in \mathbf{X}} R^2(x,z) = \bigwedge_{z \in \mathbf{X}} \{[A^{\frown}(x) \vee C] \vee [B^{\frown}(z) \vee C]\}$$

$$= [A^{\frown}(x) \vee C] \vee \{[\bigwedge_{z \in \mathbf{X}} B^{\frown}(z)] \vee C\}$$

$$= A^{\frown}(x) \vee C \vee C = A^{\frown}(x) \vee C.$$

This completes the proof. ■

In other words, this theorem ensures that being decomposable is a hereditary property with respect to the powers of a fuzzy relation $R \in \Re_d$. The same conclusion holds for a fuzzy relation $R \in \Re_D$, but we first prove the following result [21].

THEOREM 6.14. *If* $S \in \Re_D$, *then* $(R \odot S) \in \Re_D$ *and* $(S \odot R) \in \Re_D$ *for any* $R \in \mathbf{F}(\mathbf{X} \times \mathbf{X})$.

PROOF. We prove that $(R \odot S) \in \Re_D$ (the proof that $(S \odot R) \in \Re_D$ is analogous). Put $R \odot S = T$ and assume that $T(x',y) > T(x,y)$ and $T(x,y') > T(x,y)$ for some $x,x',y,y' \in \mathbf{X}$. Further, there exist $z,z' \in \mathbf{X}$ such that $T(x',y) = S(x',z) \wedge R(z,y)$ and $T(x,y') = S(x,z') \wedge R(z',y')$. Hence $S(x',z) > T(x,y)$, $S(x,z') > T(x,y)$ and $R(z,y) > T(x,y)$.

Since $S \in \Re_D$, we have $S(x,z) \geq S(x,z') \wedge S(x',z)$. Then $S(x,z) \wedge R(z,y) \geq S(x,z') \wedge S(x',z) \wedge R(z,y) > T(x,y) \geq S(x,z) \wedge R(z,y)$, a contradiction. ■

COROLLARY 6.15. *If* $R \in \Re_D$, *then* $R^2 \in \Re_D$.

If $S \neq R$, a similar result to Thm.6.14 does not hold generally in $\Re_d$, as is shown in the following example of [21]:

EXAMPLE 6.4. Let $\mathbf{X} = \{x_1, x_2, x_3\}$, $R,S \in \mathbf{F}(\mathbf{X} \times \mathbf{X})$ given by

$$\begin{Vmatrix} 0.6 & 0.6 & 0.8 \\ 0.5 & 0.5 & 0.8 \\ 0.6 & 0.6 & 0.8 \end{Vmatrix} = S \neq R = \begin{Vmatrix} 1.0 & 0.3 & 0.5 \\ 0.8 & 1.0 & 0.4 \\ 0.5 & 0.2 & 0.2 \end{Vmatrix}.$$

We have:

$$R \odot S = \begin{Vmatrix} 0.6 & 0.6 & 0.5 \\ 0.5 & 0.5 & 0.5 \\ 0.6 & 0.6 & 0.5 \end{Vmatrix}.$$

Of course $S \in \Re_d$ but $(R \odot S) \notin \Re_d$ since there is no row of $(R \odot S)$ whose entries are equal to $\sup\{R(x,y): x,y \in \mathbf{X}\}$ (this conclusion holds for the elements of $\Re_d$, enunciating of course a theorem similar to Thm.6.7).

In Ch.7 another decomposition problem will be solved.

## References

[1] J. C. Bezdek and J. D. Harris, Fuzzy partitions and relations: an axiomatic basis for

clustering, *Fuzzy Sets and Systems* 1(1978), 111-127.

[2] M. K. Chakraborty and M. Das, Studies in fuzzy relations over fuzzy subsets, *Fuzzy Sets and Systems* 9(1983), 79-89.

[3] M. K. Chakraborty and M. Das, On fuzzy equivalence I, *Fuzzy Sets and Systems* 11(1983), 185-193.

[4] A. Di Nola, W. Pedrycz and S. Sessa, Fuzzy relation equations and algorithms of inference mechanism in expert systems, in: *Approximate Reasoning in Expert Systems* (M. M. Gupta, A. Kandel, W. Bandler and J. B. Kiszka, Eds.), Elsevier Science Publishers B. V. (North Holland), Amsterdam (1985), pp.355-367.

[5] A. Di Nola, W. Pedrycz and S. Sessa, When is a fuzzy relation decomposable in two fuzzy sets?, *Fuzzy Sets and Systems* 16(1985), 87-90.

[6] H. Hashimoto, Reduction of a fuzzy retrieval model, *Inform. Sciences* 27(1982), 133-140.

[7] H. Hashimoto, Reduction of a nilpotent fuzzy matrix, *Inform. Sciences* 27(1982), 233-243.

[8] H. Hashimoto, Convergence of powers of a fuzzy transitive matrix, *Fuzzy Sets and Systems* 9(1983), 153-160.

[9] H. Hashimoto, Szpilrajn's theorem on fuzzy orderings, *Fuzzy Sets and Systems* 10(1983), 101-108.

[10] H. Hashimoto, Canonical form of a transitive fuzzy matrix, *Fuzzy Sets and Systems* 11(1983), 157-162.

[11] H. Hashimoto, Transitive reduction of a nilpotent Boolean matrix, *Discrete Appl. Math.* 8(1984), 51-61.

[12] H. Hashimoto, Transitive reduction of a rectangular Boolean matrix, *Discrete Appl. Math.* 8(1984), 153-161.

[13] H. Hashimoto, Decomposition of fuzzy matrices, *SIAM J. Alg. Disc. Math.* 6(1985), 33-38.

[14] H. Hashimoto, Transitivity of generalized fuzzy matrices, *Fuzzy Sets and Systems* 17(1985), 83-90.

[15] H. Hashimoto, Properties of negatively transitive fuzzy relations, *Preprints of II IFSA Congress* , Tokyo, 1987 July 20-25, 540-543.

[16] W. Kolodziejczyk, Orlovsky's concept of decision-making with fuzzy preference relation - Further results, *Fuzzy Sets and Systems* 19(1986), 11-20.

[17] W. Kolodziejczyk, To what extent "decomposable" means "transitive" for a fuzzy relation?, *Fuzzy Sets and Systems*, to appear.

[18] W. Kolodziejczyk, Canonical form of a strongly transitive fuzzy matrix, *Fuzzy Sets and Systems* 22(1987), 297-302.

[19] W. Kolodziejczyk, Convergence of powers of s-transitive fuzzy matrix, *Fuzzy Sets and Systems* 26(1988), 127-130.

[20] W. Kolodziejczyk, More on reduction of transitive fuzzy matrices, *Technical Univ. of Wroclaw*, Report nº 35, 1988.

[21] W. Kolodziejczyk, Decomposition problem of fuzzy relations - further results,

*Internat. J. Gen. Syst.*, to appear.

[22] J. P. Olivier and D. Serrato, Approach to an axiomatic study on the fuzzy relations on finite sets, in: *Fuzzy Information and Decision Processes* (M. M. Gupta and E. Sanchez, Eds.), North Holland Publ. Co., Amsterdam, (1982), pp.111-116.

[23] S. A. Orlovsky, Decision-making with a fuzzy preference relation, *Fuzzy Sets and Systems* 1(1978), 155-167.

[24] S. A. Orlovsky, On formalization of a general fuzzy mathematical problem, *Fuzzy Sets and Systems* 3(1980), 311-321.

[25] S. V. Ovchinnikov, Structure of fuzzy binary relations, *Fuzzy Sets and Systems* 6(1981), 169-195.

[26] S. V. Ovchinnikov, Representations of transitive fuzzy relations, in: *Aspects of Vagueness* (H. J. Skala, S. Termini and E. Trillas, Eds.), D. Reidel Publ. Co., Dordrecht (1984), pp.105-118.

[27] S. V. Ovchinnikov, On the transitivity property, *Fuzzy Sets and Systems* 20(1986), 241-243.

[28] S. V. Ovchinnikov and V. M. Ozernoy, Using fuzzy binary relations for identifying noninferior decision alternatives, *Fuzzy Sets and Systems* 25(1988), 21-23.

[29] A. Rosenfeld, Fuzzy graphs, in: *Fuzzy Sets and Their Applications to Cognitive and Decision Processes* (L. A. Zadeh, K. S. Fu, K. Tanaka and M. Shimura, Eds.), Academic Press, New York (1975), pp.77-95.

[30] M. Roubens and P. Vincke, On families of semiorders and interval orders imbedded in a valued structure of preference: a survey, *Inform. Sciences* 34(1984), 187-198.

[31] V. Tahani, A fuzzy model of document retrieval systems, *Inform. Process. Managem.* 12(1976), 177-187.

[32] S. Tamura, S. Higuchi and K. Tanaka, Pattern classification on fuzzy relations, *IEEE Trans. Syst. Man Cybern.* SMC-1(1971), 61-66.

[33] T. Tanino, Fuzzy preference orderings in group decision-making, *Fuzzy Sets and Systems* 12 (1984), 117-131.

[34] M. G. Thomason, Convergence of powers of a fuzzy matrix, *J. Math. Anal. Appl.* 57(1977), 476-480.

[35] R. T. Yeh and S. Y. Bang, Fuzzy relations, fuzzy graphs and their applications, in: *Fuzzy Sets and Their Applications to Cognitive and Decision Processes* (L. A. Zadeh, K. S. Fu, K. Tanaka and M. Shimura, Eds.), Academic Press, New York (1975), pp.77-95.

[36] L. A. Zadeh, Similarity relations and fuzzy orderings, *Inform. Sciences* 3(1971), 177-200.

# CHAPTER 7

## MAX-MIN DECOMPOSITION PROBLEM OF A FUZZY RELATION IN LINEAR LATTICES



*In this Ch., we solve a decomposition problem, more complicated than the problem mentioned in Sec.6.5, and precisely, we present a numerical algorithm, illustrated by a flowchart, which assures the existence of a fuzzy relation $Z\in \mathbf{F}(X\times X)$ such that $Z\odot Z=R$, where $R\in \mathbf{F}(X\times X)$ is an assigned fuzzy relation defined on a referential set X and assuming values in a linear lattice **L** with universal bounds 0 and 1. Connections with results already existing in recent literature are also given.*

### 7.1. Preliminaries

Let **L** be a linear lattice (the completeness of **L** is not necessary here) with universal bounds 0 and 1, $\mathbf{X}=\{x_1,x_2,\ldots,x_n\}$ be a finite set and R, $Z\in \mathbf{F}(\mathbf{X}\times\mathbf{X})$ such that

$$R = Z\odot Z. \tag{7.1}$$

By putting $R(x_i,x_j)=R_{ij}$ for any $i,j\in \mathbf{I}_n=\{1,2,\ldots,n\}$, the set of first n natural integers, Eq.(7.1) is reformulated in the following manner:

$$R_{ij} = \bigvee_{p=1}^{n} (Z_{ip} \wedge Z_{pj}) \tag{7.2}$$

where $i,p,j\in \mathbf{I}_n$ and "∧" and "∨" are the usual lattice operations of min and max in **L**, "≤" is its total ordering.

Given $R\in \mathbf{F}(\mathbf{X}\times\mathbf{X})$, let $\mathfrak{R}_0=\mathfrak{R}_0(R)$ be the set of the solutions $Z\in \mathbf{F}(\mathbf{X}\times\mathbf{X})$ satisfying Eq.(7.2).

Now let $h,q,k\in \mathbf{I}_n$ be three arbitrary indices which define a fuzzy relation $M^{(h,q,k)}\in \mathbf{F}(\mathbf{X}\times\mathbf{X})$ with membership function given by

$$[M^{(h,q,k)}]_{ij} = \begin{cases} R_{hk} & \text{if } i=h,\ j=q, \\ R_{hk} & \text{if } i=q,\ j=k, \\ 0 & \text{otherwise,} \end{cases} \tag{7.3}$$

for all $i,j \in \mathbf{I}_n$. Of course, we have $n^3$ fuzzy relations of type (7.3) as is shown in the following example (without further mention, we shall assume $\mathbf{L}=([0,1],\wedge,\vee,\leq)$ in all the numerical examples of this Ch.):

EXAMPLE 7.1. Let n=3 and $R \in \mathbf{F}(\mathbf{X}\times\mathbf{X})$ be defined as

$$R = \begin{Vmatrix} 0.6 & 0.4 & 0.5 \\ 0.4 & 0.6 & 0.5 \\ 0.3 & 0.3 & 0.7 \end{Vmatrix}.$$

Since we have 27 triples $(h,q,k) \in \mathbf{I}_3$, we can write 27 fuzzy relations which are the following:

$$M^{(1,1,1)} = \begin{Vmatrix} 0.6 & 0.0 & 0.0 \\ 0.0 & 0.0 & 0.0 \\ 0.0 & 0.0 & 0.0 \end{Vmatrix} \quad M^{(1,2,1)} = \begin{Vmatrix} 0.0 & 0.6 & 0.0 \\ 0.6 & 0.0 & 0.0 \\ 0.0 & 0.0 & 0.0 \end{Vmatrix} \quad M^{(1,3,1)} = \begin{Vmatrix} 0.0 & 0.0 & 0.6 \\ 0.0 & 0.0 & 0.0 \\ 0.6 & 0.0 & 0.0 \end{Vmatrix}$$

$$M^{(1,1,2)} = \begin{Vmatrix} 0.4 & 0.4 & 0.0 \\ 0.0 & 0.0 & 0.0 \\ 0.0 & 0.0 & 0.0 \end{Vmatrix} \quad M^{(1,2,2)} = \begin{Vmatrix} 0.0 & 0.4 & 0.0 \\ 0.0 & 0.4 & 0.0 \\ 0.0 & 0.0 & 0.0 \end{Vmatrix} \quad M^{(1,3,2)} = \begin{Vmatrix} 0.0 & 0.0 & 0.4 \\ 0.0 & 0.0 & 0.0 \\ 0.0 & 0.4 & 0.0 \end{Vmatrix}$$

$$M^{(1,1,3)} = \begin{Vmatrix} 0.5 & 0.0 & 0.5 \\ 0.0 & 0.0 & 0.0 \\ 0.0 & 0.0 & 0.0 \end{Vmatrix} \quad M^{(1,2,3)} = \begin{Vmatrix} 0.0 & 0.5 & 0.0 \\ 0.0 & 0.0 & 0.5 \\ 0.0 & 0.0 & 0.0 \end{Vmatrix} \quad M^{(1,3,3)} = \begin{Vmatrix} 0.0 & 0.0 & 0.5 \\ 0.0 & 0.0 & 0.0 \\ 0.0 & 0.0 & 0.5 \end{Vmatrix}$$

$$M^{(2,1,1)} = \begin{Vmatrix} 0.4 & 0.0 & 0.0 \\ 0.4 & 0.0 & 0.0 \\ 0.0 & 0.0 & 0.0 \end{Vmatrix} \quad M^{(2,2,1)} = \begin{Vmatrix} 0.0 & 0.0 & 0.0 \\ 0.4 & 0.4 & 0.0 \\ 0.0 & 0.0 & 0.0 \end{Vmatrix} \quad M^{(2,3,1)} = \begin{Vmatrix} 0.0 & 0.0 & 0.0 \\ 0.0 & 0.0 & 0.4 \\ 0.4 & 0.0 & 0.0 \end{Vmatrix}$$

$$M^{(2,1,2)} = \begin{Vmatrix} 0.0 & 0.6 & 0.0 \\ 0.6 & 0.0 & 0.0 \\ 0.0 & 0.0 & 0.0 \end{Vmatrix} \quad M^{(2,2,2)} = \begin{Vmatrix} 0.0 & 0.0 & 0.0 \\ 0.0 & 0.6 & 0.0 \\ 0.0 & 0.0 & 0.0 \end{Vmatrix} \quad M^{(2,3,2)} = \begin{Vmatrix} 0.0 & 0.0 & 0.0 \\ 0.0 & 0.0 & 0.6 \\ 0.0 & 0.6 & 0.0 \end{Vmatrix}$$

$$M^{(2,1,3)} = \begin{Vmatrix} 0.0 & 0.0 & 0.5 \\ 0.5 & 0.0 & 0.0 \\ 0.0 & 0.0 & 0.0 \end{Vmatrix} \quad M^{(2,2,3)} = \begin{Vmatrix} 0.0 & 0.0 & 0.0 \\ 0.0 & 0.5 & 0.5 \\ 0.0 & 0.0 & 0.0 \end{Vmatrix} \quad M^{(2,3,3)} = \begin{Vmatrix} 0.0 & 0.0 & 0.0 \\ 0.0 & 0.0 & 0.5 \\ 0.0 & 0.0 & 0.5 \end{Vmatrix}$$

$$M^{(3,1,1)} = \begin{Vmatrix} 0.3 & 0.0 & 0.0 \\ 0.0 & 0.0 & 0.0 \\ 0.3 & 0.0 & 0.0 \end{Vmatrix} \quad M^{(3,2,1)} = \begin{Vmatrix} 0.0 & 0.0 & 0.0 \\ 0.3 & 0.0 & 0.0 \\ 0.0 & 0.3 & 0.0 \end{Vmatrix} \quad M^{(3,3,1)} = \begin{Vmatrix} 0.0 & 0.0 & 0.0 \\ 0.0 & 0.0 & 0.0 \\ 0.3 & 0.0 & 0.3 \end{Vmatrix}$$

$$M^{(3,1,2)} = \begin{Vmatrix} 0.0 & 0.3 & 0.0 \\ 0.0 & 0.0 & 0.0 \\ 0.3 & 0.0 & 0.0 \end{Vmatrix} \quad M^{(3,2,2)} = \begin{Vmatrix} 0.0 & 0.0 & 0.0 \\ 0.0 & 0.3 & 0.0 \\ 0.0 & 0.3 & 0.0 \end{Vmatrix} \quad M^{(3,3,2)} = \begin{Vmatrix} 0.0 & 0.0 & 0.0 \\ 0.0 & 0.0 & 0.0 \\ 0.0 & 0.3 & 0.3 \end{Vmatrix}$$

$$M^{(3,1,3)} = \begin{Vmatrix} 0.0 & 0.0 & 0.7 \\ 0.0 & 0.0 & 0.0 \\ 0.7 & 0.0 & 0.0 \end{Vmatrix} \quad M^{(3,2,3)} = \begin{Vmatrix} 0.0 & 0.0 & 0.0 \\ 0.0 & 0.0 & 0.7 \\ 0.0 & 0.7 & 0.0 \end{Vmatrix} \quad M^{(3,3,3)} = \begin{Vmatrix} 0.0 & 0.0 & 0.0 \\ 0.0 & 0.0 & 0.0 \\ 0.0 & 0.0 & 0.7 \end{Vmatrix}$$

The following theorem holds:

THEOREM 7.1. *We have for all* $h,q,k \in \mathbf{I}_n$:

$$R_{hk} = \bigvee_{p=1}^{n} \{[M^{(h,q,k)}]_{hp} \wedge [M^{(h,q,k)}]_{pk}\}$$

PROOF. It follows immediately from formula (7.3) since

$$[M^{(h,q,k)}]_{hp} = [M^{(h,q,k)}]_{pk} = 0$$

for any $p \in \mathbf{I}_n - \{q\}$ and

$$[M^{(h,q,k)}]_{hq} = [M^{(h,q,k)}]_{qk} = R_{hk}. \qquad \blacksquare$$

Now we set:

$$m = k + (h-1) \cdot n \tag{7.4}$$

for all $h,k \in \mathbf{I}_n$. Thus $1 \leq m \leq n^2$ and arbitrarily choosing the index $i_m \in \mathbf{I}_n$, with m given from (7.4), we consider the fuzzy relation $S \in \mathbf{F}(\mathbf{X} \times \mathbf{X})$ defined as

$$S = \bigvee_{m=1}^{n^2} S^{(m)}, \tag{7.5}$$

where $S^{(m)}=M^{(h,i_m,k)}$.

THEOREM 7.2. *We have:* $S \odot S \geq R$.

PROOF. By (7.5), we have:

$$S_{ij} \geq [S^{(m)}]_{ij}$$

for all $i,j \in \mathbf{I}_n$ and for any $m \in \mathbf{I}_{n^2}$. Using Thm.7.1, we deduce that

$$(S \odot S)_{hk} = \bigvee_{p=1}^{n} (S_{hp} \wedge S_{pk}) \geq \bigvee_{p=1}^{n} \{[S^{(m)}]_{hp} \wedge [S^{(m)}]_{pk}\}$$

$$= \bigvee_{p=1}^{n} \{[M^{(h,i_m,k)}]_{hp} \wedge [M^{(h,i_m,k)}]_{pk}\} = R_{hk}$$

for all $h,k \in \mathbf{I}_n$, i.e. the thesis. ■

In general, formula (7.5) defines fuzzy relations $S \in \mathbf{F}(\mathbf{X} \times \mathbf{X})$ either satisfying the inequality $S \odot S > R$ or Eq.(7.1), as is shown in the following example:

EXAMPLE 7.2. Let $R \in \mathbf{F}(\mathbf{X} \times \mathbf{X})$ be as in Ex.7.1, and considering the following 9-tuples defined by

$$i_1=3,\ i_2=3,\ i_3=2,\ i_4=1,\ i_5=3,\ i_6=1,\ i_7=3,\ i_8=3,\ i_9=2,$$

$$i_1'=1,\ i_2'=1,\ i_3'=1,\ i_4'=1,\ i_5'=2,\ i_6'=2,\ i_7'=1,\ i_8'=1,\ i_9'=3,$$

we have:

$$S = \begin{Vmatrix} 0.4 & 0.5 & 0.6 \\ 0.5 & 0.0 & 0.7 \\ 0.6 & 0.7 & 0.3 \end{Vmatrix}, \qquad S' = \begin{Vmatrix} 0.6 & 0.4 & 0.5 \\ 0.4 & 0.6 & 0.5 \\ 0.3 & 0.0 & 0.7 \end{Vmatrix},$$

since S and S' are given by

$$S = M^{(1,3,1)} \vee M^{(1,3,2)} \vee M^{(1,2,3)} \vee M^{(2,1,1)} \vee M^{(2,3,2)} \vee M^{(2,1,3)} \vee M^{(3,3,1)} \vee M^{(3,3,2)} \vee M^{(3,2,3)}.$$

$$S' = M^{(1,1,1)} \vee M^{(1,1,2)} \vee M^{(1,1,3)} \vee M^{(2,1,1)} \vee M^{(2,2,2)} \vee M^{(2,2,3)} \vee M^{(3,1,1)} \vee M^{(3,1,2)} \vee M^{(3,3,3)}.$$

It is easily seen that $S \odot S > R$ and $S' \odot S' = R$.

Let $\mathbf{S}_0 = \mathbf{S}_0(R)$ be the subset of $\mathbf{F}(\mathbf{X}\times\mathbf{X})$ whose fuzzy relations S, given from formula (7.5), satisfy Eq.(7.1). We have:

THEOREM 7.3. $\mathbf{S}_0 \neq \emptyset$ *iff* $\mathcal{R}_0 \neq \emptyset$.

PROOF. Since $\mathbf{S}_0$ is a subset of $\mathcal{R}_0$, we show only the nontrivial implication. Let Z be an arbitrary element of $\mathcal{R}_0$. Since Eq.(7.1) holds, then, for all $h,k \in \mathbf{I}_n$, there exists an index $q \in \mathbf{I}_n$ such that

$$R_{hk} = Z_{hq} \wedge Z_{qk} = \bigvee_{p=1}^{n} (Z_{hp} \wedge Z_{ph}) \tag{7.6}$$

and this implies either

$$Z_{hq} = R_{hk} \leq Z_{qk} \tag{7.7}$$

or

$$Z_{qk} = R_{hk} \leq Z_{hq}. \tag{7.8}$$

If (7.7) holds, then we obtain:

$$Z_{hq} = R_{hk} = [M^{(h,q,k)}]_{hq} \qquad \text{and} \quad Z_{qk} \geq R_{hk} = [M^{(h,q,k)}]_{qk}.$$

If (7.8) holds, then we deduce that

$$Z_{hq} \geq R_{hk} = [M^{(h,q,k)}]_{hq} \qquad \text{and} \quad Z_{qk} = R_{hk} = [M^{(h,q,k)}]_{qk}.$$

Since $Z_{ij} \geq 0 = [M^{(h,q,k)}]_{ij}$ for any pair $(i,j) \in \{\mathbf{I}_n \times \mathbf{I}_n\} - \{(h,q),(q,k)\}$ in both cases, then we have determined, for all $h,k \in \mathbf{I}_n$, a fuzzy relation $M^{(h,q,k)}$, with q satisfying (7.6) such that

$M^{(h,q,k)} \leq Z$. By putting $i_m=q$ with m defined from (7.4), we have:

$$M^{(h,i_m,q)} \leq Z$$

for all $h,k \in \mathbf{I}_n$. Thus we have determined a $n^2$-tuple $(i_1,\ldots,i_{n^2})$ such that the fuzzy relation S, defined from (7.5), verifies the inequality $S \leq Z$. By Thm.7.2, we deduce that

$$R \leq S \odot S \leq Z \odot Z = R,$$

therefore $S \odot S=R$, i.e. $S \in \mathbf{S}_0$. This concludes the proof. ■

## 7.2. Solution of the Decomposition Problem

In other words, Thm.7.3 assures that a fuzzy relation is decomposable in accordance with Eq.(7.1) iff there exists a fuzzy relation $S \in \mathbf{S}_0$. In terms of the "$\alpha$" operator (cfr. Sec.2.1), the previous remark is translated into the following theorem:

THEOREM 7.4. *A fuzzy relation* $R \in \mathbf{F}(\mathbf{X} \times \mathbf{X})$ *has a decomposition of type* (7.1) *iff there exists a fuzzy relation* $S \in \mathbf{F}(\mathbf{X} \times \mathbf{X})$, *defined from* (7.5), *such that* $S \leq S^{-1} \alpha R$.

PROOF. If R has a decomposition of type (7.1), then there exists an element $S \in \mathbf{S}_0$ by Thm.7.3. Thus $S \leq S^{-1} \alpha R$ by the fundamental Thm.2.3. Vice versa, using Lemma 2.2 and Thm.7.2, we have:

$$R \leq S \odot S \leq (S^{-1} \alpha R) \odot S \leq R,$$

which implies $S \odot S=R$.

Now we use Thm.7.4 in the following flowchart (Fig.7.1) to build such a fuzzy relation $S \in \mathbf{S}_0$.

## 7.3. Two Numerical Examples

Following the flowchart of Fig.7.1, we present here two examples describing in detail, within a finite sequence of steps, the construction of a fuzzy relation $S \in \mathbf{S}_0$ satisfying the inequality $S \leq S^{-1} \alpha R$ and thus leading to the decomposition of the assigned fuzzy relation R or to the conclusion that R is not decomposable.

EXAMPLE 7.3. Recalling the fuzzy relation $R \in \mathbf{F}(\mathbf{X} \times \mathbf{X})$ of Ex.7.1, we have the following steps:

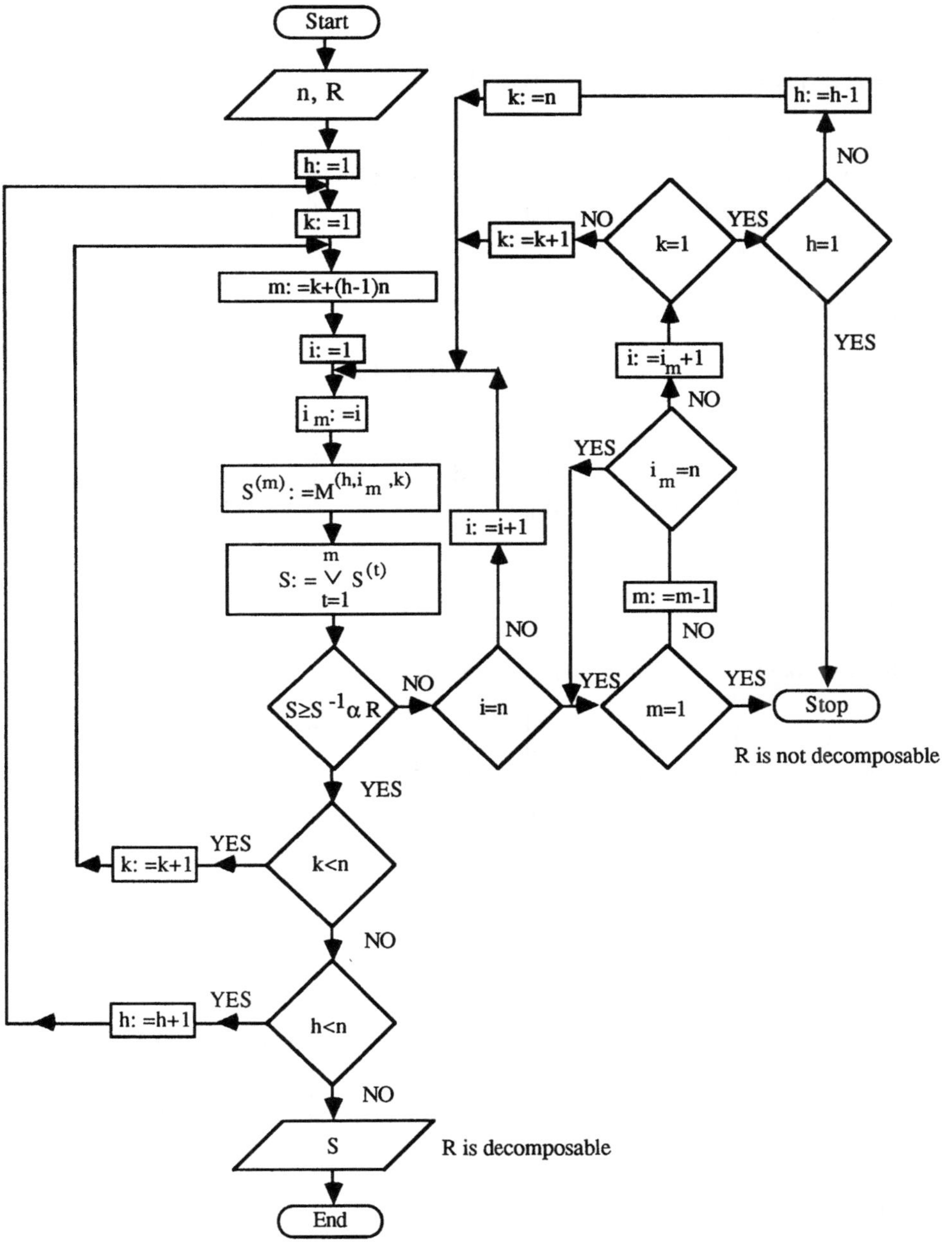


Fig. 1

Algorithm of decomposition of a fuzzy relation R

Step 1.

$h=1, k=1, m=1, i=1, i_1=1, S^{(1)}=M^{(1,1,1)}$.

$$S = S^{(1)} = \begin{Vmatrix} 0.6 & 0.0 & 0.0 \\ 0.0 & 0.0 & 0.0 \\ 0.0 & 0.0 & 0.0 \end{Vmatrix} \quad \text{and} \quad S^{-1}\alpha\, R = \begin{Vmatrix} 1.0 & 0.4 & 0.5 \\ 1.0 & 1.0 & 1.0 \\ 1.0 & 1.0 & 1.0 \end{Vmatrix} \geq S.$$

Step 2.

$h=1, k: =k+1=1+1=2, m=2, i=1, i_2=1, S^{(2)}=M^{(1,1,2)}$.

$$S = \bigvee_{t=1}^{2} S^{(t)} = \begin{Vmatrix} 0.6 & 0.4 & 0.0 \\ 0.0 & 0.0 & 0.0 \\ 0.0 & 0.0 & 0.0 \end{Vmatrix} \quad \text{and} \quad S^{-1}\alpha\, R = \begin{Vmatrix} 1.0 & 0.4 & 0.5 \\ 1.0 & 1.0 & 1.0 \\ 1.0 & 1.0 & 1.0 \end{Vmatrix} \geq S.$$

Step 3.

$h=1, k: =k+1=2+1=3, m=3, i=1, i_3=1, S^{(3)}=M^{(1,1,3)}$.

$$S = \bigvee_{t=1}^{3} S^{(t)} = \begin{Vmatrix} 0.6 & 0.4 & 0.5 \\ 0.0 & 0.0 & 0.0 \\ 0.0 & 0.0 & 0.0 \end{Vmatrix} \quad \text{and} \quad S^{-1}\alpha\, R = \begin{Vmatrix} 1.0 & 0.4 & 0.5 \\ 1.0 & 1.0 & 1.0 \\ 1.0 & 1.0 & 1.0 \end{Vmatrix} \geq S.$$

Step 4.

$h: =h+1=1+1=2, k=1, m=4, i=1, i_4=1, S^{(4)}=M^{(2,1,1)}$.

$$S = \bigvee_{t=1}^{4} S^{(t)} = \begin{Vmatrix} 0.6 & 0.4 & 0.5 \\ 0.4 & 0.0 & 0.0 \\ 0.0 & 0.0 & 0.0 \end{Vmatrix} \quad \text{and} \quad S^{-1}\alpha\, R = \begin{Vmatrix} 1.0 & 0.4 & 0.5 \\ 1.0 & 1.0 & 1.0 \\ 1.0 & 0.4 & 1.0 \end{Vmatrix} \geq S.$$

Step 5.

$h=2, k: =k+1=1+1=2, m=5, i=1, i_5=1, S^{(5)}=M^{(2,1,2)}$.

$$S = \bigvee_{t=1}^{5} S^{(t)} = \begin{Vmatrix} 0.6 & 0.6 & 0.5 \\ 0.6 & 0.0 & 0.0 \\ 0.0 & 0.0 & 0.0 \end{Vmatrix} \quad \text{and} \quad S^{-1}\alpha R = \begin{Vmatrix} 0.4 & 0.4 & 0.5 \\ 1.0 & 0.4 & 0.5 \\ 1.0 & 0.4 & 1.0 \end{Vmatrix} \not\geq S.$$

Step 6.

h=2, k=2, m=5, i: =i+1=1+1=2, $i_5$=2, $S^{(5)}=M^{(2,2,2)}$.

$$S = \bigvee_{t=1}^{5} S^{(t)} = \begin{Vmatrix} 0.6 & 0.4 & 0.5 \\ 0.4 & 0.6 & 0.0 \\ 0.0 & 0.0 & 0.0 \end{Vmatrix} \quad \text{and} \quad S^{-1}\alpha R = \begin{Vmatrix} 1.0 & 0.4 & 0.5 \\ 0.4 & 1.0 & 0.5 \\ 1.0 & 0.4 & 1.0 \end{Vmatrix} \geq S.$$

Step 7.

h=2, k: =k+1=2+1=3, m=6, i=1, $i_6$=1, $S^{(6)}=M^{(2,1,3)}$.

$$S = \bigvee_{t=1}^{6} S^{(t)} = \begin{Vmatrix} 0.6 & 0.4 & 0.5 \\ 0.5 & 0.6 & 0.0 \\ 0.0 & 0.0 & 0.0 \end{Vmatrix} \quad \text{and} \quad S^{-1}\alpha R = \begin{Vmatrix} 0.4 & 0.4 & 0.5 \\ 0.4 & 1.0 & 0.5 \\ 1.0 & 0.4 & 0.5 \end{Vmatrix} \not\geq S.$$

Step 8.

h=2, k=3, m=6, i: =i+1=1+1=2, $i_6$=2, $S^{(6)}=M^{(2,2,3)}$.

$$S = \bigvee_{t=1}^{6} S^{(t)} = \begin{Vmatrix} 0.6 & 0.4 & 0.5 \\ 0.4 & 0.6 & 0.5 \\ 0.0 & 0.0 & 0.0 \end{Vmatrix} \quad \text{and} \quad S^{-1}\alpha R = \begin{Vmatrix} 1.0 & 0.4 & 0.5 \\ 0.4 & 1.0 & 0.5 \\ 0.4 & 0.4 & 1.0 \end{Vmatrix} \geq S.$$

Step 9.

h: =h+1=2+1=3, k=1, m=7, i=1, $i_7$=1, $S^{(7)}=M^{(3,1,1)}$.

$$S = \bigvee_{t=1}^{7} S^{(t)} = \begin{Vmatrix} 0.6 & 0.4 & 0.5 \\ 0.4 & 0.6 & 0.5 \\ 0.3 & 0.0 & 0.0 \end{Vmatrix} \quad \text{and} \quad S^{-1}\alpha R = \begin{Vmatrix} 1.0 & 0.4 & 0.5 \\ 0.4 & 1.0 & 0.5 \\ 0.4 & 0.4 & 1.0 \end{Vmatrix} \geq S.$$

Step 10.

$h=3,\ k:=k+1=1+1=2,\ m=8,\ i=1,\ i_8=1,\ S^{(8)}=M^{(3,1,2)}$.

$$S = \bigvee_{t=1}^{8} S^{(t)} = \bigvee_{t=1}^{7} S^{(t)}, \text{ idem as in step 9.}$$

Step 11.

$h=3,\ k:=k+1=2+1=3,\ m=9,\ i=1,\ i_9=1,\ S^{(9)}=M^{(3,1,3)}$.

$$S = \bigvee_{t=1}^{9} S^{(t)} = \begin{Vmatrix} 0.6 & 0.4 & 0.7 \\ 0.4 & 0.6 & 0.5 \\ 0.7 & 0.0 & 0.0 \end{Vmatrix} \quad \text{and} \quad S^{-1}\alpha R = \begin{Vmatrix} 0.3 & 0.3 & 0.5 \\ 0.4 & 1.0 & 0.5 \\ 0.4 & 0.4 & 0.5 \end{Vmatrix} \not\geq S.$$

Step 12.

$h=3,\ k=3,\ m=9,\ i:=i+1=1+1=2,\ i_9=2,\ S^{(9)}=M^{(3,2,3)}$.

$$S = \bigvee_{t=1}^{9} S^{(t)} = \begin{Vmatrix} 0.6 & 0.4 & 0.5 \\ 0.4 & 0.6 & 0.7 \\ 0.3 & 0.7 & 0.0 \end{Vmatrix} \quad \text{and} \quad S^{-1}\alpha R = \begin{Vmatrix} 1.0 & 0.4 & 0.5 \\ 0.3 & 0.3 & 0.5 \\ 0.3 & 0.4 & 0.5 \end{Vmatrix} \not\geq S.$$

Step 13.

$h=3,\ k=3,\ m=9,\ i:=i+1=2+1=3,\ i_9=3,\ S^{(9)}=M^{(3,3,3)}$.

$$S = \bigvee_{t=1}^{9} S^{(t)} = \begin{Vmatrix} 0.6 & 0.4 & 0.5 \\ 0.4 & 0.6 & 0.5 \\ 0.3 & 0.0 & 0.7 \end{Vmatrix} \quad \text{and} \quad S^{-1}\alpha R = \begin{Vmatrix} 1.0 & 0.4 & 0.5 \\ 0.4 & 1.0 & 0.5 \\ 0.3 & 0.3 & 1.0 \end{Vmatrix} \geq S.$$

END because $S \odot S = R$ and therefore the given fuzzy relation R of Ex.7.1 is decomposable.

EXAMPLE 7.4. Let $\mathbf{X}=\{x_1,x_2\}$ and $R \in \mathbf{F}(\mathbf{X}\times\mathbf{X})$ be given by

$$R = \begin{Vmatrix} 1.0 & 0.2 \\ 0.8 & 0.1 \end{Vmatrix}.$$

Consequently, we have:

$$M^{(1,1,1)} = \begin{Vmatrix} 1.0 & 0.0 \\ 0.7 & 0.0 \end{Vmatrix}, \quad M^{(1,2,1)} = \begin{Vmatrix} 0.0 & 1.0 \\ 1.0 & 0.0 \end{Vmatrix}, \quad M^{(1,1,2)} = \begin{Vmatrix} 0.2 & 0.2 \\ 0.0 & 0.0 \end{Vmatrix},$$

$$M^{(1,2,2)} = \begin{Vmatrix} 0.0 & 0.2 \\ 0.0 & 0.2 \end{Vmatrix}, \quad M^{(2,1,1)} = \begin{Vmatrix} 0.8 & 0.0 \\ 0.8 & 0.0 \end{Vmatrix}, \quad M^{(2,2,1)} = \begin{Vmatrix} 0.0 & 0.0 \\ 0.8 & 0.8 \end{Vmatrix},$$

$$M^{(2,1,2)} = \begin{Vmatrix} 0.0 & 0.1 \\ 0.1 & 0.0 \end{Vmatrix}, \quad M^{(2,2,2)} = \begin{Vmatrix} 0.0 & 0.0 \\ 0.0 & 0.1 \end{Vmatrix}.$$

Using the algorithm of Fig.1, we have the following steps:

Step 1.

$h=1$, $k=1$, $m=1$, $i=1$, $i_1=1$, $S^{(1)}=M^{(1,1,1)}$.

$$S = S^{(1)} = \begin{Vmatrix} 1.0 & 0.0 \\ 0.0 & 0.0 \end{Vmatrix} \quad \text{and} \quad S^{-1}\alpha R = \begin{Vmatrix} 1.0 & 0.2 \\ 1.0 & 1.0 \end{Vmatrix} \geq S.$$

Step 2.

$h=1$, $k: =k+1=1+1=2$, $m=2$, $i=1$, $i_2=1$, $S^{(2)}=M^{(1,1,2)}$.

$$S = \bigvee_{t=1}^{2} S^{(t)} = \begin{Vmatrix} 1.0 & 0.2 \\ 0.0 & 0.0 \end{Vmatrix} \quad \text{and} \quad S^{-1}\alpha R = \begin{Vmatrix} 1.0 & 0.2 \\ 1.0 & 1.0 \end{Vmatrix} \geq S.$$

Step 3.

$h: =h+1=1+1=2$, $k=1$, $m=3$, $i=1$, $i_3=1$, $S^{(3)}=M^{(2,1,1)}$.

$$S = \bigvee_{t=1}^{3} S^{(t)} = \begin{Vmatrix} 1.0 & 0.2 \\ 0.8 & 0.0 \end{Vmatrix} \quad \text{and} \quad S^{-1}\alpha\, R = \begin{Vmatrix} 1.0 & 0.1 \\ 1.0 & 1.0 \end{Vmatrix} \ngeq S.$$

Step 4.

h=2, k=1, m=3, i: =i+1=1+1=2, $i_3$=1, $S^{(3)}=M^{(2,2,1)}$.

$$S = \bigvee_{t=1}^{3} S^{(t)} = \begin{Vmatrix} 1.0 & 0.2 \\ 0.8 & 0.8 \end{Vmatrix} \quad \text{and} \quad S^{-1}\alpha\, R = \begin{Vmatrix} 1.0 & 0.1 \\ 1.0 & 0.1 \end{Vmatrix} \ngeq S.$$

i=2? (YES),

m=1? (NO),

m: =m-1=3-1=2,

$i_2$=2? (NO),

i: =$i_2$+1=1+1=2,

k=1? (YES),

h=1? (NO),

h: =h-1=2-1=1,

k=2.

Step 5.

h=1, k=2, m=2, i=2, $i_2$=2, $S^{(2)}=M^{(1,2,2)}$.

$$S = \bigvee_{t=1}^{2} S^{(t)} = \begin{Vmatrix} 1.0 & 0.2 \\ 0.0 & 0.2 \end{Vmatrix} \quad \text{and} \quad S^{-1}\alpha\, R = \begin{Vmatrix} 1.0 & 0.2 \\ 1.0 & 0.1 \end{Vmatrix} \ngeq S.$$

i=2? (YES),

m=1? (NO),

m: =m-1=2-1=1,

$i_1$=2? (NO),

i: =$i_1$+1=1+1=2,

k=1? (NO),

k: =k-1=2-1=1.

Step 6.

h=1, k=1, m=1, i=2, $i_1$=2, $S^{(1)}=M^{(1,2,1)}$.

$$S = S^{(1)} = \begin{Vmatrix} 0.0 & 1.0 \\ 1.0 & 0.0 \end{Vmatrix} \quad \text{and} \quad S^{-1}\alpha R = \begin{Vmatrix} 0.8 & 0.1 \\ 1.0 & 0.2 \end{Vmatrix} \ngeq S.$$

i=2? (YES),

m=1? (YES),

STOP.

Then the assigned fuzzy relation R is not decomposable.

The results established here can be found in [1]. An algebraic characterization of the minimal and maximal elements of the set $\mathfrak{R}_0$ has been studied in [2], but this question is not treated here.

### 7.4. Max-min Decomposition of Transitive Fuzzy Relations

Here we assume **L**=[0,1] with the usual lattice operations "$\wedge$" and "$\vee$". As in [21, Ch.6] we need to extend the notion of strong transitivity introduced in Sec.6.5, defining the set $\mathfrak{R}_s$ of the fuzzy relations R$\in$**F(X×X)** such that if R(x,y)>R(y,z) and R(y,z)>R(z,y) with x≠y, y≠z, x≠z, then we have R(x,z)≥R(z,x).

Of course, $\mathfrak{R}_\Delta \subset \mathfrak{R}_s$ and hence, by Thm.6.9, the following hierarchy of strict inclusions holds:

$$\mathfrak{R}_{\square} \subset \mathfrak{R}_D \subset \mathfrak{R}_{\wedge} \subset \mathfrak{R}_{\Delta} \subset \mathfrak{R}_s.$$

A fuzzy relation $R \in [\mathfrak{R}_D \cup (\mathfrak{R}_{\wedge} - \mathfrak{R}_D)]$ is not necessarily an element of the set $\mathfrak{R}_0$ as is easily seen in the following example of Kolodziejczyk [21, Ch.6]:

EXAMPLE 7.5. Let $\mathbf{X}=\{x_1,x_2, x_3\}$, $R_1$, $R_2 \in \mathbf{F}(\mathbf{X}\times\mathbf{X})$ given by

$$R_1 = \begin{Vmatrix} 0.3 & 0.5 & 0.1 \\ 0.2 & 0.2 & 0.1 \\ 0.2 & 0.2 & 0.1 \end{Vmatrix}, \qquad R_2 = \begin{Vmatrix} 0.7 & 0.4 & 0.5 \\ 0.0 & 0.2 & 0.3 \\ 0.0 & 0.0 & 0.0 \end{Vmatrix}.$$

Indeed, $R_1 \in \mathfrak{R}_D$ and $R_2 \in (\mathfrak{R}_{\wedge} - \mathfrak{R}_D)$ but it is seen that $R_1, R_2 \notin \mathfrak{R}_0$ using the flowchart of Fig.7.1.

The following results of Kolodziejczyk [21, Ch.6] hold:

THEOREM 7.5. *If* $R \in \mathfrak{R}_D \cap \mathfrak{R}_0$, *then there exists a fuzzy relation* $S \in \mathfrak{R}_{\wedge}$ *such that* $R=S \odot S$.

PROOF. Since $R \in \mathfrak{R}_0$, then there exists a fuzzy relation $Q \in \mathbf{F}(\mathbf{X}\times\mathbf{X})$ such that $Q^2=R$. Let $S=Q \vee R$, so that $S \odot S=(Q \vee R)\odot(Q \vee R) \geq Q \odot Q=R$ and assume that $S^2>R$, i.e.

$$S^2(x,y) = [Q(x,z) \vee R(x,z)] \wedge [Q(z,y) \vee R(z,y)] > R(x,y)$$

for some $x,y,z \in \mathbf{X}$. We can distinguish four cases:

(i) $Q(x,z) \vee R(x,z) = R(x,z)$, $Q(z,y) \vee R(z,y) = R(z,y)$.

In this case, we would have $R(x,z) \wedge R(z,y) > R(x,y)$, a contradiction since $R \in \mathfrak{R}_{\wedge}$ by Thm.6.9.

(ii) $Q(x,z) \vee R(x,z) = Q(x,z)$, $Q(z,y) \vee R(z,y) = Q(z,y)$.

In this case, since $Q \odot Q=R$, we would have:

$$Q(x,y) \wedge Q(y,z) > R(x,y) \geq Q(x,y) \wedge Q(y,z),$$

a contradiction.

(iii) $Q(x,z) \vee R(x,z) = Q(x,z)$, $Q(z,y) \vee R(z,y) = R(z,y)$.

In this case, we would have $Q(x,z) \wedge R(z,y) > R(x,y)$ which implies $R(z,y)>R(x,y)$, i.e.

$$R(x,y) = \bigvee_{w \in \mathbf{X}} R(x,w),$$

since $R \in \mathfrak{R}_D$. Further, $R(z,y)=Q(z,x_o) \wedge Q(x_o,y)$ for some $x_o \in \mathbf{X}$ and then

$$R(x,x_o) \geq Q(x,z) \wedge Q(z,x_o) \geq Q(x,z) \wedge Q(z,x_o) \wedge Q(x_o,y)$$

$$= Q(x,z) \wedge R(z,y) > R(x,y) \geq R(x,x_o),$$

a contradiction.

(iv) $Q(x,z) \vee R(x,z) = R(x,z)$, $Q(z,y) \vee R(z,y) = Q(z,y)$.

It suffices to reason as in the case (iii) in order to obtain another contradiction.

Therefore $S^2=R$, further we have $S \in \mathfrak{R}_\wedge$ since $S^2=R \leq Q \vee R=S$. This concludes the proof. ■

REMARK 7.1. It is not yet known if from $R \in \mathfrak{R}_d \cap \mathfrak{R}_0$ it follows that a fuzzy relation $S \in \mathfrak{R}_c$ exists such that $S^2=R$. However, this thesis is valid if one considers the min-max composition of fuzzy matrices (cfr. Sec.2.8). In this case, a result similar to Thm.7.5 can be proved using the fuzzy relation $S=Q \wedge R$.

THEOREM 7.6. *If* $R \in \mathfrak{R}_\wedge \cap \mathfrak{R}_0$, *then there exists a fuzzy relation* $S \in \mathfrak{R}_s$ *such that* $R=S \odot S$.

PROOF. Let $R \in \mathfrak{R}_\wedge \cap \mathfrak{R}_0$. Hence there exists a fuzzy relation $Q \in \mathbf{F}(\mathbf{X} \times \mathbf{X})$ such that $Q^2=R$. If $Q \in \mathfrak{R}_s$, the thesis is proved.

If $Q \notin \mathfrak{R}_s$, then $Q(x_o,y_o)>Q(y_o,x_o)$, $Q(y_o,z_o)>Q(z_o,y_o)$, $Q(z_o,x_o)>Q(x_o,z_o)$ for some $x_o,y_o,z_o \in \mathbf{X}$ with $x_o \neq y_o$, $y_o \neq z_o$, $x_o \neq z_o$. Without loss of generality, assume that

$$Q(y_o,x_o) = Q(y_o,x_o) \wedge Q(z_o,y_o) \wedge Q(x_o,z_o)$$

and by putting:

$$Z(y_o,x_o) = \wedge \{Q(x,y) : Q(x,y) > Q(y_o,x_o)\},$$

we define the following fuzzy relation $Q' \in \mathbf{F}(\mathbf{X} \times \mathbf{Y})$ as

$$Q'(x,y) = Z(y_o,x_o) \quad \text{if } x=y_o \text{ and } y=x_o, \qquad (7.9)$$
$$= Q(x,y) \quad \text{otherwise,}$$

for all $x,y \in \mathbf{X}$. Of course, $Q' \geq Q$ and then $Q'^2 \geq Q^2 = R$ and suppose that $Q'^2 > R$, i.e. $Q'^2(x',y') > R(x',y')$ for some $x',y' \in \mathbf{X}$, hence $Q'(x',z') \wedge Q'(z',y') > R(x',y')$ for some $z' \in \mathbf{X}$. We must have either $z'=x_o$ or $y'=x_o$, otherwise if $z' \neq x_o$ and $y' \neq x_o$, we would obtain from (7.9):

$$Q(x',z') \wedge Q(z',y') = Q'(x',z') \wedge Q'(z',y') > R(x',y') \geq Q(x',z') \wedge Q(z',y'),$$

a contradiction. Suppose that $z'=x_o$ (the other case $y'=x_o$ can be treated analogously) and then $Q'(x',x_o) \wedge Q'(x_o,y') > R(x',y')$. We must have $x'=y_o$ otherwise we would deduce, from (7.9), that

$$Q(x',x_o) \wedge Q(x_o,y') = Q'(x',x_o) \wedge Q'(x_o,y') > R(x',y') \geq Q(x',x_o) \wedge Q(x_o,y'),$$

a contradiction. Therefore

$$Q'(y_o,x_o) \wedge Q(x_o,y') = Z(y_o,x_o) \wedge Q(x_o,y') > R(y_o,y') \geq Q(y_o,x_o) \wedge Q(x_o,y') \qquad (7.10)$$

and this implies that

$$Q(x_o,y') \geq Z(y_o,x_o) \wedge Q(x_o,y') > Q(y_o,x_o) \wedge Q(x_o,y'),$$

i.e. $Q(y_o,x_o) \wedge Q(x_o,y') = Q(y_o,x_o) < Q(x_o,y')$. Then

$$R(z_o,y') \geq Q(z_o,x_o) \wedge Q(x_o,y') > Q(y_o,x_o), \qquad (7.11)$$

since $Q(z_o,x_o) > Q(x_o,z_o) \geq Q(y_o,x_o)$. On the other hand,

$$R(x_o,z_o) \geq Q(x_o,y_o) \wedge Q(y_o,z_o) > Q(y_o,x_o) \wedge Q(z_o,y_o)$$
$$= Q(y_o,x_o) \qquad (7.12)$$

and

$$R(y_o,x_o) \geq Q(y_o,z_o) \wedge Q(z_o,x_o) > Q(z_o,y_o) \wedge Q(x_o,z_o)$$
$$\geq Q(y_o,x_o). \qquad (7.13)$$

Since $R\in\mathfrak{R}_\wedge$, one has:

$$R(y_o,y') \geq R(y_o,x_o)\wedge R(x_o,y') \geq R(y_o,x_o) \wedge R(x_o,z_o)\wedge R(z_o,y') > Q(y_o,x_o)$$

by (7.11), (7.12) and (7.13). Since $R(y_o,y') = Q(y_o,z'') \wedge Q(z'',y')$ for some $z''\in \mathbf{X}$, we have $Q(y_o,z'')>Q(y_o,x_o)$ and $Q(z'',y')>Q(y_o,x_o)$, i.e.

$$Z(y_o,x_o) \leq Q(y_o,z'') \wedge Q(z'',y') = R(y_o,y'),$$

a contradiction because (7.10) implies that

$$Z(y_o,x_o) \wedge Q(x_o,y') = Z(y_o,x_o) > R(y_o,y').$$

Thus we have $Q'^2=R$ and obviously, using the transformation (7.9) a finite number of times, we build a fuzzy relation $Q'\in\mathfrak{R}_s$ such that $Q'^2=R$. This concludes the proof. ■

Concluding as in [21, Ch.6], we can say that a max-min transitive fuzzy relation $R\in\mathfrak{R}_0$ can be further decomposed by using fuzzy relations $R\in\mathfrak{R}_s$ which are transitive in a wider sense than R.

## References


[1] A. Di Nola, W. Pedrycz and S. Sessa, Decomposition problem of fuzzy relations, *Internat. J. Gen. Syst.* 10 (1985), 123-133.

[2] A. Di Nola, W. Pedrycz, S. Sessa and M. Higashi, Minimal and maximal solutions of a decomposition problem of fuzzy relations, *Internat. J. Gen. Syst.* 11 (1985), 103-116.

# CHAPTER 8

## FUZZY RELATION EQUATIONS WITH LOWER AND UPPER SEMICONTINUOUS TRIANGULAR NORMS



*In this Ch. we present another generalization of the fuzzy relation equations which have been considered in the previous chapters. Here we focus our attention on a broad class of logical connectives applied in fuzzy set theory, i.e. triangular norms (for short, t-norms) and conorms (for short, s-norms) which in turn enable us to consider max-t and min-s compositions. In Sec.8.1 we present essentially the concept of the logical connectives modelled by t-norms and related s-norms. In Sec.8.2 we analyze max-t fuzzy relation equations and related dual min-s fuzzy relation equations, assuming that t (resp s) is lower (resp. upper) semicontinuous. In Sec.8.3 we pay attention to equations of complex structure and a related adjoint fuzzy relation equation is studied in Sec.8.4. In the last Sec.8.5, max-t fuzzy relation equations under upper semicontinuous t-norms are studied. All the equations considered in this Ch. are assigned on finite referential sets.*

### 8.1. Triangular Norms as Logical Connectives in Fuzzy Set Theory

Triangular norms were introduced by Menger [26] with regard to so-called statistical metric spaces.

DEFINITION 8.1. *A t-norm is an associative binary operation on* $[0,1]$, *commutative, nondecreasing in each argument and such that* $0tx=0$ *and* $xt1=x$ *for any* $x\in[0,1]$.

From an algebraic point of view, each t-norm defines a semigroup on [0,1] with 1 as unit and 0 as annihilator with order-preserving and commutative operation. We can distinguish some special classes of t-norms ([36]÷[39]).

DEFINITION 8.2. *An s-norm is an associative binary operation on* [0,1], *commutative, nondecreasing in each argument and such that* 0sx=x *and* 1sx=1 *for any* x∈[0,1].

Given any t-norm, the corresponding s-norm (called also conorm s), dual of t, is defined by setting:

$$xsy = 1 - (1-x)\,t\,(1-y)$$

for all x,y∈[0,1]. An extensive list of t-norms and s-norms can be found in literature, e.g., see [1]÷[5], [11], [12], [24], [25], [28], [40]÷[42], [44]÷[50], [17, Ch.4].

Some examples of t-norms and related dual s-norms are reported below.

Yager [44]:

$a t_1^{(p)} b = 1-\min\{1, [(1-a)^p + (1-b)^p]^{1/p}\}$, $a s_1^{(p)} b = \min\{1, (a^p+b^p)^{1/p}\}$, $p \geq 1$,

Frank [18]:

$a t_2^{(p)} b = \log_p \{1+ [(p^a-1)(p^b-1)]/(p-1)\}$, $a s_2^{(p)} b = 1-\log_p \{1+ [(p^{1-a}-1)(p^{1-b}-1)]/(p-1)\}$,

$0<p$, $p \neq 1$,

Zadeh [8, Ch.1]:

$a t_3 b = a \cdot b$, $a s_3 b = a+b-ab$,

Hamacher [24]:

$a t_4^{(p)} b = ab/[p+(1-p)(a+b-ab)]$, $a s_4^{(p)} b = [ab(p-2) + a + b]/[ab(p-1)+1]$, $p \geq 0$,

Sugeno [41]:

$a t_5^{(p)} b = \max\{0, (p+1)(a+b-1) - pab\}$, $a s_5^{(p)} b = \min\{1, a+b+pab\}$, $p \geq -1$,

Drastic product:

$$a t_6 b = \begin{cases} a & \text{if } b=1, \\ b & \text{if } a=1, \\ 0 & \text{otherwise,} \end{cases} \qquad a s_6 b = \begin{cases} a & \text{if } b=0, \\ b & \text{if } a=0, \\ 1 & \text{otherwise,} \end{cases}$$

Schweizer and Sklar [39]:

$at_7^{(p)}b = [\max(0, a^p + b^p - 1)]^{1/p}$, $as_7^{(p)}b = 1 - \{\max[0, (1-a)^p + (1-b)^p - 1]\}^{1/p}$, $p \neq 0$.

There are several ways to build a triangular norm: some methods are based on the theory of functional equations ([1], [18], [19]). We also recall the method of ordinal sums ([38], [39]) to construct a new t-norm (resp. s-norm) from a family of t-norms (resp. s-norms). For any t-norm and s-norm, the following inequalities hold for all $a,b \in [0,1]$:

$$at_6b \leq atb \leq \min\{a,b\},$$
$$\max\{a,b\} \leq asb \leq as_6b.$$

Due to the associativity of the t-norms and s-norms, we may also consider the quantities $(a_1\, t\, a_2\, t \ldots t\, a_n)$ and $(a_1\, s\, a_2\, s \ldots s\, a_n)$, $a_i \in [0,1]$, defined recursively by

$$\mathop{t}_{i=1}^{n} a_i = a_1\, t\, a_2\, t \ldots t\, a_n = (a_1\, t\, a_2\, t \ldots t\, a_{n-1})\, t\, a_n,$$

$$\mathop{s}_{i=1}^{n} a_i = a_1\, s\, a_2\, s \ldots s\, a_n = (a_1\, s\, a_2\, s \ldots s\, a_{n-1})\, s\, a_n.$$

It is instructive to present some limit cases of the t-norms summarized above.

$$\lim_{p\to+\infty} at_1^{(p)}b = \lim_{p\to 0} at_2^{(p)}b = \lim_{p\to-\infty} at_7^{(p)}b = \min\{a, b\}.$$

If $p=1$, then $at_1^{(p)}b = \max\{0, a+b-1\}$.

If $p=0$, then $at_1^{(p)}b = at_6b$.

$$\lim_{p\to+\infty} at_2^{(p)}b = \max\{0, a+b-1\}.$$

$$\lim_{p\to 1} at_2^{(p)}b = \lim_{p\to 0} at_7^{(p)}b = at_3b.$$

If $p=1$, then $at_4^{(p)}b = at_3b$.

$$\lim_{p\to+\infty} at_4^{(p)}b = \lim_{p\to+\infty} at_5^{(p)}b = \lim_{p\to+\infty} at_7^{(p)}b = at_6b.$$

If p=-1, then $at_5^{(p)}b = at_3b$.

If p=0, then $at_5^{(p)}b = \max\{0, a+b-1\}$.

Of course, similar properties hold for s-norms.

The triangular norms can be used for modelling the connectives used in fuzzy set theory. In his pioneering work [8, Ch.1], Zadeh suggested two models of intersection and union such as $t_3$ and $s_3$. Only few authors considered fundamental properties of these operators ([5], [6], [20]). It was shown, in practice ([35], [43], [50]) that in some situations the minimum operator is not an appropriate operator for the intersection of fuzzy sets. In [7], local properties of logical connectives applied to fuzzy sets are pointed out, however the main idea is that the operators for intersection and union should be chosen with respect to the problem under consideration [46]. This fact was also confirmed by further research, when compensative operators were proposed and discussed in [17] and [50]. See also Alsina [2]÷[4], Dubois and Prade [13]÷[16], Valverde [48], Yu [47] about the usage of t-norms in different contexts.

In the applications, one might find that t-norms and s-norms parametrized by a real number p, such as, e.g., $t_1^{(p)}$, $t_2^{(p)}$, $s_1^{(p)}$, $s_2^{(p)}$, are adjustable for available empirical data. Particular attention should be paid to an aspect of modelling different types of "**and**" and "**or**" connectives.

It is obvious that some concepts are strongly related to each other while there exist also cases where they are discussed as almost unrelated. For instance, the nature of and-ing the two following statements:

"*red* **and** *new car*"

"*large* **and** *expensive house*"

is quite different: in the first case, the properties of the car are obviously unrelated, while in the second case one might expect the properties of the house to be to a certain extent related to each other. Hence in the second case "**and**" might be interpreted by means of, e.g., $t_3$, while in the first case the minimum is preferable. Thus, by adjusting the parameter p of the t-norm and s-norm, one could easily model the situation in which the fuzziness is manifested.

## 8.2. Fuzzy Relation Equations with Lower Semicontinuous Triangular Norms and Upper Semicontinuous Conorms

The main results of this Sec. are contained in the papers of Pedrycz [29]÷[34].

Let **X** and **Y** be finite sets, **L**=[0,1] with the usual "$\wedge$" and "$\vee$" lattice operations.

DEFINITION 8.3. *Let* $A \in \mathbf{F}(\mathbf{X})$ *and* $R \in \mathbf{F}(\mathbf{X} \times \mathbf{Y})$. *We define* max-t (*resp.* min-s) *composition of* R *and* A *to be the fuzzy set* $B \in \mathbf{F(Y)}$, *in symbols* B=RtA (*resp.* B=RsA), *whose membership function is given by*

$$B(y) = (R\ t\ A)\ (y) = \bigvee_{x \in \mathbf{X}} [A(x)\ t\ R(x,y)] \tag{8.1}$$

*resp.*

$$B(y) = (R\ s\ A)\ (y) = \bigwedge_{x \in \mathbf{X}} [A(x)\ s\ R(x,y)] \tag{8.2}$$

*for any* $y \in \mathbf{Y}$.

By considering Eqs.(8.1) and (8.2), we put the following problems (cfr. Sec.2.2):

(p1) *determine R if A and B are given,*

(p2) *determine A if B and R are given.*

In order to solve these problems, we introduce the following definition [30]:

DEFINITION 8.4. *An operator* $\varphi_t$: $[0,1]^2 \to [0,1]$ *(resp.* $\beta_s$*) is associated with a t-norm (resp. s-norm) if*

(i) $a\ \varphi_t \max\{b,c\} = \max\{a\varphi b, a\varphi c\}$ (resp. $a\ \beta_s \min\{b,c\} = \min\{a\beta_s b, a\beta_s c\}$),

(ii) $a\ t\ (a\varphi_t b) \le b$ (resp. $a\ s\ (a\beta_s b) \ge b$),

(iii) $a\ \varphi_t\ (atb) \ge b$ (resp. $a\ \beta_s\ (asb) \le b$),

*for all* $a,b,c \in [0,1]$.

Note that if t=min, the above properties reduce to the properties (i), (ii), (iii) of the operator "α" given in Sec.2.1 and if s=max, then $\beta_s = \varepsilon$ (cfr. Sec.2.5). Since no misunderstanding can arise, from now on, we write φ (resp.β) instead of $\varphi_t$ (resp.$\beta_s$).

The following result provides a connection between operators φ and t-norms:

THEOREM 8.1. *If* φ *is associated with a t-norm, then*

$$a\varphi b = \sup\{c \in [0,1]: atc \le b\}. \tag{8.3}$$

PROOF. By property (ii) of Def.8.4, aφb does not exceed the supremum. If $a\varphi b < \sup\{c \in [0,1]: atc \le b\}$, then there exists some $c_0 \in [0,1]$ such that $a\varphi b < c_0$ and $atc_0 \le b$.

Hence $c_0 \leq a\varphi(atc_0) \leq a\varphi b < c_0$, a contradiction. ■

Conversely, the following result holds (cfr. [21]÷[23]):

THEOREM 8.2. *An operator φ is associated with a t-norm iff t is lower semicontinuous.*

PROOF. We first notice that the lower semicontinuity of the t-norm, due to its monotonicity, is equivalent to the left continuity of t with respect to both arguments. This can be seen in the equivalent condition:

$$a\,t\,(\sup_{i\in I} b_i) = \sup_{i\in I} (atb_i),$$

i.e. this equality holds for any family $(b_i)_{i\in I}$ of real numbers of [0,1].

Define φ by using (8.3). It is easily seen that property (i) of Def.8.4 holds. Moreover,

$$a\,t\,(a\varphi b) = a\,t\,\sup\,\{c\in[0,1]: atc\leq b\}$$

$$= \sup\{atc: atc \leq b\} \leq b$$

and

$$a\varphi(atb) = \sup\,\{c\in[0,1]: atc \leq atb\} \geq b,$$

i.e. properties (ii) and (iii) of Def.8.4.

Vice versa, assume that the left continuity is not preserved. Then there exist $a, b_i \in [0,1]$, $i\in I$, such that

$$b = \sup_{i\in I} (atb_i) < at\,(\sup_{i\in I} b_i).$$

Thus we observe that $atb_i \leq b$. Hence $b_i$ belongs to the set $\{c\in[0,1]: atc\leq b\}$ for any $i\in I$, then

$$\sup\{c\in[0,1]: atc \leq b\} \geq \sup_{i\in I} b_i$$

and since t is nondecreasing in the second argument, we have:

$$at(a\varphi b) = at\ \sup\{c\in[0,1]\colon atc \le b\} \ge at(\sup_{i\in I} b_i) > b,$$

a contradiction to the property (ii) of Def.8.4. ■

As noted in Remark 1 of [27], the left continuity of t in the second argument is essential in order to guarantee that $a\varphi b$, given by (8.3), belongs to the set $\{c\in[0,1]\colon atc\le b\}$.

Indeed, considering the t-norm $t_6$, we note that $(at_6x)$ is not left continuous at $x=1$. Further, we have if $1>a>b$:

$$a\varphi_6 b = 1 \notin [0,1) = \{c\in[0,1]\colon at_6c \le b\}.$$

Dually, it is proved that

THEOREM 8.3. *An operator* $\beta$ *is associated with an s-norm iff s is upper semicontinuous. Further, we have* $a\beta b=\inf\{c\in[0,1]\colon asc\ge b\}$ *for all* $a,b\in[0,1]$.

All t-norms (resp. s-norms) mentioned in Sec.8.1, except $t_6$ (resp. $s_6$), are lower (resp. upper) semicontinuous. The corresponding operators $\varphi$ and $\beta$ are listed below. Of course, $a\varphi b=1$ if $a\le b$ and $a\beta b=0$ if $a\ge b$, otherwise we have that

$$a\varphi_1^{(p)}b = 1-[(1-b)^p - (1-a)^p]^{1/p}, \qquad a\beta_1^{(p)}b = (b^p-a^p)^{1/p}, \qquad p\ge 1,$$

$$a\varphi_2^{(p)}b = \log_p\left[1+\frac{(p^b-1)\cdot(p-1)}{p^a-1}\right], \qquad a\beta_2^{(p)}b = 1-\log_p\left[1+\frac{(p^{1-b}-1)\ (p-1)}{(p^{1-a}-1)}\right],$$

$$0<p,\ p\ne 1,$$

$$a\varphi_3\, b = b/a, \qquad a\beta_3\, b = \frac{b-a}{1-a}$$

$$a\varphi_4^{(p)}b = \frac{pb+(1-p)ab}{a-b+pb+(1-p)ab}, \qquad a\beta_4^{(p)}b = \frac{b-a}{a[(p-1)\,(1-b)-1]+1}, \qquad p\ge 0,$$

$$a\varphi_5^{(p)}b = \frac{b-a+p+1-pa}{p+1-pa}, \qquad a\beta_5^{(p)}b = \frac{b-a}{1+pa}, \qquad p\ge -1,$$

$$a\varphi_7^{(p)}b = (1+b^p-a^p)^{1/p}, \qquad a\beta_7^{(p)}b = 1-[1+(1-b)^p-(1-a)^p]^{1/p}, \qquad p\ne 0.$$

Note that

$$\lim_{p\to+\infty} a\varphi_1{}^{(p)}b = a\alpha b \qquad \text{and} \qquad \lim_{p\to+\infty} a\beta_1{}^{(p)}b = a\varepsilon b$$

for all a,b∈ [0,1]. The operators $\varphi_1{}^{(p)}$, $\varphi_2{}^{(p)}$, $\beta_1{}^{(p)}$, $\beta_2{}^{(p)}$ are depicted in Figs. 8.1, 8.2, 8.3, 8.4, respectively.

DEFINITION 8.5. *Let* $A\in \mathbf{F}(\mathbf{X})$ *and* $B\in \mathbf{F}(\mathbf{Y})$. *We define the φ-composition (resp. β-composition) of* A *and* B *to be the fuzzy relation* $(A\varphi B)\in \mathbf{F}(\mathbf{X}\times\mathbf{Y})$ *(resp.* $(A\beta B)\in \mathbf{F}(\mathbf{X}\times\mathbf{Y})$*) with membership function given by*

$$(A\varphi B)\,(x,y) = A(x)\,\varphi\,B(y) \qquad (\text{resp. } (A\beta B)\,(x,y) = A(x)\,\beta\,B(y))$$

*for all* $x\in \mathbf{X}$, $y\in \mathbf{Y}$.

DEFINITION 8.6. *Let* $R\in \mathbf{F}(\mathbf{X}\times\mathbf{Y})$ *and* $B\in \mathbf{F}(\mathbf{Y})$. *We define the φ-composition (resp. β-composition) of* R *and* B *to be the fuzzy set* $(R\varphi B)\in \mathbf{F}(\mathbf{X})$ *(resp.* $(R\beta B)\in \mathbf{F}(\mathbf{Y})$*) with membership function given by*

$$(R\varphi B)\,(x) = \min_{y\in \mathbf{Y}} \,[R(x,y)\,\varphi\,B(y)],$$

resp.

$$(R\beta B)\,(x) = \max_{y\in \mathbf{Y}} \,[R(x,y)\,\beta\,B(y)],$$

*for any* $x\in \mathbf{X}$.

From now on, we assume that t is lower semicontinuous and s is upper semicontinuous. As in Thms.2.3 and 2.4, we can prove the following theorems that solve the questions (p1) and (p2) for Eq.(8.1):

THEOREM 8.4. *If* $A\in \mathbf{F}(\mathbf{X})$ *and* $B\in \mathbf{F}(\mathbf{Y})$ *are given and Eq.*(8.1) *has solutions* $R\in \mathbf{F}(\mathbf{X}\times\mathbf{Y})$, *then* $(A\varphi B)\in \mathbf{F}(\mathbf{X}\times\mathbf{Y})$ *is the greatest solution.*

THEOREM 8.5. *If* $B\in \mathbf{F}(\mathbf{Y})$ *and* $R\in \mathbf{F}(\mathbf{X}\times\mathbf{Y})$ *and Eq.*(8.1) *has solutions* $A\in \mathbf{F}(\mathbf{X})$, *then* $(R\varphi B)\in \mathbf{F}(\mathbf{X})$ *is the greatest solution.*

REMARK 8.1. Thms.8.4 and 8.5 can be considered as consequences of the basic Thms.2.3 and 2.4 respectively, observing that the structure ([0,1], ∧, ∨, t, 0, 1) is a complete residuated lattice in the sense of Sec.2.1, where t is assumed to be the binary multiplication satisfying the isotonicity condition (2.1).

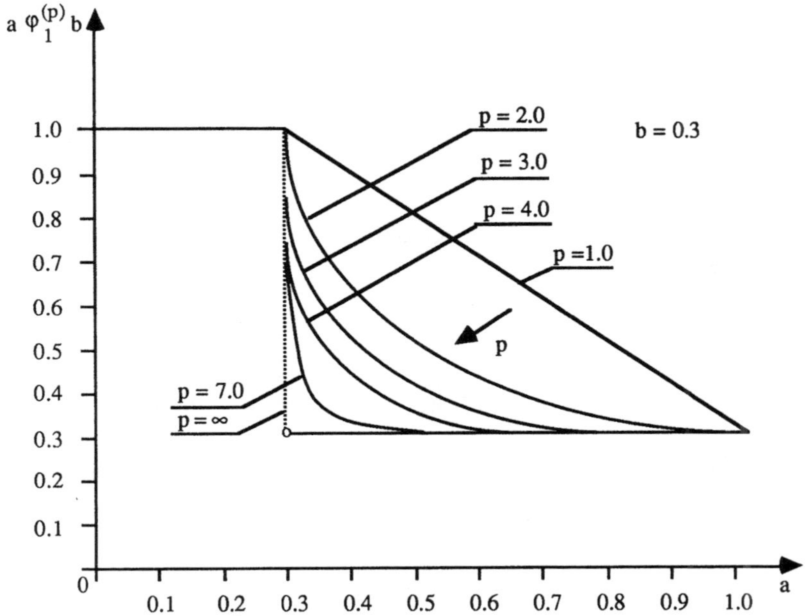


Fig. 8.1

Operator $\varphi_1^{(p)}$ for b = 0.3

We illustrate Thm.8.4 with the following example [12]:

EXAMPLE 8.1. From [10], we consider the lower semicontinuous t-norm defined as $atb=0$ if $a+b-1\leq 0$ and $atb=\min\{a,b\}$ if $a+b-1>0$, where $a,b\in[0,1]$. We have $a\varphi b=1$ if $a\leq b$, $a\varphi b=b$ if $1-a\leq b<a$ and $a\varphi b=1-a$ if $b<\min\{a,1-a\}$.

Let $\mathbf{X}=\{x_1,x_2\}$, $\mathbf{Y}=\{y_1,y_2\}$, and $A\in\mathbf{F}(\mathbf{X})$, $B\in\mathbf{F}(\mathbf{Y})$ be defined as $A(x_1)=A(x_2)=B(y_1)=0.5$ and $B(y_2)=0$. We have:

$$(A\varphi B) = \begin{Vmatrix} 1.0 & 0.5 \\ 1.0 & 0.5 \end{Vmatrix}$$

and $At(A\varphi B)=B$. We note that this equation has no lower solutions. Indeed, we have the following system of equations for any solution R:

$$\begin{cases} (0.5 \text{ t } R_{11}) \vee (0.5 \text{ t } R_{21}) = 0.5 \\ (0.5 \text{ t } R_{12}) \vee (0.5 \text{ t } R_{22}) = 0, \end{cases}$$

where $R_{ij}=R(x_i,y_j)$ for i,j=1,2. This system is satisfied iff

$$\begin{cases} 0.5 \text{ t } R_{11} = 0.5 \\ 0.5 \text{ t } R_{21} \le 0.5 \\ 0.5 \text{ t } R_{12} = 0 \\ 0.5 \text{ t } R_{22} = 0 \end{cases} \quad \text{or} \quad \begin{cases} 0.5 \text{ t } R_{11} \le 0.5 \\ 0.5 \text{ t } R_{21} = 0.5 \\ 0.5 \text{ t } R_{12} = 0 \\ 0.5 \text{ t } R_{22} = 0 , \end{cases}$$

i.e. either $R_{11} \in (0.5,1]$, $R_{21} \in [0,1]$ or $R_{11} \in [0,1]$, $R_{21} \in (0.5,1]$ and $R_{12}$, $R_{22} \in [0,0.5]$. For $a \in [0.5,1]$, we put:

$$R_a' = \begin{Vmatrix} a & 0.0 \\ 0.0 & 0.0 \end{Vmatrix}, \qquad R_a'' = \begin{Vmatrix} 0.0 & 0.0 \\ a & 0.0 \end{Vmatrix}.$$

Then the fuzzy relations $R_a'$ and $R_a''$ are solutions of the given equations if a>0.5, while $R_{0.5}'$ and $R_{0.5}''$ are not solutions.

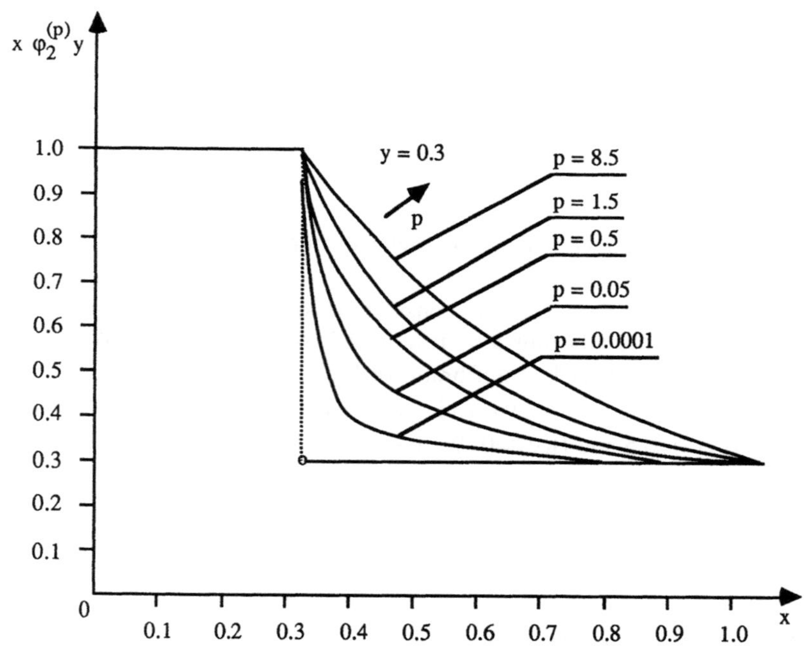


Fig. 8.2

Operator $\varphi_2^{(p)}$ for y = 0.3

In other words, we can say that if t is lower semicontinuous, Eq.(8.1) has a unique greatest solution by Thm.8.4 but, in general as Ex.8.1 shows, it does not have lower solutions. However, the membership functions of A and B may be such that some lower solutions may exist, as is easily seen in the following example:

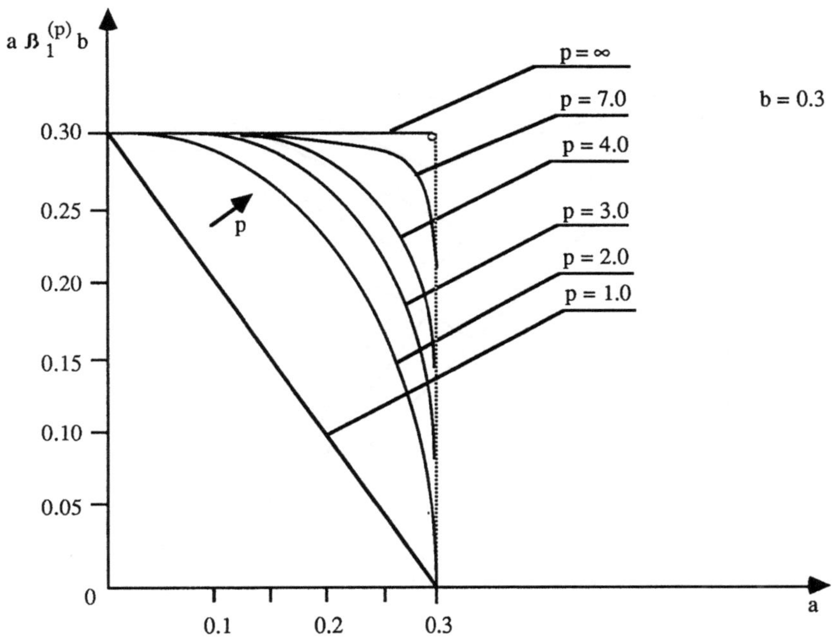


Fig. 8.3

Operator $\beta_1^{(p)}$ for b = 0.3

EXAMPLE 8.2. Let t, **X**, **Y**, B be as in Ex.8.1 and A∈ **F**(**X**) be defined as $A(x_1)=0.6$ and $A(x_2)=0.5$. We have:

$$(A\varphi B) = \begin{Vmatrix} 0.5 & 0.4 \\ 1.0 & 0.5 \end{Vmatrix}$$

and At(AφB)=B. Reasoning as in Ex.8.1, it is proved that the following fuzzy relations:

$$M = \begin{Vmatrix} 0.5 & 0.0 \\ 0.0 & 0.0 \end{Vmatrix}, \qquad R_x = \begin{Vmatrix} 0.0 & 0.0 \\ x & 0.0 \end{Vmatrix},$$

where $x \in [0.5,1]$, are solutions of the given equation if $x>0.5$. Furthermore, M is a lower solution but $R_{0.5}$ is not a solution.

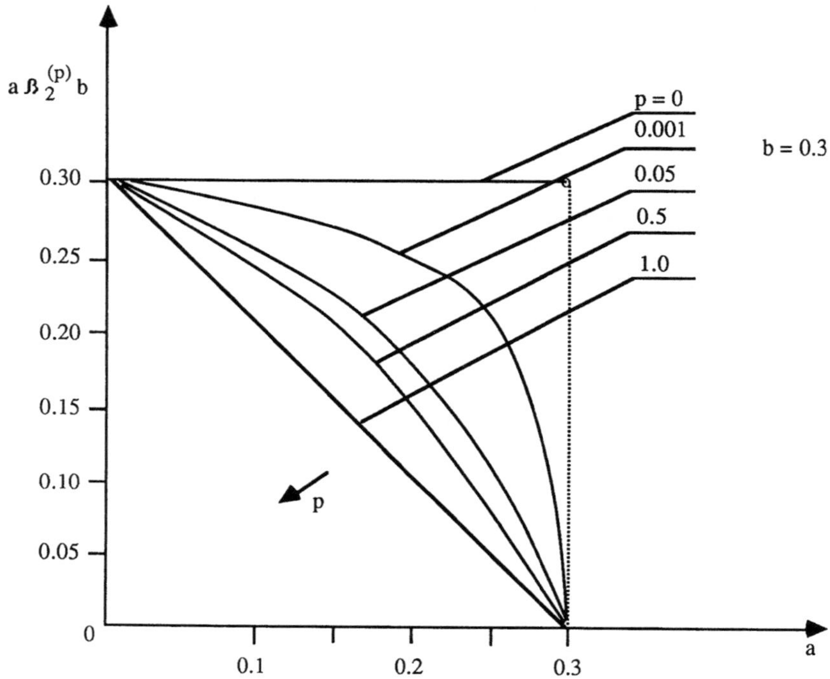


Fig. 8.4

Operator $\mathcal{B}_2^{(p)}$ for b = 0.3

We now consider a system of max-t fuzzy relation equations and, as in Thm.3.10, one can establish that

THEOREM 8.6. *If the system* $RtA_h=B_h$, $h=1,2,\ldots,m$, *has solutions* $R \in \mathbf{F}(\mathbf{X}\times\mathbf{Y})$, *where* $A_h \in \mathbf{F}(\mathbf{X})$ *and* $B_h \in \mathbf{F}(\mathbf{Y})$, *then the fuzzy relation* $R^{\frown} \in \mathbf{F}(\mathbf{X}\times\mathbf{Y})$ *defined by*

$$R^{\frown} = \bigwedge_{h=1}^{m} (A_h \,\varphi\, B_h)$$

*is the greatest solution of the given system.*

If the system has no solutions, we can look for approximate solutions of fuzzy relation equations, where the term *approximate* is defined appropriately to the concrete problem under consideration (this is discussed extensively in Ch.10).

We now enunciate, as Thms.2.13 and 2.14, two results which solve the question (p1) and (p2) for Eq.(8.2).

THEOREM 8.7. *If* $A \in \mathbf{F}(\mathbf{X})$ *and* $B \in \mathbf{F}(\mathbf{Y})$ *are given and Eq.*(8.2) *has solutions* $R \in \mathbf{F}(\mathbf{X}\times\mathbf{Y})$, *then* $(A \beta B) \in \mathbf{F}(\mathbf{X}\times\mathbf{Y})$ *is the smallest solution.*

THEOREM 8.8. *If* $B \in \mathbf{F}(\mathbf{Y})$ *and* $R \in \mathbf{F}(\mathbf{X}\times\mathbf{Y})$ *are given and Eq.*(8.2) *has solutions* $A \in \mathbf{F}(\mathbf{X})$, *then* $(R \beta B) \in \mathbf{F}(\mathbf{X})$ *is the smallest solution.*

## 8.3. Fuzzy Relation Equations of Complex Structure

Up to now we have solved fuzzy relation equations of simple structure. We extend our results to equations having the following form:

$$B = R \text{ t } A_1 \text{ t } A_2 \text{ t } \ldots \text{ t } A_n , \tag{8.4}$$

$$B = R \text{ s } A_1 \text{ s } A_2 \text{ s } \ldots \text{ s } A_n , \tag{8.5}$$

where $A_i \in \mathbf{F}(\mathbf{X}_i)$, $i=1,\ldots,n$, $B \in \mathbf{F}(\mathbf{Y})$ and $R \in \mathbf{F}(\mathop{\times}\limits_{i=1}^{n} \mathbf{X}_i \times \mathbf{Y})$, $\mathbf{X}_i$ being n finite sets.

In terms of membership functions, Eqs.(8.4) and (8.5) are written as follows:

$$B(y) = \max_{\substack{x_i \in \mathbf{X}_i \\ i=1,\ldots,n}} [A_1(x_1) \text{ t } \ldots \text{ t } A_n(x_n) \text{ t } R(x_1,\ldots,x_n,y)] ,$$

$$B(y) = \min_{\substack{x_i \in \mathbf{X}_i \\ i=1,\ldots,n}} [A_1(x_1) \text{ s } \ldots \text{ s } A_n(x_n) \text{ s } R(x_1,\ldots,x_n,y)]$$

for any $y \in \mathbf{Y}$, respectively. The resolution of these equations can be stated in a similar fashion as for the equations discussed before. Thus we put the following questions (cfr. Sec.8.2):

(c1) *determine R if* $A_1,\ldots, A_n$ *and B are given,*

(c2) *determine* $A_j$, $j \in \{1,\ldots,n\}$, *if* $A_1,\ldots, A_{j-1}, A_{j+1},\ldots, A_n$, *B and R are given.*

Considering (c1), we note that Eq.(8.4) can be rewritten as B=RtA, where $A \in \mathbf{F}(\mathbf{X}_1 \times \ldots \times \mathbf{X}_n)$ is a fuzzy relation given by

$$A = A_1 \, t \, A_2 \, t \ldots t \, A_n.$$

By Thm.8.4, the greatest fuzzy relation $R^\frown$ satisfying Eq.(8.4) is computed as

$$R^\frown = A \, \varphi \, B,$$

i.e.

$$R^\frown(\mathbf{x}, y) = A(\mathbf{x}) \, \varphi \, B(y)$$

for all $\mathbf{x}=(x_1,\ldots, x_n) \in \times_{i=1}^{n} \mathbf{X}_i$, $y \in \mathbf{Y}$.

Concerning (c2), we rewrite Eq.(8.4) in the following form:

$$B = G \, t \, A_j \,,$$

where

$$G=R \, t \, A_1 \, t \ldots t \, A_{j-1} \, t \, A_{j+1} \, t \ldots t \, A_n.$$

By Thm.8.5, the greatest fuzzy set $A_j \in \mathbf{F}(\mathbf{X}_j)$ satisfying Eq.(8.4) is equal to

$$A_j^\frown = G \, \varphi \, B,$$

i.e.

$$A_j^\frown(x_j) = \min_{y \in \mathbf{Y}} \, [G(x_j,y) \, \varphi \, B(y)]$$

for any $x_j \in \mathbf{X}_j$. By using Thms.8.7 and 8.8, it is clear that similar results can be established for Eq.(8.5).

We now consider composite fuzzy relation equations:

$$T = S \, t \, Q \,, \tag{8.6}$$

$$T = S \, s \, Q \,, \tag{8.7}$$

where $Q \in \mathbf{F}(\mathbf{X} \times \mathbf{Y})$, $S \in \mathbf{F}(\mathbf{Y} \times \mathbf{Z})$ and $T \in \mathbf{F}(\mathbf{X} \times \mathbf{Z})$, $\mathbf{X}$, $\mathbf{Y}$ and $\mathbf{Z}$ being finite referential sets.

In terms of membership functions, Eqs. (8.6) and (8.7) are read as

$$T(x,z) = \bigvee_{y \in \mathbf{Y}} [Q(x,y) \; t \; S(y,z)] \, ,$$

$$T(x,z) = \bigwedge_{y \in \mathbf{Y}} [Q(x,y) \; s \; S(y,z)]$$

for all $x \in \mathbf{X}$, $z \in \mathbf{Z}$.

Bearing in mind Remark 8.1, a direct application of Thms.2.3 and 2.4 yields the following results:

THEOREM 8.9. *If* $Q \in \mathbf{F}(\mathbf{X} \times \mathbf{Y})$ *and* $T \in \mathbf{F}(\mathbf{X} \times \mathbf{Z})$ *are given and Eq.*(8.6) *has solutions* $S \in \mathbf{F}(\mathbf{Y} \times \mathbf{Z})$, *then the fuzzy relation* $(Q^{-1} \varphi T) \in \mathbf{F}(\mathbf{Y} \times \mathbf{Z})$ *defined as*

$$(Q^{-1} \varphi T)(y,z) = \min_{x \in \mathbf{X}} [Q^{-1}(y,x) \varphi T(x,z)]$$

*for all* $y \in \mathbf{Y}$, $z \in \mathbf{Z}$, *is the greatest solution.*

THEOREM 8.10. *If* $S \in \mathbf{F}(\mathbf{Y} \times \mathbf{Z})$ *and* $T \in \mathbf{F}(\mathbf{X} \times \mathbf{Z})$ *are given and Eq.*(8.6) *has solutions* $Q \in \mathbf{F}(\mathbf{X} \times \mathbf{Y})$, *then the fuzzy relation* $(S \varphi T^{-1})^{-1} \in \mathbf{F}(\mathbf{X} \times \mathbf{Y})$ *defined as*

$$(S \varphi T^{-1})^{-1}(y,x) = \min_{z \in \mathbf{Z}} [S(y,z) \varphi T^{-1}(z,x)]$$

*for all* $x \in \mathbf{X}$, $y \in \mathbf{Y}$, *is the greatest solution.*

Of course, analogous results hold for Eq.(8.7).

## 8.4. A Related Adjoint Fuzzy Relation Equation

Here we are interested in fuzzy relation equations for which max-t composition is replaced by min-$\varphi$ composition already introduced in the statements of Thms. 8.9 and 8.10. It is a so-called adjoint fuzzy relation equation studied by Miyakoshi and Shimbo [27].

Dealing with fuzzy relation equations of the form:

$$T = S \varphi Q, \tag{8.8}$$

i.e.

$$T(x,z) = \min_{y \in \mathbf{Y}} [Q(x,y) \varphi S(y,z)]$$

for all $x\in \mathbf{X}$, $z\in \mathbf{Z}$, where $Q\in \mathbf{F}(\mathbf{X}\times\mathbf{Y})$, $S\in \mathbf{F}(\mathbf{Y}\times\mathbf{Z})$, $T\in \mathbf{F}(\mathbf{X}\times\mathbf{Z})$ and $\mathbf{X}$, $\mathbf{Y}$, $\mathbf{Z}$ are finite sets, we formulate, as usual, two problems:

(d1) *determine S if Q and T are given,*

(d2) *determine Q if S and T are given.*

Reasoning as in Ch.6 where a particular Eq.(8.8) was studied for $\varphi=\alpha$ (but assuming $\mathbf{L}$ to be a complete Brouwerian lattice), it is easily seen that the questions (d1) and (d2) are resolved by means of the following theorems (cfr. Thm.6.5).

THEOREM 8.11. *If* $Q\in \mathbf{F}(\mathbf{X}\times\mathbf{Y})$ *and* $T\in \mathbf{F}(\mathbf{X}\times\mathbf{Z})$ *are given and Eq.*(8.8) *has solutions* $S\in \mathbf{F}(\mathbf{Y}\times\mathbf{Z})$, *then the fuzzy relation* $(T\ t\ Q^{-1})\in \mathbf{F}(\mathbf{Y}\times\mathbf{Z})$ *defined by*

$$(T\ t\ Q^{-1})\ (y,z) = \bigvee_{x\in \mathbf{X}} [Q^{-1}(y,x)\ t\ T(x,z)]$$

*for all* $y\in \mathbf{Y}$, $z\in \mathbf{Z}$, *is the smallest solution.*

THEOREM 8.12. *If* $S\in \mathbf{F}(\mathbf{Y}\times\mathbf{Z})$ *and* $T\in \mathbf{F}(\mathbf{X}\times\mathbf{Z})$ *are given and Eq.*(8.8) *has solutions* $Q\in \mathbf{F}(\mathbf{X}\times\mathbf{Y})$, *then the fuzzy relation* $(T\ \varphi\ S^{-1})\in \mathbf{F}(\mathbf{X}\times\mathbf{Y})$ *defined as*

$$(T\ \varphi\ S^{-1})\ (x,y) = \min_{z\in \mathbf{Z}} [T(x,z)\ \varphi\ S^{-1}(z,y)]$$

*for all* $x\in \mathbf{X}$, $y\in \mathbf{Y}$, *is the greatest solution.*

We recall an example drawn from [27] that proves that lower solutions $Q\in \mathbf{F}(\mathbf{X}\times\mathbf{Y})$ cannot exist even if Eq.(8.8) has solutions and $t=\min$, $\varphi=\alpha$.

EXAMPLE 8.3. Let $\mathbf{X}=\{x_1,x_2\}$, $\mathbf{Y}=\{y_1,y_2\}$, $\mathbf{Z}=\{z_1,z_2\}$, $t=\min$, $\varphi=\alpha$ and $S\in \mathbf{F}(\mathbf{X}\times\mathbf{Y})$, $T\in \mathbf{F}(\mathbf{X}\times\mathbf{Z})$ be given by

$$S = \begin{Vmatrix} 0.3 & 0.4 \\ 0.6 & 0.1 \end{Vmatrix}, \qquad T = \begin{Vmatrix} 1.0 & 0.1 \\ 0.3 & 0.1 \end{Vmatrix}.$$

$$(T\varphi S^{-1}) = \begin{Vmatrix} 0.3 & 0.6 \\ 1.0 & 1.0 \end{Vmatrix}$$

and $(T\varphi S^{-1})\varphi S=T$. Thus the given equation has $(T\varphi S^{-1})$ as the greatest solution, but it has no lower solutions Q.

Indeed, we have the following system of equations for any solution Q:

$$(Q_{11}\ \alpha\ 0.3) \wedge (Q_{12}\ \alpha\ 0.6) = 1, \qquad (Q_{11}\ \alpha\ 0.4) \wedge (Q_{12}\ \alpha\ 0.1) = 0.1,$$

$$(Q_{21}\ \alpha\ 0.3) \wedge (Q_{22}\ \alpha\ 0.6) = 0.3, \qquad (Q_{21}\ \alpha\ 0.4) \wedge (Q_{22}\ \alpha\ 0.1) = 0.1,$$

where $Q(x_i,y_j)=Q_{ij}$ for i,j=1,2. This system is satisfied iff $Q_{11}\in[0,0.3]$, $Q_{12}\in(0.1,0.6]$, $Q_{21}\in(0.3,1]$, $Q_{22}\in(0.1,1]$ but a solution Q cannot be minimal since $Q_{12}$, $Q_{21}$, $Q_{22}$ belong to left-open intervals.

REMARK 8.2. If Eq.(8.8) is written for fuzzy sets, i.e.

$$B = R\ \varphi\ A\,, \tag{8.9}$$

where $A\in \mathbf{F}(\mathbf{X})$, $B\in \mathbf{F}(\mathbf{Y})$ and $R\in \mathbf{F}(\mathbf{X}\times\mathbf{Y})$, then , specializing Thm.8.11 to Eq.(8.9), we see that the fuzzy relation $(A\ t\ B)\in \mathbf{F}(\mathbf{X}\times\mathbf{Y})$ defined as $(A\ t\ B)(x,y)=A(x)\ t\ B(y)$ for all $x\in\mathbf{X}$, $y\in\mathbf{Y}$, is the smallest solution of Eq.(8.9) if R is unknown. It is also easily seen that if the system $R\ \varphi\ A_h=B_h$, h=1,2,...,m, has solutions $R\in \mathbf{F}(\mathbf{X}\times\mathbf{Y})$, where $A_h\in \mathbf{F}(\mathbf{X})$ and $B_h\in \mathbf{F}(\mathbf{Y})$, then the fuzzy relation $R^{\cup}\in \mathbf{F}(\mathbf{X}\times\mathbf{Y})$ defined by

$$R^{\cup} = \bigvee_{h=1}^{m} (A_h\ t\ B_h)$$

is the smallest solution of the given system.

## 8.5. Fuzzy Relation Equations with Upper Semicontinuous Triangular Norms

The result of this Sec. are contained in [12]. The operator, given in Def. 8.4, related to an upper semicontinuous s-norm, can be associated with an upper semicontinuous t-norm by setting for all $a,b\in[0,1]$:

$$\begin{aligned} a\psi_t b &= 0 && \text{if } a < b,\\ &= \inf\{c\in[0,1]: a t c \geq b\} && \text{if } a \geq b. \end{aligned}$$

An example of t-norm which is upper (but not lower) semicontinuous is $t_6$ (cfr. Sec.8.1), whose associated operator $\psi_6$ is defined as $a\psi_6 b=0$ if $a<b$ or $b=0$, $a\psi_6 b=b$ if $1=a>b$ and $a\psi_6 b=1$ if $1>a>b$ or $a=b$, where $a,b\in[0,1]$.

REMARK 8.3. If $a \geq b$, then (cfr. property (ii) of the operator $\mathbf{B}_S$ in Def.8.4) we have

$$a\ t\ (a\psi_t b) \geq b. \tag{8.10}$$

Note that if t=min, then $a\psi_t b = a\sigma b$, where $\sigma$ is the operator defined in Sec.3.1. For brevity, from now on, we write $a\psi b$ instead of $a\psi_t b$.

We now consider Eq.(8.1) defined on the finite sets $\mathbf{X}$ and $\mathbf{Y}$. If $A \in \mathbf{F}(\mathbf{X})$, $B \in \mathbf{F}(\mathbf{Y})$ and t is assumed to be upper semicontinuous, we define the following sets:

$$\mathbf{K}_t(y) = \{x \in \mathbf{X}: A(x)\ t\ [A(x)\ \psi B(y)] = B(y)\}$$

for any $y \in \mathbf{Y}$. Further, $\mathfrak{R}_t = \mathfrak{R}_t(A,B)$ be the set of all the solutions $R \in \mathbf{F}(\mathbf{X} \times \mathbf{Y})$ of Eq.(8.1). Then the following result holds (cfr.Thms. 3.1÷3.3):

THEOREM 8.13. $\mathfrak{R}_t \neq \emptyset$ *iff* $\mathbf{K}_t(y) \neq \emptyset$ *for any* $y \in \mathbf{Y}$.

PROOF. Let $\mathfrak{R}_t \neq \emptyset$. Then, we have for all $y \in \mathbf{Y}$, $R \in \mathfrak{R}_t$:

$$B(y) = \bigvee_{x \in \mathbf{X}} [A(x)\ t\ R(x,y)] = A(x')\ t\ R(x',y)$$

for some $x' \in \mathbf{X}$. Thus $R(x',y)$ belongs to the nonempty set

$$\{c \in [0,1]: A(x')\ t\ c \geq B(y)\},$$

whose infimum is $A(x')\psi B(y)$. Hence $A(x')\psi B(y) \leq R(x',y)$ and then, by (8.10), we have

$$B(y) \leq A(x')\ t\ [A(x')\ \psi\ B(y)] \leq A(x')\ t\ R(x',y) = B(y),$$

i.e. $x' \in \mathbf{K}_t(y)$. Vice versa, let $\mathbf{K}_t(y) \neq \emptyset$ for any $y \in \mathbf{Y}$. Define a fuzzy relation $(A\psi B) \in \mathbf{F}(\mathbf{X} \times \mathbf{Y})$ with membership function given, for all $x \in \mathbf{X}$, $y \in \mathbf{Y}$, by $(A\psi B)(x,y) = A(x)\psi B(y)$ if $B(y) > 0$ and $x \in \mathbf{K}_t(y)$ and $(A\psi B)(x,y) = 0$ otherwise. Then we have

$$\bigvee_{x \in X} [A(x)\ t\ (A\psi B)\ (x,y)] = \bigvee_{x \in \mathbf{K}_t(y)} \{A(x)\ t\ [A(x)\ \psi\ B(y)]\}$$

$$\vee \{ \bigvee_{x \notin \mathbf{K}_t(y)} A(x)\ t\ 0\}$$

$$= \{ \bigvee_{x \in \mathbf{K}_t(y)} B(y)\} \vee 0 = B(y)$$

for any $y \in \mathbf{Y}$. This means that $(A\psi B)$ belongs to $\mathfrak{R}_t$, i.e. $\mathfrak{R}_t \neq \emptyset$. ∎

REMARK 8.4. Note that if $\mathfrak{R}_t \neq \emptyset$, then for any $y \in \mathbf{Y}$, there exists an element $x(\in \mathbf{K}_t(y))$ of $\mathbf{X}$ such that $A(x) \geq B(y)$. The converse implication is not generally true as is proved in the following example (cfr. Ex.5 of [12]).

EXAMPLE 8.4. Let $t=t_6$, $\mathbf{X}=\{x_1,x_2,x_3\}$, $\mathbf{Y}=\{y_1,y_2,y_3\}$ and $A \in \mathbf{F}(\mathbf{X})$, $B \in \mathbf{F}(\mathbf{Y})$ be defined by

$$A = \| 0.2 \quad 0.5 \quad 0.9 \|, \qquad\qquad B = \| 0.8 \quad 0.5 \quad 0.0 \|.$$

Then $A(x_3) \geq B(y_j)$ for $j=1,2,3$ but $\mathfrak{R}_t = \emptyset$ since $\mathbf{K}_t(y_1)=\emptyset$. Indeed, we have for $i=1,2$:

$$A(x_i)\ t_6\ [A(x_i)\ \psi_6\ B(y_1)] = A(x_i)\ t\ [A(x_i)\ \psi_6\ 0.8] = A(x_i)\ t\ 0 = 0 \neq 0.8 = B(y_1)$$

and

$$A(x_3)\ t_6\ [A(x_3)\ \psi_6\ B(y_1)] = 0.9\ t_6\ [0.9\ \psi_6\ 0.8] = 0.9\ t_6\ 1 = 0.9 \neq 0.8 = B(y_1).$$

However, as in [9, Ch.3], if one assumes:

$$\textit{for any } a \in [0,1],\ t(a,\cdot) \textit{ is continuous in } [0,1], \tag{8.11}$$

then we can say that $x \in \mathbf{K}_t(y)$ iff $A(x) \geq B(y)$ and hence if $t=\min$ $(\psi=\sigma)$, the necessity and the sufficiency of the condition of Thm.8.13 are given by Thms.3.1 and 3.2, respectively.

As already established for max-min fuzzy relation equations of type (3.1) in Thms. 3.6÷3.8, the following results, whose proof is omitted, hold:

THEOREM 8.14. *If* $\mathfrak{R}_t \neq \emptyset$, $\mathfrak{R}_t$ *has minimal elements (obtained by choosing a nonzero element in the y-th columns of* $(A\psi B)$ *for which* $B(y)>0$) *whose fuzzy union is just* $(A\psi B)$.

THEOREM 8.15. *For any* $R \in \mathfrak{R}_t$, *there exists a minimal element* $L \in \mathfrak{R}_t$ *such that* $L \leq R$.

We clarify Thm.8.14 with the following example:

EXAMPLE 8.5. Let t, $\mathbf{X}$, $\mathbf{Y}$, B as in Ex.8.4 and $A \in \mathbf{F}(\mathbf{X})$ defined as $A(x_1)=0.2$, $A(x_2)=0.5$, $A(x_3)=1$. We have:

$$(A\psi_6 B) = \begin{Vmatrix} 0.0 & 0.0 & 0.0 \\ 0.0 & 1.0 & 0.0 \\ 0.8 & 0.5 & 0.0 \end{Vmatrix}.$$

Note that $\mathbf{K}_t(y_1)=\{x_3\}$, $\mathbf{K}_t(y_2)=\{x_2,x_3\}$, $\mathbf{K}_t(y_3)=X$ and then, since $\mathfrak{R}_t \neq \emptyset$, we have the following minimal elements:

$$\begin{Vmatrix} 0.0 & 0.0 & 0.0 \\ 0.0 & 1.0 & 0.0 \\ 0.8 & 0.0 & 0.0 \end{Vmatrix}, \qquad \begin{Vmatrix} 0.0 & 0.0 & 0.0 \\ 0.0 & 0.0 & 0.0 \\ 0.8 & 0.5 & 0.0 \end{Vmatrix}.$$

It is immediately seen that $\mathfrak{R}_t$ has not the greatest element.

Summarizing, we can say that if t is upper semicontinuous, the set of all the lower solutions of Eq.(8.1) is completely determined, but in general, as Ex.8.5 shows, the greatest solution does not exist. However, it is possible to exhibit easy examples in which the membership functions of A and B can be such that the greatest solution exists.

We now consider Eq.(8.6) defined on finite sets $\mathbf{X}$, $\mathbf{Y}$, $\mathbf{Z}$ with $Q \in \mathbf{F}(\mathbf{X}\times\mathbf{Y})$ and $T \in \mathbf{F}(\mathbf{X}\times\mathbf{Z})$ assigned and t is upper semicontinuous. As already observed in Sec.3.2, Eq.(8.6) is equivalent to the following system of n(=card $\mathbf{X}$) equations of type (8.1),

$$S \; t \; Q_x = T_x, \qquad x \in \mathbf{X}, \tag{8.12}$$

where $Q_x \in \mathbf{F}(\{x\}\times\mathbf{Y})$ and $T_x \in (\{x\}\times\mathbf{Z})$ are fuzzy sets whose membership functions are the restrictions of Q and T to the specified Cartesian products.

If $\mathbf{S}_t=\mathbf{S}_t(Q,T)$ (resp. $\mathfrak{R}_{t,x}=\mathfrak{R}_{t,x}(Q_x,T_x)$) denotes the set of all the solutions $S \in \mathbf{F}(\mathbf{Y}\times\mathbf{Z})$ of Eq.(8.6) (resp. Eq.(8.12)), obviously we have:

$$\mathbf{S}_t = \bigcap_{x \in \mathbf{X}} \mathfrak{R}_{t,x}.$$

Assume that $\mathbf{S}_t \neq \emptyset$. Then $\mathfrak{R}_{t,x} \neq \emptyset$ for any $x \in \mathbf{X}$ and let $\mathfrak{L}_{t,x}=\mathfrak{L}_{t,x}(Q_x,T_x)$, (which is nonempty by Thm.8.14) be the finite set of all the minimal elements of $\mathfrak{R}_{t,x}$. If $L_x$ denotes an arbitrary element of $\mathfrak{L}_{t,x}$, we define the following finite subset $\mathfrak{L}_t$ of $\mathbf{F}(\mathbf{Y}\times\mathbf{Z})$:

$$\mathfrak{L}_t = \mathfrak{L}_t(Q,T) = \{L \in \mathbf{F}(\mathbf{Y}\times\mathbf{Z}) \colon L = \bigvee_{x \in \mathbf{X}} L_x,\ L_x \in \mathfrak{L}_{t,x}\}.$$

Now, adopting the same proof of Thm.3.11 and by using Thm.8.15, it is not difficult to prove the following result:

THEOREM 8.16. *Let* $\mathbf{S}_t \neq \emptyset$. *Then* $\mathbf{S}_t \cap \mathbf{L}_t \neq \emptyset$ *and an element* S *is minimal in* $\mathbf{S}_t$ *iff it is minimal in* $\mathbf{S}_t \cap \mathbf{L}_t$.

We illustrate Thm.8.16 with the following example of [12]:

EXAMPLE 8.6. Let $t=t_6$, $\mathbf{X}=\{x_1,x_2\}$, $\mathbf{Y}=\{y_1,y_2\}$, $\mathbf{Z}=\{z_1,z_2\}$, and $Q \in \mathbf{F}(\mathbf{X}\times\mathbf{Y})$, $T \in \mathbf{F}(\mathbf{X}\times\mathbf{Z})$ be defined as

$$Q = \begin{Vmatrix} 0.9 & 1.0 & 0.6 \\ 0.7 & 0.5 & 1.0 \end{Vmatrix}, \qquad T = \begin{Vmatrix} 0.8 & 0.6 \\ 0.2 & 0.7 \end{Vmatrix}.$$

Then, by using the same notations of Ex.3.3, we have

$$(Q_1\, \psi_6\, T_1) = \begin{Vmatrix} 0.0 & 0.0 \\ 0.8 & 0.6 \\ 0.0 & 1.0 \end{Vmatrix}, \qquad (Q_2\, \psi_6\, T_2) = \begin{Vmatrix} 0.0 & 1.0 \\ 0.0 & 0.0 \\ 0.2 & 0.7 \end{Vmatrix}.$$

Thus $\mathbf{L}_{t,x_1}=\{L_1,M_1\}$ and $\mathbf{L}_{t,x_2}=\{L_2,M_2\}$, where

$$L_1 = \begin{Vmatrix} 0.0 & 0.0 \\ 0.8 & 0.6 \\ 0.0 & 0.0 \end{Vmatrix}, \qquad M_1 = \begin{Vmatrix} 0.0 & 0.0 \\ 0.8 & 0.0 \\ 0.0 & 1.0 \end{Vmatrix},$$

$$L_2 = \begin{Vmatrix} 0.0 & 1.0 \\ 0.0 & 0.0 \\ 0.2 & 0.0 \end{Vmatrix}, \qquad M_2 = \begin{Vmatrix} 0.0 & 0.0 \\ 0.0 & 0.0 \\ 0.2 & 0.7 \end{Vmatrix}.$$

Then $\mathbf{L}_t=\{L,S,U,W\}$, where

$$L = L_1 \vee L_2 = \begin{Vmatrix} 0.0 & 1.0 \\ 0.8 & 0.6 \\ 0.2 & 0.0 \end{Vmatrix}, \qquad S = L_1 \vee M_2 = \begin{Vmatrix} 0.0 & 0.0 \\ 0.8 & 0.6 \\ 0.2 & 0.7 \end{Vmatrix},$$

$$U = M_1 \vee L_2 = \begin{Vmatrix} 0.0 & 1.0 \\ 0.8 & 0.0 \\ 0.2 & 1.0 \end{Vmatrix}, \qquad W = M_1 \vee M_2 = \begin{Vmatrix} 0.0 & 0.0 \\ 0.8 & 0.0 \\ 0.2 & 1.0 \end{Vmatrix}.$$

It is easily seen that $\mathbf{S}_t \cap \mathbf{L}_t=\{S\}$, hence S is the unique minimal element, i.e. the minimum, of $\mathbf{S}_t$. We note that $\mathbf{S}_t$ has not the greatest element.

In [9, Ch.3], fuzzy relation equations under a t-norm satisfying property (8.11) were studied. In this case, since t is both lower and upper semicontinuous, Eq.(8.1) has the greatest solution and lower solutions by Thms.8.4 and 8.14.

Additionally, in [9, Ch.3], using methods similar to those of [7, Ch.3], the solution of Eqs.(8.1) and (8.6) with the smallest value of fuzziness (measured with the use of a t-norm in the sense of Yager [45]) were characterized. Similar questions were solved for max-$t_1^{(p)}$ fuzzy relation equations in [8, Ch.3]. For further results, see also [8], [9], [10], [12], [27], [15, Ch.3].

It is obvious that analogous results can be established using lower semicontinuous s-norms. Similar considerations can be suitably made for the equations dealt in Secs.8.3 and 8.4 .

## References

[1] J. Aczel, *Lectures on Functional Equations and Their Applications*, Academic Press, New York, 1986.

[2] C. Alsina, On convex triangle functions, *Aequationes Math.* 26 (1983), 191-196.

[3] C. Alsina, On a family of connectives for fuzzy sets, *Fuzzy Sets and Systems* 16 (1985), 231-235.

[4] C. Alsina, On a functional equation characterizing two binary operations on the space of membership functions, *Fuzzy Sets and Systems* 27 (1988), 5-9.

[5] C. Alsina, E. Trillas and L. Valverde, On some logical connectives for fuzzy set theory, *J. Math. Anal. Appl.* 93 (1983), 15-26.

[6] R.E. Bellman and M. Giertz, On the analytic formalism of the theory of fuzzy sets, *Inform. Sciences* 5(1973), 149-156.

[7] R.E. Bellman and L.A. Zadeh, Local and fuzzy logics, in: *Modern Uses of Multiple Valued Logics* (J.M. Dunn and D. Epstein, Eds.), D. Reidel Publ. Co., Dordrecht (1977), pp.103-165.

[8] L. Bour and M. Lamotte, Solutions minimales d'equations de relations floues avec la composition max-norme triangulaire, *BUSEFAL* 31 (1987), 24-31.

[9] L. Bour and M. Lamotte, Equations de relations floues avec la composition conorm-norm triangulaires, *BUSEFAL* 34 (1988), 86-94.

[10] L. Bour, G. Hirsch and M. Lamotte, Determination d'un operateur de maximalization pour la resolution d'equations de relation floue, *BUSEFAL* 25 (1986), 95-106.

[11] W.F. Darsow and M.J. Frank, Associative functions and Abel-Schröder systems, *Publ. Math. Debrecen* 30 (1983), 253-272.

[12] A. Di Nola, W. Pedrycz and S. Sessa, Fuzzy relation equations under LSC and USC T-norms and its Boolean solutions, *Stochastica*, to appear.

[13] D. Dubois and H. Prade, A class of fuzzy measures based on triangular norms, *Internat. J. Gen. Systems* 8 (1982), 43-61.

[14] D. Dubois and H. Prade, A theorem on implication functions defined from triangular norms, *Stochastica* 8, no.3 (1984), 267-279.

[15] D. Dubois and H. Prade, Fuzzy-set-theoretic differences and inclusions and their use

in the analysis of fuzzy equations, *Control and Cybern.* 13 (1984), 129-146.

[16] D. Dubois and H. Prade, A review of fuzzy set aggregation connectives, *Inform. Sciences* 36 (1985), 85-121.

[17] H. Dyckhoff and W. Pedrycz, Generalized means as model of compensative connectives, *Fuzzy Sets and Systems* 14 (1984), 145-154.

[18] M.J Frank, On the simultaneous associativity of F(x,y) and x+y-F(x,y), *Aequationes Math.* 19 (1979), 194-226.

[19] L. Fuchs, *Partially Ordered Algebraic Systems*, Pergamon Press, Oxford, 1963.

[20] R. Goetschel and W. Voxman, A note on characterization of the max and min operators, *Inform. Sciences* 30 (1983), 5-10.

[21] S. Gottwald, Generalization of some results of E. Sanchez, *BUSEFAL* 16 (1983), 54-60.

[22] S. Gottwald, Characterization of the solvability of fuzzy equations, *Elektron. Inf. verarb. Kybern.* 22 (1986), n. 2/3, 67-91.

[23] S. Gottwald, Fuzzy set theory with T-norms and φ-operators, in *Topics in the Mathematics of Fuzzy Systems* (A. Di Nola and A. Ventre, Eds.), Verlag TÜV Rheinland, Köln (1986), pp.143-196.

[24] H. Hamacher, Über logische Vernüpfungen unscharfer Aussagen und deren Zugehörige Bewertungs-funktionen, in: *Progress in Cybernetics and Systems Research,* Vol.II, (R. Trappl and F. de P. Hanika, Eds.), Hemisphere Publ. Corp., New York (1975), pp.276-287.

[25] E.P. Klement, Construction of fuzzy σ-algebras using triangular norms, *J. Math. Anal. Appl.* 85 (1982), 543-565.

[26] K. Menger, Statistical metric spaces, *Proc. Nat. Acad. Sci. USA* 28 (1942), 535-537.

[27] M. Miyakoshi and M. Shimbo, Solutions of composite fuzzy relational equations with triangular norms, *Fuzzy Sets and Systems* 16 (1985), 53-63.

[28] M. Mizumoto, T-norms and their pictorial representations *BUSEFAL* 25 (1986), 67-78.

[29] W. Pedrycz, Fuzzy relational equations with triangular norms and their resolution, *BUSEFAL* 11 (1982), 24-32.

[30] W. Pedrycz, Some aspects of fuzzy decision-making, *Kybernetes* 11 (1982), 297-301.

[31] W. Pedrycz, Fuzzy relational equations with generalized connectives and their applications, *Fuzzy Sets and Systems* 10 (1983), 185-201.

[32] W. Pedrycz, Numerical and applicational aspects of fuzzy relational equations, *Fuzzy Sets and Systems* 11 (1983), 1-18.

[33] W. Pedrycz, Some applicational aspects of fuzzy relational equations in system analysis, *Internat. J. Gen. Systems* 9 (1983), 125-132.

[34] W. Pedrycz, On generalized fuzzy relational equations and their applications, *J. Math. Anal. Appl.* 107 (1985), 520-536.

[35] W. Rodder, *On "and" and "or" connectives in fuzzy set theory*, RWTH Aachen,

75/07, 1975.

[36] B. Schweizer and A.Sklar, Statistical metric spaces, *Pacific J. Math.* 10 (1960), 313-334.

[37] B. Schweizer and A.Sklar, Associative functions and statistical triangle inequalities, *Publ. Math. Debrecen* 8 (1961), 169-186.

[38] B. Schweizer and A.Sklar, Associative functions and abstract semigroups, *Publ. Math. Debrecen* 10 (1963), 69-81.

[39] B. Schweizer and A.Sklar, *Probabilistic Metric Spaces*, North-Holland, N.Y., 1983.

[40] H. Sherwood, Characterizing dominates on a family of triangular norms, *Aequationes Math.* 27 (1984), 255-273.

[41] M. Sugeno, Fuzzy measures and fuzzy integrals: a survey, in: *Fuzzy Automata and Decision Processes* (M.M. Gupta, G.N. Saridis and B.R. Gaines, Eds.), North-Holland, Amsterdam (1977), pp.89-102.

[42] M. Tardiff, On a generalized Minkowsky inequality and its relation to dominates for t-norms, *Aequationes Math.* 27 (1984), 308-316.

[43] U. Thole, H.J. Zimmermann and P. Zysno, On the suitability of minimum and product operators for intersection of fuzzy sets, *Fuzzy Sets and Systems* 2 (1979), 167-180.

[44] R.R. Yager, On a general class of fuzzy connectives, *Fuzzy Sets and Systems* 4 (1980), 235-242.

[45] R.R. Yager, Measures of fuzziness based on T-norms, *Stochastica* 3 (1982), 207-229.

[46] R.R. Yager, Some procedures for selecting fuzzy set-theoretic operators, *Internat. J. Gen. Systems* 8 (1982), 115-124.

[47] Y.D. Yu, Triangular norms and TNF-sigma algebras, *Fuzzy Sets and Systems* 16 (1985), 251-264.

[48] L. Valverde, On the structure of F-indistinguishability operators, *Fuzzy Sets and Systems* 17 (1985), 313-328.

[49] S. Weber, A general concept of fuzzy connectives, negations and implications based on t-norms and t-conorms, *Fuzzy Sets and Systems* 11 (1983), 115-134.

[50] H.J. Zimmermann and P. Zysno, Latent connectives in human decision-making, *Fuzzy Sets and Systems* 4 (1980), 37-51.

## CHAPTER 9

## FUZZY RELATION EQUATIONS WITH EQUALITY AND DIFFERENCE COMPOSITION OPERATORS



*In this Ch. we will introduce other composition operators which have a plausible logical interpretation. They will be called equality and difference operator, respectively. We first define a notion of equality and difference of any two grades of membership of a fuzzy set and of a fuzzy relation. In the sequel this concept will be utilized to form the respective composition operators and the related fuzzy equations. Afterwards, we provide a method of resolution of these equations, characterizing completely the set of the solutions. All the fuzzy sets involved are defined on finite sets and with membership values in [0,1].*

### 9.1. Equality and Difference Operators

One of the main problems that arise in fuzzy set theory is how a property of equivalence of a fuzzy set and of a fuzzy relation can be evaluated.

When one discusses the equivalence of two propositions "**p**" and "**q**", one has the following expression:

$$(p\&q) = (p\rightarrow q) \wedge (q\rightarrow p),$$

which in terms of the truth values $|p|$ and $|q|$ of the mentioned propositions ($|p|$ and $|q|$ belong to [0,1]) is read as

$$|p\&q| = (|p| \,\alpha\, |q|) \wedge (|q| \,\alpha\, |p|), \tag{9.1}$$

where "$\alpha$" is the implication operator defined in Sec.2.1 and $|p\&q|$ represents the degree of equivalence of the propositions "**p**" and "**q**".

Formalizing the formula (9.1) to the case of a fuzzy set $A\in \mathbf{F}(\mathbf{X})$ and a fuzzy relation $R\in \mathbf{F}(\mathbf{X}\times\mathbf{Y})$, we define as degrees of equality of A and R the following expressions:

$$(R\&A)\,(y) = \bigvee_{x\in \mathbf{X}} [A(x)\&R(x,y)] = B(y), \qquad (9.2)$$

$$(R\&'A)\,(y) = \bigwedge_{x\in \mathbf{X}} [A(x)\&R(x,y)] = B(y), \qquad (9.2')$$

for any $y\in \mathbf{Y}$, where $\&:[0,1]^2\to[0,1]$ is the "equality" operator defined as $(a\&b)=(a\alpha b)\wedge(b\alpha a)$ for any $a,b\in[0,1]$. Giving an interpretation of the expression (9.2) (resp. (9.2')), we can say that B(y), for y fixed, supplies an *optimistic* (resp. *pessimistic*) *value* of the degree of equality between A and R.

Let $\varepsilon$: $[0,1]^2\to[0,1]$ be the operator, dual of "$\alpha$", defined in Sec.2.5. We consider additionally the following expressions:

$$(R\&^{\frown}A)\,(y) = \bigwedge_{x\in \mathbf{X}} [A(x)\&^{\frown}R(x,y)] = B(y), \qquad (9.3)$$

$$(R\&'^{\frown}A)\,(y) = \bigvee_{x\in \mathbf{X}} [A(x)\&^{\frown}R(x,y)] = B(y), \qquad (9.3')$$

for any $y\in \mathbf{Y}$, where $\&^{\frown}$: $[0,1]^2\to[0,1]$ is the "difference" operator defined as $(a\&^{\frown}b)=(a\varepsilon b)\vee(b\varepsilon a)$ for any $a,b\in[0,1]$. Interpreting the expression (9.3) (resp. (9.3')), we can say that B(y), for y fixed, is a *pessimistic* (resp. *optimistic*) *value* of the degree of difference between A and R.

Here we assume that **X** and **Y** are finite sets and we denote briefly Eqs. (9.2), (9.2'), (9.3), (9.3') by the formulas:

$$R\&A = B, \qquad (9.4)$$

$$R\&'A = B, \qquad (9.4')$$

$$R\&^{\frown}A = B, \qquad (9.5)$$

$$R\&'^{\frown}A = B, \qquad (9.5')$$

respectively, where A and B are given and R is unknown. We establish the theory for Eqs.(9.4) and (9.4'), the results for Eqs.(9.5) and (9.5') being dually obtained.

## 9.2. Basic Lemmas

We have already seen in Sec.2.2 that, in particular, for the max-min composition, the monotonicity property holds, i.e. if $R \leq R'$, then $R \odot A \leq R' \odot A$, where $A \in \mathbf{F}(\mathbf{X})$ and $R, R' \in \mathbf{F}(\mathbf{X} \times \mathbf{Y})$. This property is crucial in the proof of Thm.2.3 and it is easily seen that, in general, if $R \leq R'$, then $R \& A \leq R' \& A$ and $R \&' A \leq R' \&' A$. However we shall prove results similar to the fundamental Thm.2.3 and to achieve this aim, we need to define the operator $\xi: [0,1]^2 \rightarrow [0,1]$ as $a \xi b = a \alpha b$ if $b<1$ and $a \xi b = a$ if $b=1$, where $a, b \in [0,1]$.

A straightforward computation proves that

(i) $a \; \xi \max \{b, c\} = \max \{a \xi b, a \xi c\}$ if $\max \{b, c\} < 1$,

(ii) $a \; \xi \; (a \& b) \geq b$,

(iii) $(a \& b) \leq (a \& c)$ if $a \& b < 1$ and $b \leq c$,

(iv) $a \& (a \xi b) \leq b$,

(v) $\min\{a, (a \& b)\} = \min\{a, b\}$,

(vi) $a \& \min\{a, b\} = a \alpha b \geq b$,

where $a, b, c \in [0,1]$.

REMARK 9.1. Property (i) is false if $\max \{b, c\} = 1$. Indeed, we have for $a \leq c < b = 1$:

$$a \; \xi \max \{b, c\} = a \; \xi 1 = a < 1 = \max\{a, 1\} = \max \{a \xi b, a \xi c\}.$$

REMARK 9.2. Property (iii) is false if $a \& b = 1$, i.e. $a = b$. Indeed, we have for $a = b < c$:

$$a \& c = (a \alpha c) \wedge (c \alpha a) = 1 \wedge a = a < 1 = a \& b.$$

Let $A \in \mathbf{F}(\mathbf{X})$ and $B \in \mathbf{F}(\mathbf{Y})$. We define the $\xi$-composition of A and B to be the fuzzy relation $(A \xi B) \in \mathbf{F}(\mathbf{X} \times \mathbf{Y})$ defined by

$$(A \xi B)(x, y) = A(x) \; \xi \; B(y)$$

for all $x \in X$, $y \in Y$. Then we can prove the following basic Lemmas for Eq.(9.4):

LEMMA 9.1. *For any* $A \in \mathbf{F}(\mathbf{X})$ *and* $B \in \mathbf{F}(\mathbf{Y})$, *we have:*

$$(A\xi B) \,\&\, A \leq B.$$

PROOF. In terms of membership functions, the property (iv) implies that

$$[(A\xi B) \,\&\, A]\,(y) = \bigvee_{x\in \mathbf{X}} \{A(x) \,\&\, [A(x)\,\xi\, B(y)]\} \leq \bigvee_{x\in \mathbf{X}} B(y) = B(y)$$

for any $y\in \mathbf{Y}$ and therefore the thesis. ■

LEMMA 9.2. *Let* $A\in \mathbf{F}(\mathbf{X})$, $R\in \mathbf{F}(\mathbf{X}\times\mathbf{Y})$ *and* $y\in \mathbf{Y}$ *be such that* $(R\&A)(y)<1$. *Then we have:*

$$R(x,y) \leq [A\,\xi\,(R\&A)]\,(x,y)$$

*for any* $x\in \mathbf{X}$.

PROOF. By properties (i) and (ii), we have:

$$\begin{aligned}[A\xi(R\&A)]\,(x,y) &= A(x)\,\xi\,\{[(R\&A)]\,(y)\}\\ &= A(x)\,\xi\,\{\bigvee_{x'\in \mathbf{X}} [A(x') \,\&\, R(x',y)]\}\\ &= A(x)\,\xi\,\{\bigvee_{x'\neq x} [A(x') \,\&\, R(x',y)] \vee [A(x) \,\&\, R(x,y)]\}\\ &\geq A(x)\,\xi\,[A(x) \,\&\, R(x,y)] \geq R(x,y)\end{aligned}$$

for any $x\in \mathbf{X}$, i.e. the thesis. ■

LEMMA 9.3. *If* $R(x,y)\leq R'(x,y)$ *for any* $x\in \mathbf{X}$ *and for some* $y\in \mathbf{Y}$, *then we have:*

$$(R\&A)(y) \leq (R'\&A)(y),$$

*provided that* $(R\&A)(y)<1$.

PROOF. Since $\mathbf{X}$ is finite, we have:

$$1>(R\&A)(y) = \bigvee_{x\in \mathbf{X}} [A(x) \,\&\, R(x,y)] = A(x') \,\&\, R(x',y)$$

for some $x'\in \mathbf{X}$. By property (iii), we deduce that

$$[A(x') \,\&\, R(x',y)] \leq [A(x') \,\&\, R'(x',y)]$$

and hence:

$$(R\&A)(y) = \bigvee_{x\in \mathbf{X}} [A(x) \,\&\, R(x,y)] = A(x') \,\&\, R(x',y)$$

$$\leq A(x')\&R'(x',y) \leq \bigvee_{x\in \mathbf{X}} [A(x) \,\&\, R'(x,y)] = (R'\&A)(y),$$

i.e. the thesis. ■

Analogously, by using properties (vi), (v) and (iii), we can show the following basic Lemmas for Eq.(9.4'):

LEMMA 9.1'. *For all* $A\in \mathbf{F}(\mathbf{X})$ *and* $B\in \mathbf{F}(\mathbf{Y})$, *we have:*

$$(A\times B) \,\&'\, A \geq B,$$

*where* $(A\times B)\in \mathbf{F}(\mathbf{X}\times\mathbf{Y})$ *is defined in Sec.*6.1.

LEMMA 9.2'. *For all* $A\in \mathbf{F}(\mathbf{X})$ *and* $R\in \mathbf{F}(\mathbf{X}\times\mathbf{Y})$,*we have:*

$$R \geq [A \times (R\&'A)].$$

LEMMA 9.3'. *If* $R(x,y)\leq R'(x,y)$ *for any* $x\in \mathbf{X}$ *and for some* $y\in \mathbf{Y}$, *then we have*

$$(R\&'A)(y) \leq (R'\&'A)(y),$$

*provided that* $(R\&'A)(y)<1$.

### 9.3. Resolution of the Fuzzy Equation (9.4)

Let $\mathcal{Z}=\mathcal{Z}(A,B)$ be the set of all fuzzy relations $R\in \mathbf{F}(\mathbf{X}\times\mathbf{Y})$ satisfying Eq.(9.4). The following fundamental theorem holds:

THEOREM 9.4. $\mathcal{Z}\neq\emptyset$ iff $(A\xi B)\in\mathcal{Z}$. *Furthermore,* $R\leq(A\xi B)$ *for any* $R\in\mathcal{Z}$ *if* B *is not normal, i.e.* $\mathbf{B}^{-1}(1)=\{y\in \mathbf{Y}\colon B(y)=1\}=\emptyset$.

PROOF. Let $\mathcal{Z}\neq\emptyset$ and R be arbitrarily chosen in $\mathcal{Z}$. Let $y\in \mathbf{Y}$ such that

$(R\&A)(y)=B(y)<1$ and then by Lemma 9.2:

$$R(x,y) \leq (A \,\xi\, B)(x,y) \tag{9.6}$$

for any $x \in \mathbf{X}$. By Lemma 9.3, we deduce that

$$B(y) = (R\&A)(y) \leq [(A \,\xi\, B) \,\&\, A]\,(y) \tag{9.7}$$

and by Lemma 9.1:

$$[(A \,\xi\, B) \,\&\, A](y) \leq B(y). \tag{9.8}$$

The inequalities (9.7) and (9.8) imply that

$$[(A \,\xi\, B) \,\&\, A](y) = B(y). \tag{9.9}$$

Now let $y \in \mathbf{B}^{-1}(1)$. Then we have:

$$[(A\xi B) \,\&\, A]\,(y) = \bigvee_{x \in \mathbf{X}} \{A(x) \,\&\, [(A \,\xi\, B)(x,y)]\} = \bigvee_{x \in \mathbf{X}} [A(x) \,\&\, A(x)] = 1 = B(y). \tag{9.10}$$

The equalities (9.9) and (9.10) ensure that $(A\xi B) \in \mathcal{Z}$. The converse implication is trivial. Of course, the second part of the thesis follows from (9.6). ■

If $B(y)=1$ for some $y \in \mathbf{Y}$, $(A\xi B)$ is not the greatest element of $\mathcal{Z}$ as is shown in the following example:

EXAMPLE 9.1. Let $\mathbf{X}=\{x_1,x_2,x_3\}$, $\mathbf{Y}=\{y_1,y_2\}$ and $A \in \mathbf{F}(\mathbf{X})$, $B \in \mathbf{F}(\mathbf{Y})$ defined as

$$A = \| \; 0.3 \quad 1.0 \quad 0.6 \; \|, \qquad B = \| \; 0.3 \quad 1.0 \; \|.$$

We have:

$$\left\| \begin{matrix} 1.0 & 0.3 \\ 0.3 & 1.0 \\ 0.3 & 0.6 \end{matrix} \right\| = (A \,\xi\, B) < (A \,\alpha\, B) = \left\| \begin{matrix} 1.0 & 1.0 \\ 0.3 & 1.0 \\ 0.3 & 1.0 \end{matrix} \right\|$$

and it is seen that $(A\alpha B)\&A=(A\xi B)\&A=B$.

Indeed, in this case the following theorem holds:

THEOREM 9.5. *Let* $\mathbf{Z}\neq\emptyset$. *If* A *and* B *are normal, i.e.* $\mathbf{A}^{-1}(1)=\{x\in\mathbf{X}: A(x)=1\}\neq\emptyset$ *and* $\mathbf{B}^{-1}(1)\neq\emptyset$, *then the fuzzy relation* $(A\alpha B)\in\mathbf{F}(\mathbf{X}\times\mathbf{Y})$ (cfr. Def.2.6) *is the greatest element of* $\mathbf{Z}$.

PROOF. Let R be arbitrarily chosen in $\mathbf{Z}$. For any $x\in\mathbf{X}$ and for any $y\in\mathbf{Y}\text{-}\mathbf{B}^{-1}(1)$, we have by (9.6):

$$R(x,y) \leq (A \xi B)(x, y) = A(x)\, \alpha\, B(y) \tag{9.11}$$

while, for any $x\in\mathbf{X}$ and for any $y\in\mathbf{B}^{-1}(1)$, obviously we deduce that

$$R(x,y) \leq 1 = A(x)\, \alpha\, B(y). \tag{9.12}$$

Thus (9.11) and (9.12) ensure that $R\leq(A\alpha B)$ for any $R\in\mathbf{Z}$. The thesis is proved if we show that $(A\alpha B)$ belongs to $\mathbf{Z}$. Indeed, if $B(y)<1$ we have (cfr. (9.9)):

$$\begin{aligned}[(A\alpha B)) \,\&\, A]\,(y) &= \bigvee_{x\in\mathbf{X}} \{A(x) \,\&\, [A(x)\, \alpha\, B(y)]\} \\ &= \bigvee_{x\in\mathbf{X}} \{A(x) \,\&\, [A(x)\, \xi\, B(y)]\} \\ &= [(A \,\xi\, B) \,\&\, A](y) = B(y).\end{aligned} \tag{9.13}$$

If $B(y) = 1$, then we obtain that

$$\begin{aligned}[(A\alpha B) \,\&\, A]\,(y) &= \bigvee_{x\in\mathbf{X}} \{A(x) \,\&\, [A(x)\, \alpha\, B(y)]\} \\ &= \bigvee_{x\in\mathbf{X}} [A(x) \,\&\, 1] \\ &= \{\bigvee_{x\notin\mathbf{A}^{-1}(1)} [A(x) \,\&\, 1]\} \vee \{\bigvee_{x\in\mathbf{A}^{-1}(1)} [A(x) \,\&\, 1]\} \\ &= \{\bigvee_{x\notin\mathbf{A}^{-1}(1)} A(x)\} \vee 1 = 1 = B(y).\end{aligned} \tag{9.14}$$

Thus (9.13) and (9.14) assure that $(A\alpha B)\in\mathbf{Z}$. ∎

Ex. (9.1) verifies Thm.9.5. It remains to study the question when $\mathbf{A}^{-1}(1)=\emptyset$, i.e. $A(x)<1$ for any $x\in\mathbf{X}$ and $B(y)=1$ for some $y\in\mathbf{Y}$. In this case, there exist in $\mathbf{Z}$ maximal elements as is proved in the following theorem:

THEOREM 9.6. *Let* $\mathcal{Z}=\emptyset$. *If* B *is normal and* A *is not normal, the fuzzy relation* $R_Z \in \mathbf{F}(\mathbf{X}\times\mathbf{Y})$, *where* z *is a fixed point of* $\mathbf{X}$, *pointwise defined as*

$$R_Z(x, y) = \begin{cases} A(z) & \text{if } x = z \text{ and } B(y) = 1, \\ A(x)\ \alpha\ B(y) & \text{otherwise,} \end{cases}$$

*for all* $x\in\mathbf{X}$, $y\in\mathbf{Y}$, *belongs to* $\mathcal{Z}$. *Varying* z *in* $\mathbf{X}$, *the set of such relations is the set of the maximal elements of* $\mathcal{Z}$.

PROOF: If $B(y)<1$, we have:

$$(R_Z \,\&\, A)(y) = \bigvee_{x\in\mathbf{X}} [A(x) \,\&\, R_Z(x, y)] = \bigvee_{x\in\mathbf{X}} \{A(x) \,\&\, [A(x)\ \alpha\ B(y)]\}$$

$$= \bigvee_{x\in\mathbf{X}} \{A(x) \,\&\, [A(x)\ \xi\ B(y)]\}$$

$$= [(A\ \xi\ B) \,\&\, A](y) = B(y), \tag{9.15}$$

since $(A\ \xi\ B)\in\mathcal{Z}$ by Thm.9.4 (cfr. (9.9)).

If $B(y)=1$, we have:

$$(R_Z \,\&\, A)(y) = \bigvee_{x\in\mathbf{X}} [A(x) \,\&\, R_Z(x, y)]$$

$$= \{\bigvee_{x\neq z} [A(x) \,\&\, R_Z(x, y)]\} \vee [A(z) \,\&\, A(z)]$$

$$= \{\bigvee_{x\neq z} A(x)\} \vee 1 = 1 = B(y). \tag{9.16}$$

Thus (9.15) and (9.16) ensure that $R_Z$ belongs to $\mathcal{Z}$ for any $z\in\mathbf{X}$. Now we show that $R_Z$ is maximal in $\mathcal{Z}$ for any $z\in\mathbf{X}$, i.e. if R belongs to $\mathcal{Z}$ and if $R\geq R_Z$ then $R=R_Z$.

If $B(y)<1$, we have by (9.6):

$$R(x, y) \leq (A\ \xi\ B)(x, y) = A(x)\ \alpha\ B(y) = R_Z(x, y) \tag{9.17}$$

for any $x\in\mathbf{X}$. Since the opposite inequality holds, we deduce that

$$R(x, y) = R_Z(x, y) \tag{9.18}$$

for any $x \in \mathbf{X}$ and for any $y \in \mathbf{Y}-\mathbf{B}^{-1}(1)$. Furthermore, by the same definition of $R_Z$, we have:

$$R_Z(x, y) = A(x)\ \alpha\ B(y) = A(x)\ \alpha\ 1 = 1 = R(x, y) \tag{9.19}$$

for any $x \in \mathbf{X}-\{z\}$ and for any $y \in \mathbf{B}^{-1}(1)$. We must only prove that

$$R(z, y) = R_Z(z, y) = A(z) \tag{9.20}$$

for any $y \in \mathbf{B}^{-1}(1)$. Indeed, assume that $R(z,y)>A(z)$ for some $y \in \mathbf{B}^{-1}(1)$. Then (9.19) implies that

$$\begin{aligned} 1 = B(y) &= \bigvee_{x \in \mathbf{X}} [A(x)\ \&\ R(x, y)] \\ &= \{ \bigvee_{x \neq z} [A(x)\ \&\ 1]\} \vee [A(z)\ \&\ R(z, y)] \\ &= \{ \bigvee_{x \neq z} A(x) \} \vee A(z), \end{aligned}$$

which is a contradiction since $A(x)<1$ for any $x \in \mathbf{X}$ by hypothesis. Thus the equality (9.20) holds and this implies, together with (9.18) and (9.19), that $R=R_Z$.

In order to characterize entirely the set of all maximal elements of $\mathcal{Z}$, we must show that if R is maximal in $\mathcal{Z}$, we have necessarily $R=R_Z$ for some $z \in \mathbf{X}$.

Indeed, let $y \in \mathbf{B}^{-1}(1)$. Since $R \in \mathcal{Z}$, there exists an element $z \in \mathbf{X}$ such that

$$1 = \bigvee_{x \in \mathbf{X}} [A(x)\ \&\ R(x, y)] = A(z)\ \&\ R(z, y)$$

and hence $R(z,y)=A(z)$ since $A(z)<1$ by hypothesis. Let us consider the fuzzy relation $R_Z$ as defined in the statement. Then (9.17) and (9.20) hold. Further, we have:

$$R(x, y) \leq 1 = R_Z(x, y) \tag{9.21}$$

for any $x \in \mathbf{X}-\{z\}$ and for any $y \in \mathbf{B}^{-1}(1)$.

The inequalities (9.17), (9.20) and (9.21) imply that $R \leq R_Z$ and therefore $R=R_Z$ since R is maximal in $\mathcal{Z}$.

This completes the proof. ■

We illustrate Thm.9.6 with the following example:

EXAMPLE 9.2. Let **X**, **Y** as in Ex. 9.1 and A∈**F**(**X**), B∈**F**(**Y**) be defined as

$$A = \| 0.3 \quad 0.6 \quad 0.6 \|, \qquad B = \| 0.3 \quad 1.0 \|.$$

Then $\mathcal{Z}$ has three maximal elements given by

$$\begin{Vmatrix} 1.0 & 0.3 \\ 0.3 & 1.0 \\ 0.3 & 1.0 \end{Vmatrix}, \quad \begin{Vmatrix} 1.0 & 1.0 \\ 0.3 & 0.6 \\ 0.3 & 1.0 \end{Vmatrix}, \quad \begin{Vmatrix} 1.0 & 1.0 \\ 0.3 & 1.0 \\ 0.3 & 0.6 \end{Vmatrix}.$$

## 9.4. Resolution of the Fuzzy Equation (9.4')

Let $\mathcal{Z}'=\mathcal{Z}'(A,B)$ be the set of all the solutions R∈**F**(**X**×**Y**) satisfying Eq.(9.4'). For any y∈**Y**, we define the following sets:

$$\mathbf{C}(y) = \{x\in \mathbf{X}: A(x)>B(y)\},$$

$$\mathbf{D}(y) = \{x\in \mathbf{X}: A(x)=B(y)\},$$

$$\mathbf{E}(y) = \{x\in \mathbf{X}: A(x)<B(y)\}.$$

Of course $\mathbf{X}=\mathbf{C}(y)\cup\mathbf{D}(y)\cup\mathbf{E}(y)$ for any y∈**Y** and the following theorems hold:

THEOREM 9.7. *Let* $\mathcal{Z}'\neq\emptyset$. *If* $\mathbf{Y}=\mathbf{B}^{-1}(1)$, *then* $\mathcal{Z}'=\{(A\times B)\}$.

PROOF. Obvious. ∎

This theorem is not invertible as is proved in the following example:

EXAMPLE 9.3. Let **X**, **Y** as in Ex.9.1 and A∈**F**(**X**), B∈**F**(**Y**) be defined as

$$A = \| 0.2 \quad 0.1 \quad 0.6 \|, \qquad B = \| 0.3 \quad 1.0 \|.$$

We have:

$$A\times B = \begin{Vmatrix} 0.2 & 0.2 \\ 0.1 & 0.1 \\ 0.3 & 0.6 \end{Vmatrix}.$$

It is easily seen that $\mathcal{Z}'=\{(A\times B)\}$ but $B(y_1)=0.3$.

THEOREM 9.8. *Let* $\mathbf{Y}\neq\mathbf{B}^{-1}(1)$. *Then* $\mathcal{Z}'\neq\emptyset$ *iff either* $\mathbf{C}(y)\neq\emptyset$ *or* $\mathbf{D}(y)\neq\emptyset$ *for any* $y\in\mathbf{Y}-\mathbf{B}^{-1}(1)$.

PROOF. Let $\mathcal{Z}'\neq\emptyset$ and $\mathbf{C}(y)=\mathbf{D}(y)=\emptyset$ for some $y\in\mathbf{Y}-\mathbf{B}^{-1}(1)$. Let $x'\in\mathbf{E}(y)$ be such that

$$1>B(y) = \bigwedge_{x\in\mathbf{X}} [A(x) \,\&\, R(x,y)] = A(x') \,\&\, R(x',y).$$

Then $A(x')\neq R(x',y)$ and hence

$$B(y) = A(x') \,\&\, R(x',y) = A(x') \wedge R(x',y) \leq A(x'),$$

a contradiction. Vice versa, let either $\mathbf{C}(y)\neq\emptyset$ or $\mathbf{D}(y)\neq\emptyset$ for any $y\in\mathbf{Y}-\mathbf{B}^{-1}(1)$. Then we define the following fuzzy relation $T\in\mathbf{F}(\mathbf{X}\times\mathbf{Y})$ as $T(x,y)=B(y)$ for any $x\in\mathbf{C}(y)$ if $B(y)<1$ and $\mathbf{C}(y)\neq\emptyset$, $T(x,y)=1$ for any $x\in\mathbf{D}(y)$ if $B(y)<1$ and $\mathbf{D}(y)\neq\emptyset$, otherwise $T(x,y)=A(x)$, where $x\in\mathbf{X}$ and $y\in\mathbf{Y}$.

If $B(y)=1$, we have:

$$(T \,\&'\, A)(y) = \bigwedge_{x\in\mathbf{X}} [A(x) \,\&\, T(x,y)] = \bigwedge_{x\in\mathbf{X}} [A(x) \,\&\, A(x)]$$

$$= 1 = B(y).$$

For any $y\in\mathbf{Y}-\mathbf{B}^{-1}(1)$, we have (if $\mathbf{C}(y)$, $\mathbf{D}(y)$, $\mathbf{E}(y)$ are nonempty otherwise one of these sets does not appear in the equality below):

$$(T \,\&'\, A)(y) = \bigwedge_{x\in\mathbf{X}} [A(x) \,\&\, T(x,y)] = \{\bigwedge_{x\in\mathbf{C}(y)} [A(x) \,\&\, B(y)]\}$$

$$\wedge \{\bigwedge_{x\in\mathbf{D}(y)} [A(x) \,\&\, 1]\} \wedge \{\bigwedge_{x\in\mathbf{E}(y)} [A(x) \,\&\, A(x)]\}$$

$$= B(y) \wedge B(y) \wedge 1 = B(y).$$

The above equalities imply that T is in $\mathcal{Z}'$, i.e. $\mathcal{Z}'\neq\emptyset$. ■

THEOREM 9.9. *Let* $\mathcal{Z}'\neq\emptyset$ *and* $\mathbf{Y}\neq\mathbf{B}^{-1}(1)$. *Then* $(A\times B)\in\mathcal{Z}'$ *iff* $C(y)\neq\emptyset$ *for any* $y\in\mathbf{Y}-\mathbf{B}^{-1}(1)$. *Furthermore,* $R\geq(A\times B)$ *for any* $R\in\mathcal{Z}'$.

PROOF. Let $R\in\mathcal{Z}'$. By Lemma 9.2', we have:

$$R \geq A \times (R \,\&'\, A) = A \times B. \tag{9.22}$$

Assume that $(A\times B)\in \mathcal{Z}'$ and $\mathbf{C}(y)=\emptyset$ for some $y\in \mathbf{Y}-\mathbf{B}^{-1}(1)$, i.e. $A(x)\leq B(y)$ for any $x\in \mathbf{X}$. Then we have:

$$1 > B(y) = \bigwedge_{x\in \mathbf{X}} \{A(x) \,\&\, [A(x) \wedge B(y)]\} = \bigwedge_{x\in \mathbf{X}} [A(x) \,\&\, A(x)] = 1,$$

a contradiction. Vice versa, let $y\in \mathbf{Y}-\mathbf{B}^{-1}(1)$ and assume that

$$[(A\times B) \,\&'\, A]\,(y) = \bigwedge_{x\in \mathbf{X}} \{A(x) \,\&\, [A(x) \wedge B(y)]\} = 1.$$

Thus $A(x)=A(x)\wedge B(y)$, i.e. $A(x)\leq B(y)$ for any $x\in \mathbf{X}$. This means that $\mathbf{C}(y)=\emptyset$, a contradiction. Then $[(A\times B)\&'A]\,(y)<1$ and by Lemmas 9.1', 9.3' and (9.22), we have:

$$B(y) = (R \,\&'\, A)\,(y) \geq [(A \times B) \,\&'\, A]\,(y) \geq B(y),$$

i.e. $[(A\times B)\&'A](y)=B(y)$ for any $y\in \mathbf{Y}-\mathbf{B}^{-1}(1)$. Let $y\in \mathbf{B}^{-1}(1)$ and hence:

$$[(A \times B) \,\&'\, A]\,(y) = \bigwedge_{x\in \mathbf{X}} \{A(x) \,\&\, [A(x) \wedge 1]\}$$

$$= \bigwedge_{x\in \mathbf{X}} [A(x) \,\&\, A(x)] = 1 = B(y).$$

Then $(A\times B)\in \mathcal{Z}'$ and this concludes the proof. ■

REMARK 9.3. If $\mathbf{C}(y)=\emptyset$ for some $y\in \mathbf{Y}-\mathbf{B}^{-1}(1)$ (but $\mathbf{D}(y)\neq\emptyset$ by Thm. 9.8), then Thm.9.9 fails and (9.22) holds even if $(A\times B)\notin \mathcal{Z}'$ as is shown in the following example:

EXAMPLE 9.4. Let $\mathbf{X}=\{x_1,x_2,x_3\}$, $\mathbf{Y}=\{y_1,y_2,y_3\}$ and $A\in \mathbf{F}(\mathbf{X})$, $B\in \mathbf{F}(\mathbf{Y})$ be defined as

$$A = \begin{Vmatrix} 0.1 & 0.3 & 0.5 \end{Vmatrix}, \qquad B = \begin{Vmatrix} 0.5 & 0.4 & 1.0 \end{Vmatrix}.$$

We have:

$$A\times B = \begin{Vmatrix} 0.1 & 0.1 & 0.1 \\ 0.3 & 0.3 & 0.3 \\ 0.5 & 0.4 & 0.5 \end{Vmatrix}.$$

Since $[(A\times B)\ \&'A](y_1)=1>B(y_1)$, then $(A\times B)\notin \mathcal{Z}'$. It is easily seen that $\mathcal{Z}'$ is constituted by the fuzzy relations $R_x\in \mathbf{F}(\mathbf{X}\times\mathbf{Y})$ such that

$$\begin{Vmatrix} 0.1 & 0.1 & 0.1 \\ 0.3 & 0.3 & 0.3 \\ x & 0.4 & 0.5 \end{Vmatrix}.$$

where $x\in(0.5,1]$. Thus $(A\times B)<R_x$.

About the existence of the maximal elements of $\mathcal{Z}'$, we have the following results.

THEOREM 9.10. *Let* $\mathbf{Y}\neq\mathbf{B}^{-1}(1)$ *and* $\mathbf{D}(y)\neq\emptyset$ *for any* $y\in\mathbf{Y}-\mathbf{B}^{-1}(1)$. *Then* $\mathcal{Z}'\neq\emptyset$ *and the fuzzy relation* $U\in\mathbf{F}(\mathbf{X}\times\mathbf{Y})$ *defined as*

$$\begin{aligned} U(x,y) &= 1 \quad \textit{for any } x\in\mathbf{C}(y)\cup\mathbf{D}(y) \ \textit{ if } B(y)<1, \\ &= A(x) \ \ \textit{otherwise}, \end{aligned} \tag{9.23}$$

*where* $x\in\mathbf{X}$, $y\in\mathbf{Y}$, *is the greatest element of* $\mathcal{Z}'$, *i.e.* $R\leq U$ *for any* $R\in\mathcal{Z}'$.

PROOF. $\mathcal{Z}'$ is nonempty by Thm.9.8 and let R be an arbitrary element of $\mathcal{Z}'$. The thesis is clearly proved if we show that $R(x,y)\leq U(x,y)$ for any $x\in\mathbf{E}(y)$ such that $\mathbf{E}(y)\neq\emptyset$. Indeed, let $x'\in\mathbf{E}(y)$ such that $R(x',y)>U(x',y)=A(x')$. Then $A(x')=R(x',y)\&A(x')\geq B(y)$, i.e. $x'\in\mathbf{C}(y)\cup\mathbf{D}(y)$, a contradiction. ■

REMARK 9.4. Ex.9.4 proves that Thm.9.10 is not invertible, i.e. $\mathbf{D}(y)$ could be empty for some $y\in\mathbf{Y}-\mathbf{B}^{-1}(1)$ but $\mathcal{Z}'$ has the greatest element. In Ex.9.4, $R_x$, with $x=1$, is the greatest element of $\mathcal{Z}'$ but $D(y_2)=\emptyset$.

If $\mathbf{D}(y)=\emptyset$ for some $y\in\mathbf{Y}-\mathbf{B}^{-1}(1)$, it is seen, using techniques similar to those already exhibited in the proofs of the previous results, that the following theorem holds:

THEOREM 9.11. *Let* $\mathcal{Z}'\neq\emptyset$, $\mathbf{Y}\neq\mathbf{B}^{-1}(1)$ *and* $\mathbf{D}(y)=\emptyset$ *for some* $y\in\mathbf{Y}-\mathbf{B}^{-1}(1)$. *Then the fuzzy relation* $V\in\mathbf{F}(\mathbf{X}\times\mathbf{Y})$ *defined as* $V(x,y)=A(x)$ *for any* $x\in\mathbf{E}(y)$ *if* $\mathbf{E}(y)\neq\emptyset$ *or* $B(y)=1$, *otherwise* $V(x,y)=1$ *for any* $x\in\mathbf{C}(y)\cup\mathbf{D}(y)$ *if* $\mathbf{C}(y)\neq\emptyset$ *and* $\mathbf{D}(y)\neq\emptyset$, $V(z,y)=B(y)$ *for some* $z\in\mathbf{C}(y)$ *and* $V(x,y)=1$ *for any* $x\in\mathbf{C}(y)-\{z\}$ *if* $\mathbf{C}(y)\neq\emptyset$ *and* $\mathbf{D}(y)=\emptyset$, $V(x,y)=1$ *for any* $x\in\mathbf{D}(y)$ *if* $\mathbf{C}(y)=\emptyset$ *and* $\mathbf{D}(y)\neq\emptyset$, *where* $x\in\mathbf{X}$, $y\in\mathbf{Y}$, *is a maximal element of* $\mathcal{Z}'$. *Any maximal element of* $\mathcal{Z}'$ *is a fuzzy relation* V *defined as above.*

REMARK 9.5. Let $\mathbf{Y}\neq\mathbf{B}^{-1}(1)$, $\mathcal{Z}'\neq\emptyset$ and $\mathbf{W}=\{y\in\mathbf{Y}-\mathbf{B}^{-1}(1):\ \mathbf{D}(y)=\emptyset\}$. We have $\mathbf{C}(y)\neq\emptyset$ for any $y\in\mathbf{W}$ by Thm.9.8 and it is obvious that the number q of the maximal elements of $\mathcal{Z}'$ is given by

$$q = \prod_{y \in \mathbf{W}} \operatorname{card} \mathbf{C}(y).$$

Of course q=1 if $\mathbf{W}=\emptyset$ (cfr. Thm.9.10).

We illustrate Thm.9.11 with the following example:

EXAMPLE 9.5. Let $\mathbf{X}=\{x_1,x_2,x_3,x_4\}$, $\mathbf{Y}=\{y_1,y_2,y_3,y_4\}$ and $A\in \mathbf{F}(\mathbf{X})$, $B\in \mathbf{F}(\mathbf{Y})$ be defined as

$$A=\|0.3 \quad 0.8 \quad 0.9 \quad 0.6\|, \qquad B=\|0.7 \quad 1.0 \quad 0.5 \quad 0.6\|.$$

We have $\mathbf{B}^{-1}(1)=\{y_2\}$, $\mathbf{C}(y_1)=\{x_2,x_3\}$, $\mathbf{C}(y_3)=\{x_2,x_3,x_4\}$, $\mathbf{D}(y_1)=\mathbf{D}(y_3)=\emptyset$ and $\mathbf{D}(y_4)=\{x_4\}$. Hence $\mathcal{Z}'$ is nonempty by Thm.9.8 and q=6 since $\mathbf{W}=\{y_1,y_3\}$. The maximal elements of $\mathcal{Z}'$ are the following fuzzy relations:

$$\begin{Vmatrix} 0.3 & 0.3 & 0.3 & 0.3 \\ 1.0 & 0.8 & 1.0 & 1.0 \\ 0.7 & 0.9 & 0.5 & 1.0 \\ 0.6 & 0.6 & 1.0 & 1.0 \end{Vmatrix}, \qquad \begin{Vmatrix} 0.3 & 0.3 & 0.3 & 0.3 \\ 1.0 & 0.8 & 1.0 & 1.0 \\ 0.7 & 0.9 & 1.0 & 1.0 \\ 0.6 & 0.6 & 0.5 & 1.0 \end{Vmatrix},$$

$$\begin{Vmatrix} 0.3 & 0.3 & 0.3 & 0.3 \\ 1.0 & 0.8 & 0.5 & 1.0 \\ 0.7 & 0.9 & 1.0 & 1.0 \\ 0.6 & 0.6 & 1.0 & 1.0 \end{Vmatrix}, \qquad \begin{Vmatrix} 0.3 & 0.3 & 0.3 & 0.3 \\ 0.7 & 0.8 & 1.0 & 1.0 \\ 1.0 & 0.9 & 0.5 & 1.0 \\ 0.6 & 0.6 & 1.0 & 1.0 \end{Vmatrix},$$

$$\begin{Vmatrix} 0.3 & 0.3 & 0.3 & 0.3 \\ 0.7 & 0.8 & 1.0 & 1.0 \\ 1.0 & 0.9 & 1.0 & 1.0 \\ 0.6 & 0.6 & 0.5 & 1.0 \end{Vmatrix}, \qquad \begin{Vmatrix} 0.3 & 0.3 & 0.3 & 0.3 \\ 0.7 & 0.8 & 0.5 & 1.0 \\ 1.0 & 0.9 & 1.0 & 1.0 \\ 0.6 & 0.6 & 1.0 & 1.0 \end{Vmatrix}.$$

Summarizing all the results of this Sec., we can say that Eq.(9.4'), except the trivial case given in Thm.9.7, if $\mathcal{Z}'\neq\emptyset$ and $\mathbf{Y}\neq\mathbf{B}^{-1}(1)$, has the greatest solution, defined by (9.23), if $\mathbf{W}=\emptyset$ and q upper solutions (cfr. Thm.9.11) if $\mathbf{W}\neq\emptyset$. Furthermore, it has the smallest solution (A×B) if $\mathbf{C}(y)\neq\emptyset$ for any $y\in \mathbf{Y}\text{-}\mathbf{B}^{-1}(1)$ (cfr. Thm.9.9).

## 9.5. Resolution of Fuzzy Equations with Difference Operator

In this Sec., we give the resolution of Eq.(9.5) (resp. (9.5')), denoting by $\mathcal{Z}^{\cap}=\mathcal{Z}^{\cap}(A,B)$ (resp. $\mathcal{Z}'^{\cap}=\mathcal{Z}'^{\cap}(A,B)$) the set of its solutions.

We first introduce the operator $\xi^{\cap}:[0,1]^2\rightarrow[0,1]$ defined as $a\xi^{\cap}b=a\varepsilon b$ if b>0 and

$a\xi^{\cap}b=a$ if $b=0$. Extending this operation to two fuzzy sets $A\in\mathbf{F}(\mathbf{X})$ and $B\in\mathbf{F}(\mathbf{Y})$, we define the fuzzy relation $(A\xi^{\cap}B)\in\mathbf{F}(\mathbf{X}\times\mathbf{Y})$ as

$$(A\xi^{\cap}B)\,(x,y) = A(x)\ \xi^{\cap}\ B(y)$$

for all $x\in\mathbf{X}$, $y\in\mathbf{Y}$.

The following results, dual of the theorems of Sec.9.3, hold:

THEOREM 9.12. *$\mathbf{Z}^{\cap}\neq\emptyset$ iff $(A\xi^{\cap}B)\in\mathbf{Z}^{\cap}$. Furthermore, $(A\xi^{\cap}B)\leq R$ for any $R\in\mathbf{Z}^{\cap}$ if $B(y)>0$ for any $y\in\mathbf{Y}$.*

THEOREM 9.13. *Let $\mathbf{Z}^{\cap}\neq\emptyset$. If $A(x)=B(y)=0$ for some $x\in\mathbf{X}$, $y\in\mathbf{Y}$, then the fuzzy relation $(A\varepsilon B)\in\mathbf{F}(\mathbf{X}\times\mathbf{Y})$, defined as $(A\varepsilon B)(x,y)=A(x)\varepsilon B(y)$ for all $x\in\mathbf{X}$, $y\in\mathbf{Y}$, is the smallest element of $\mathbf{Z}^{\cap}$.*

THEOREM 9.14. *Let $\mathbf{Z}^{\cap}\neq\emptyset$. If $B(y)=0$ for some $y\in\mathbf{Y}$ and $A(x)>0$ for any $x\in\mathbf{X}$, the fuzzy relation $R_z^{\cap}\in\mathbf{F}(\mathbf{X}\times\mathbf{Y})$, where z is a fixed point of $\mathbf{X}$, pointwise defined as*

$$\begin{aligned} R_z^{\cap}(x,y) &= A(z) && \textit{if } x=z \textit{ and } B(y)=0,\\ &= A(x)\ \varepsilon\ B(y) && \textit{otherwise,} \end{aligned}$$

*for all $x\in\mathbf{X}$, $y\in\mathbf{Y}$, belongs to $\mathbf{Z}^{\cap}$. Varying z in $\mathbf{X}$, the set of such relations is the set of the minimal elements of $\mathbf{Z}^{\cap}$.*

Now we define the fuzzy relation $(A\vee B)\in\mathbf{F}(\mathbf{X}\times\mathbf{Y})$ as

$$(A\vee B)\,(x,y) = A(x)\vee B(y)$$

for all $x\in\mathbf{X}$, $y\in\mathbf{Y}$. Except the trivial case that $\mathbf{Y}=\mathbf{B}^{-1}(0)=\{y\in\mathbf{Y}: B(y)=0\}$ for which we have $\mathbf{Z}'^{\cap}=\{(A\vee B)\}$, we can say that, if $\mathbf{Z}'^{\cap}\neq\emptyset$ and $\mathbf{Y}\neq\mathbf{B}^{-1}(0)$, Eq.(9.5') has the smallest solution $M\in\mathbf{F}(\mathbf{X}\times\mathbf{Y})$ given by

$$\begin{aligned} M(x,y) &= A(x) && \text{if } x\in\mathbf{C}(y)\neq\emptyset \text{ or } B(y)=0,\\ &= 0 && \text{otherwise,} \end{aligned}$$

where $x\in\mathbf{X}$, $y\in\mathbf{Y}$, if $\mathbf{W}'=\{y\in\mathbf{Y}-\mathbf{B}^{-1}(0): \mathbf{D}(y)=\emptyset\}=\emptyset$ and $\mathbf{W}'\neq\emptyset$, only q' lower solutions $N\in\mathbf{F}(\mathbf{X}\times\mathbf{Y})$ pointwise defined as $N(x,y)=A(x)$ for any $x\in\mathbf{C}(y)$ if $\mathbf{C}(y)\neq\emptyset$ or $B(y)=0$, otherwise $N(x,y)=0$ for any $x\in\mathbf{D}(y)\cup\mathbf{E}(y)$ if $\mathbf{D}(y)\neq\emptyset$ and $\mathbf{E}(y)\neq\emptyset$, $N(z,y)=B(y)$ for some $z\in\mathbf{E}(y)$ and $N(x,y)=0$ for any $x\in\mathbf{E}(y)-\{z\}$ if $\mathbf{D}(y)=\emptyset$ and $\mathbf{E}(y)\neq\emptyset$, $N(x,y)=0$ for any $x\in\mathbf{D}(y)$ if $\mathbf{D}(y)\neq\emptyset$ and $\mathbf{E}(y)=\emptyset$, where $x\in\mathbf{X}$, $y\in\mathbf{Y}$, and

$$q' = \prod_{y\in\mathbf{W}'} \text{card}\,\mathbf{E}(y).$$

Furthermore, if $\mathbf{Y} \neq \mathbf{B}^{-1}(0)$, then $\mathbf{Z}'^{\cap} \neq \emptyset$ iff either $\mathbf{D}(y) \neq \emptyset$ or $\mathbf{E}(y) \neq \emptyset$ for any $y \in \mathbf{Y}-\mathbf{B}^{-1}(0)$. Eq.(9.5') has the greatest solution $(A \vee B)$ if $\mathbf{E}(y) \neq \emptyset$ for any $y \in \mathbf{Y}-\mathbf{B}^{-1}(0)$.

Here we have supposed that fuzzy sets and relations have membership values in [0,1], but it is easily seen that all the results hold assuming that membership values lie in a linear lattice. The results presented can be found in [1] and [3].

## 9.6. Concluding Remarks

The results derived for these equations are not as strong as for the equations discussed in the previous Chs. This is due to the structure of the composition which, in fact, is much more complicated here. This implies, for instance, that the resolution of a system of Eqs.(9.2) or (9.3):

$$R \,\&\, A_h = B_h\,,$$

or

$$R \,\&^{\cap}\, A_h = B_h\,,$$

where h=1,2,...,m, is possible if one is able to determine the greatest or the smallest solution. Further details about this class of equations can be found in [2] and Ch.13.

## References


[1] A. Di Nola, W. Pedrycz and S. Sessa, Fuzzy relation equations with equality and difference composition operators, *Fuzzy Sets and Systems* 25 (1988), 205-215.

[2] A. Di Nola, W. Pedrycz and S. Sessa, Modus ponens for fuzzy data realized via equations with equality operators, *Internat. J. Intelligent Systems*, to appear.

[3] A. Di Nola, W. Pedrycz and S.Sessa, On some finite fuzzy relation equations, *Inform. Sciences*, to appear.

# CHAPTER 10

## APPROXIMATE SOLUTIONS OF FUZZY RELATION EQUATIONS



*In the previous Chs. we have discussed various problems concerning solutions of fuzzy relation equations. Obviously an underlying assumption is that there exists a nonempty set of solutions. A situation, may occurr, and indeed it is quite common, in which no solution exists. Nevertheless even in this case one might be interested to obtain an approximate solution and know to which extent it can be viewed as a solution. This stream of investigations is particularly interesting for applicational purposes. Contrary to the topics already discussed in the previous Chs., this is a field of research which has not been developed enough so far. It concerns studies on solvability properties of fuzzy relation equations. In this Ch., we shall try to answer, by using several techniques, the following basic question: how difficult is it to attain a situation in which the system of equations has solutions and then how to measure this property?*

### 10.1. Preliminaries

In [33, Ch.8], a numerical approach was suggested to solve fuzzy relation equations; the optimization procedures proposed there enable us to find a fuzzy relation which leads to the minimum of a specified performance index.

Further, in [2] and [22, Ch.8], a solvability index has been proposed and it measures to what extent the system of equations or a single equation is solvable. This is useful not only to detect a situation in which no solution exists but to measure how *easily* the equation(s) can be solved. Nevertheless, it is a *passive* instrument in the sense that it evaluates the certain property (i.e. solvability) but it does not intervene in the structure of the given fuzzy sets and relations. Of course, it does not provide the user with any possible suggestion how the data may be modified to make the equation(s) solvable or to make the value of the index higher.

Several proposals, concerning systems of equations, can be found in [4] and [9]. They rely, respectively, on a special modification of the entire data set (for instance, omission of some elements in the data set for purpose, e.g., of model identification). The use of auxiliary probabilistic information has been discussed in [6], [7] and [8].

The aim of this Ch. is to present various approaches to the determination of approximate solutions of a fuzzy relation equation (or a system of equations) and their related characterization. At the very beginning, we start with expressions of a property of solvability of these equations, afterwards moving to some technical details. In the whole Ch., the sets involved are finite and we assume that **L**=[0,1] with the usual "$\wedge$" and "$\vee$" lattice operations.

## 10.2. On Expressing a Property of Solvability of Fuzzy Relation Equations

Let t be a lower semicontinuous t-norm and $\varphi$ be the related associated operator (cfr. Thm.8.2). Making use of the equality index of two fuzzy sets $A, A' \in \mathbf{F}(\mathbf{X})$, defined in the same finite space **X**, given by

$$[A \equiv A'] = \bigwedge_{x \in \mathbf{X}} \{[A(x)\ \varphi\ A'(x)]\ t\ [A'(x)\ \varphi\ A(x)]\}, \tag{10.1}$$

we shall discuss the single fuzzy Eq. (8.1) and the system of equations:

$$R\ t\ A_h = B_h, \tag{10.2}$$

where h=1,2,...,m, $A_h \in \mathbf{F}(\mathbf{X})$, $B_h \in \mathbf{F}(\mathbf{X})$, **Y** being another finite space.

Looking for a fuzzy relation R fulfilling Eq.(8.1) or the system (10.2) and referring to the contents of Ch.8, it is evident that to give a simple and easy-to-test condition for the existence of such a solution is extremely difficult. Instead of discussing directly the problem of solvability of the above equations, we introduce a solvability index defined as follows [22, Ch.8]:

DEFINITION 10.1. *By a solvability index* of Eq.(8.1), *we mean a number* $\xi \in [0,1]$ *equal to*

$$\xi = \max\{[R\ t\ A \equiv B] : R \in \mathbf{F}(\mathbf{X} \times \mathbf{Y})\}. \tag{10.3}$$

DEFINITION 10.2. *By a solvability index of the system* (10.2), *we mean a number* $\xi \in [0,1]$ *equal to*

$$\xi = \max\{ \mathop{t}_{h=1}^{m} [R \, t \, A_h \equiv B_h] : R \in \mathbf{F}(\mathbf{X} \times \mathbf{Y})\}, \tag{10.4}$$

*where the indicated recursive formula is defined as in* Sec.8.1.

Note that Eq.(8.1) or the system (10.2) is solvable iff $\xi=1$. In [22, Ch.8], Gottwald proved that the solvability index (10.3) can be expressed, for lower semicontinuous t-norms, as

$$\xi = \bigwedge_{y \in \mathbf{Y}} \{B(y) \, \varphi \, \{ \bigvee_{x \in \mathbf{X}} \{A(x) \, t \, [A(x) \, \varphi \, B(y)]\}\}\}. \tag{10.5}$$

In [22, Ch.8], Gottwald also determined lower and upper bounds for the solvability index (10.4). Still more transparent results are derived for continuous t-norms. Indeed, the following results of [4] hold:

LEMMA 10.1. *If* t *is continuous, then* $b\varphi(a \, t \, (a\varphi b))=b\varphi \, a$ *for any* $a,b \in [0,1]$.

PROOF. (This Lemma contains results already known in Brouwerian semilattice theory, but we give this proof for sake of completeness.) We first prove that

$$a \, t \, (a\varphi b) = a \wedge b \tag{10.6}$$

for any $a,b \in [0,1]$. If $a \leq b$, then (cfr. formula (8.3)) $a\varphi b=1$ and hence $at(a\varphi b)=at1=a=a\wedge b$. Now let $a>b$. Since t is continuous, there exists $x \in [0,1]$ such that $atx=b$. Hence $x \leq a\varphi b$ and then $b=atx \leq at(a\varphi b) \leq b$ by property (ii) of Def.8.4, i.e. $at(a\varphi b)=b=a\wedge b$.

The thesis follows immediately from (10.6) since $b\varphi(at(a\varphi b))=b\varphi(a\wedge b)=b\varphi a$ which is obvious for $a \leq b$ while we have $b\varphi b=1=b\varphi a$ for $a>b$. ■

THEOREM 10.2. *If* t *is continuous, then the solvability index* (10.3) *is given by*

$$\xi = \mathrm{hgt}(B) \, \varphi \, \mathrm{hgt}(A). \tag{10.7}$$

(We recall that the height, abbreviated as hgt, of a fuzzy set is the supremum of its membership function).

PROOF. By Lemma (10.1) and property (i) of Def.8.4, we have by using (10.5):

$$\xi = \bigwedge_{y\in \mathbf{Y}} \bigvee_{x\in \mathbf{X}} \{B(y)\ \varphi\ \{A(x)\ t\ [A(x)\ \varphi\ B(y)]\}\}$$

$$= \bigwedge_{y\in \mathbf{Y}} \{B(y)\ \varphi\ [\bigvee_{x\in \mathbf{X}} A(x)]\}.$$

We now observe that the formula (8.3), which defines the operator φ, implies $\min\{a\varphi c, b\varphi c\} = \max\{a,b\}\varphi c$ for any $a,b,c \in [0,1]$. Then the thesis follows since we deduce from the previous equality, **Y** being finite,

$$\xi = [\bigvee_{y\in \mathbf{Y}} B(y)]\ \varphi\ [\bigvee_{x\in \mathbf{X}} A(x)].$$ ■

Unfortunately, for the system (10.2), one cannot specify a direct expression of the value of the solvability index (10.4). However, some bounds have been obtained in [4]. For instance, we have the following result here only enunciated (for a proof, cfr. Prop.5 and inequality (10) of [4]):

THEOREM 10.3. *If* t *is continuous, then the solvability index* (10.4) *has a lower bound*, i.e.

$$\xi \geq \mathop{t}_{h=1}^{m} \bigwedge_{y\in \mathbf{Y}} \{B_h(y)\ \varphi\ \{hgt(A_h)\ t\ [\bigwedge_{h=1}^{m} B_h(y)]\}\}. \tag{10.8}$$

*If* t=*min, then we have:*

$$\xi = \bigwedge_{h=1}^{m} \bigwedge_{y\in \mathbf{Y}} \{B_h(y)\ \alpha\ \{\bigvee_{x\in \mathbf{X}} \{A_h(x) \wedge \{\bigwedge_{h=1}^{m} [A_h(x)\ \alpha\ B_h(y)]\}\}\}\}, \tag{10.9}$$

*where "α" is the operator associated with min* (cfr. Sec. 2.1).

Returning to Eq.(8.1), the formula (10.7) has a clear interpretation. If the fuzzy set A is normal (i.e., hgt(A)=1), Eq.(8.1) has always a solution since ξ=1, despite the values of B. If hgt(A)<1, then the value of the solvability index depends on the height of B; as soon as it is smaller than the height of A, we still have ξ=1, further ξ=hgt(A) if hgt(B)=1. Thus for a single equation, it is quite easy to check whether there exists a solution. The following is an easy Corollary to Thm.10.2.

COROLLARY 10.4. *If for any* $y\in \mathbf{Y}$, *there exists* $x\in \mathbf{X}$ *such that* $A(x)\geq B(y)$, *then* Eq.(8.1) *has solutions if* t *is continuous.*

Note that the same result was obtained for t=min in Thm.3.3.

Let $\mathfrak{R}_h=\mathfrak{R}_h(A_h,B_h)$ be the set of the solutions of the h-th equation of the system (10.2). Thm.8.6 assures that if $\cap_{i=1}\mathfrak{R}_h\neq\emptyset$, then $R^\cap$ is the greatest solution of the system (10.2).

Concluding this Sec., we consider two extreme situations illustrating two related results.

Assume that for some $h,j\in\{1,2,\ldots,m\}$, $m\geq 2$:

(1) $A_h=A_j$ while the corresponding fuzzy sets $B_h$, $B_j$ vary *significantly*,

(2) $A_h$, $A_j$ vary *significantly* and $B_h$, $B_j$ are *similar* to each other (in the sequel, we shall specify a measure of similarity or dissimilarity of two fuzzy sets).

Concerning situation (1), we see that if $\mathfrak{R}_h,\mathfrak{R}_j\neq\emptyset$ and $B_h\neq B_j$, then $\mathfrak{R}_h\cap\mathfrak{R}_j=\emptyset$, otherwise for some $R\in\mathfrak{R}_h\cap\mathfrak{R}_j$ we would deduce $B_h=RtA_h=RtA_j=B_j$. Consequently, $\cap_{h=1}\mathfrak{R}_h=\emptyset$.

Considering situation (2), we have this result of [3], where, for simplicity, we supposed t=min, but here we assume that t is lower semicontinuous.

THEOREM 10.5. *For all* $h,j\in\{1,2,\ldots,m\}$, *let* $\mathfrak{R}_h,\mathfrak{R}_j\neq\emptyset$ *and* $\mathbf{supp}\ A_h\cap\mathbf{supp}A_j=\emptyset$ ($\mathbf{supp}\ A_h=\{x\in\mathbf{X}: A_h(x)>0\}$), $h\neq j$. *Then* $\cap_{h=1}\mathfrak{R}_h\neq\emptyset$.

PROOF. Since $\mathfrak{R}_h\neq\emptyset$ for any $h\in\{1,2,\ldots,m\}$, $(A_h\ \varphi\ B_h)\in\mathfrak{R}_h$ by Thm.8.4. Let

$$R^\cap=\bigwedge_{h=1}^{m}(A_h\ \varphi\ B_h)$$

and $h,j\in\{1,2,\ldots,m\}$ such that $h\neq j$. The hypothesis ensures that $A_j(x)=0$ for any $j\neq h$ and $x\in\mathbf{supp}\ A_h$, so that

$$\bigwedge_{j\neq h}[A_j(x)\ \varphi\ B_j(y)]=\bigwedge_{j\neq h}[0\ \varphi\ B_j(y)]=1.$$

For any $h\in\{1,2,\ldots,m\}$, then we have:

$$(R^\cap\ t\ A_h)(y)=\bigvee_{x\in\mathbf{X}}\{A_h(x)\ t\ \{[A_h(x)\ \varphi\ B_h(y)]\wedge\{\bigwedge_{j\neq h}[A_j(x)\ \varphi\ B_j(y)]\}\}\}$$

$$=\bigvee_{x\in\mathbf{supp}\ A_h}\{A_h(x)\ t\ [A_h(x)\ \varphi\ B_h(y)]\}$$

$$= \bigvee_{x \in \mathbf{X}} \{A_h(x)\ t\ [A_h(x)\ \varphi\ B_h(y)]\} = B_h(y)$$

for any $y \in \mathbf{Y}$. This means that $R^{\frown} \in \Re_h$ for any $h \in \{1,2,\ldots,m\}$, i.e. $R^{\frown} \in \cap_{h=1} \Re_h$. ■

Not all the data $(A_h,B_h)$, $h=1,\ldots,m$, satisfy Eqs.(10.2) for several reasons. However one can modify the data by imposing some threshold levels, as we shall prove later.

## 10.3. On Solving Fuzzy Relation Equations via Modification of Fuzzy Sets: a Role of γ-Level Fuzzy Sets

One of the possible ways leading to increment the value of the solvability index of Eq.(8.1) (resp. system (10.2)) is to modify the value of the membership functions of A (resp. $A_h$) and/or B (resp. $B_h$). Of course, this can be done in many ways, but we try here a more systematic approach.

Since the values of the membership functions are not precise, one might expect the elements with the highest grades of membership to be evaluated or given correctly, free of error, but when moving down some errors might occur. To avoid their influence, we introduce a fuzzy set $A^\gamma \in \mathbf{F}(\mathbf{X})$ that is built on the basis of A following the formula:

$$A^\gamma(x) = \max\{A(x),\gamma\}$$

for any $x \in \mathbf{X}$, where $\gamma \in [0,1]$.

A threshold level $\gamma$ underlines the fact that the membership grades smaller than $\gamma$ are less reliable than those greater than $\gamma$ and should be discarded. In both extreme cases, we have the original fuzzy set A if $\gamma=0$ or the fuzzy set identically equal to 1 if $\gamma=1$ (cfr. Fig.10.1).

We now investigate the influence of $\gamma$ on the values of the solvability index (10.3) (resp.(10.4)). In accordance with our discussion, we change the data A,B (resp. $A_h$, $B_h$) into $A^\gamma$, $B^\gamma$ (resp. $A_h^\gamma$, $B_h^\gamma$) in Eq.(8.1) (resp. system (10.2)) and we denote by $\xi(\gamma)$ the corresponding solvability index (10.3) (resp.(10.4)) of the modified equation (resp. system):

$$R\ t\ A^\gamma = B^\gamma, \tag{10.10}$$

$$R\ t\ A_h^\gamma = B_h^\gamma, \quad h=1,2,\ldots,m. \tag{10.11}$$

Restricting our attention to the case of continuous t-norms, we have the following results of [4]:

THEOREM 10.6. *Let* t *be continuous and* $\gamma, \gamma_1, \gamma_2 \in [0,1]$. *Then we have for the solvability indexes* (10.3) *related to* Eqs.(8.1) *and* (10.10):

(i) $\xi(0)=\xi$, $\xi(1)=1$,

(ii) $\gamma \leq \xi(\gamma) \leq 1$,

(iii) $\xi \leq \xi(\gamma)$,

(iv) *If* $\gamma_1 \leq \gamma_2$, *then* $\xi(\gamma_1) \leq \xi(\gamma_2)$.

PROOF. (i) is obvious.

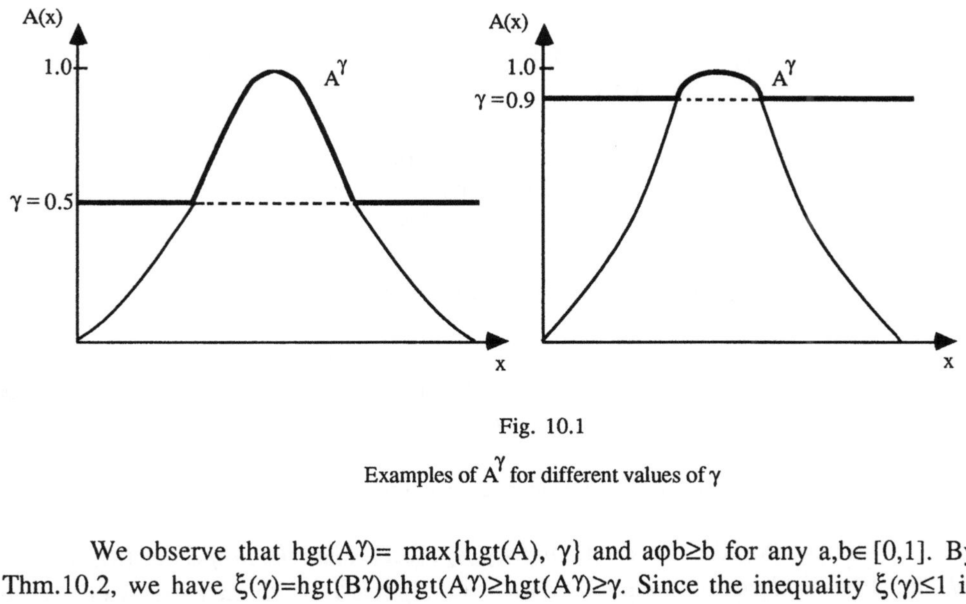


Fig. 10.1

Examples of $A^\gamma$ for different values of $\gamma$

We observe that $hgt(A^\gamma)= \max\{hgt(A), \gamma\}$ and $a\varphi b \geq b$ for any $a,b \in [0,1]$. By Thm.10.2, we have $\xi(\gamma)=hgt(B^\gamma)\varphi hgt(A^\gamma) \geq hgt(A^\gamma) \geq \gamma$. Since the inequality $\xi(\gamma) \leq 1$ is obvious, the property (ii) is proved. We now prove (iii). If $hgt(B) \leq hgt(A)$, then $hgt(B^\gamma) \leq hgt(A^\gamma)$ and if $hgt(B) \leq \gamma$, $hgt(A) \leq \gamma$, then $hgt(B^\gamma)=hgt(A^\gamma)$. Hence $\xi(\gamma)=1$ in both cases. If $hgt(B)>\max\{hgt(A), \gamma\}$, then $hgt(B^\gamma)=hgt(B)$ and then $\xi(\gamma)=hgt(B^\gamma)\varphi hgt(A^\gamma)$ $=hgt(B)\varphi hgt(A^\gamma) \geq hgt(B)\varphi hgt(A)=\xi$ since $\varphi$ is nondecreasing in the second argument. Thus (iii) is proved and (iv) follows from the fact that $(A^{\gamma_1})^{\gamma_2}=A^{\gamma_2}$ for any $A \in \mathbf{F}(\mathbf{X})$ if $\gamma_1 \leq \gamma_2$, hereafter one applies property (iii). ■

THEOREM 10.7. *Let* t *be continuous. Then we have for the solvability indices* (10.4) *related to the systems* (10.2) *and* (10.11):

(i) $\xi(0)=\xi,\ \xi(1)=1,$

(ii) $\mathop{t}\limits_{h=1}^{m}(\gamma\, t\, \gamma)\leq\xi(\gamma)\leq 1,$

(iii) *if all the fuzzy sets* $A_h$ *are normal, then* $\mathop{t}\limits_{h=1}^{m}\gamma\leq\xi(\gamma).$

PROOF. (i) and the inequality $\xi(\gamma)\leq 1$ in (ii) are obvious.

Since $\varphi$ is decreasing in the first argument and t is monotone in both arguments, we have on using (10.8):

$$\xi(\gamma)\geq \mathop{t}\limits_{h=1}^{m}\ \bigwedge_{y\in \mathbf{Y}}\ \{B_h^{\gamma}(y)\ \varphi\ \{hgt(A_h^{\gamma})\ t\ [\ \bigwedge_{h=1}^{m} B_h^{\gamma}(y)]\}\}$$

$$\geq \mathop{t}\limits_{h=1}^{m}\ \bigwedge_{y\in \mathbf{Y}}\ [1\ \varphi\ (\gamma\, t\, \gamma)]=\ \mathop{t}\limits_{h=1}^{m}\ (\gamma\, t\, \gamma),$$

hence property (ii) is proved. (iii) follows again from (10.8), since we have:

$$\xi(\gamma)\geq \mathop{t}\limits_{h=1}^{m}\ \bigwedge_{y\in \mathbf{Y}}\ [1\varphi\ (1 t\, \gamma)] =\ \mathop{t}\limits_{h=1}^{m}\ \gamma.$$ ■

If t=min, we have still more properties as is seen in the following theorem:

THEOREM 10.8. *Let* t=min *and* $\gamma,\gamma_1,\gamma_2\in[0,1]$. *Then the properties* (ii), (iii), (iv) *of* Thm.10.6 *hold for the solvability indices* (10.4) *relatively to the systems* (10.2) *and* (10.11).

PROOF. (ii) follows immediately from Thm.10.7 (ii). We now observe that

$$A_h(x)\ \alpha\ B_h(y)\leq A_h^{\gamma}(x)\ \alpha\ B_h^{\gamma}(y) \qquad (10.12)$$

always holds for all $x\in \mathbf{X}$, $y\in \mathbf{Y}$, $\alpha$ being the $\varphi$-operator associated with min. Indeed, if $A_h(x)\leq\gamma$, we have the value 1 on the right-side of (10.12). If $A_h(x)>\gamma$, we have $A_h^{\gamma}(x)=A_h(x)$ and then (10.12) follows since $B_h(y)\leq B_h^{\gamma}(y)$.

Now (10.12) implies that

$$\max\{\bigvee_{x\in \mathbf{X}}\{A_h(x)\wedge\{\bigwedge_{h=1}^{m}[A_h(x)\,\alpha\,B_h(y)]\}\},\gamma\}\leq \bigvee_{x\in \mathbf{X}}\{A_h^{\gamma}(x)\wedge\{\bigwedge_{h=1}^{m}[A_h^{\gamma}(x)\,\alpha\,B_h^{\gamma}(y)]\}\}$$

for all $x\in \mathbf{X}$, $y\in \mathbf{Y}$. Then, by the same reasoning used to establish (10.12), we get for any $h\in\{1,2,\ldots,m\}$:

$$B_h(y)\,\alpha\,\{\bigvee_{x\in \mathbf{X}}\{A_h(x)\wedge\{\bigwedge_{h=1}^{m}[A_h(x)\,\alpha\,B_h(y)]\}\}\}$$

$$\leq B_h^{\gamma}(y)\,\alpha\,\{\bigvee_{x\in \mathbf{X}}\{A_h^{\gamma}(x)\wedge\{\bigwedge_{h=1}^{m}[A_h^{\gamma}(x)\,\alpha\,B_h^{\gamma}(y)]\}\}\}$$

for all $x\in \mathbf{X}$, $y\in \mathbf{Y}$. Now, using (10.9), we deduce (iii). Reasoning as in Thm.10.6, one deduces (iv) from (iii). ■

In order to solve fuzzy relation equations by means of the approach given here, the choice of a suitable value of the parameter $\gamma$ constitutes a critical point. Note that too high values of $\gamma$ may cause the result obtained to be completely useless. In a limit case, by putting $\gamma=1$ for which $\xi(1)=1$, we get a fuzzy relation identically equal to 1 and obviously one does not discover any relationship in the data set.

Therefore a choice of $\gamma$ should form a certain compromise between a value of the solvability index and the ability of the modified fuzzy sets $A^{\gamma}$, $B^{\gamma}$, $A_h^{\gamma}$, $B_h^{\gamma}$ to determine the fuzzy relation. Some plots of the performance of $\xi(\gamma)$ vs. the values of $\gamma$ are visualized in Fig.10.2.

They present different forms of changes of $\xi(\gamma)$ with respect to $\gamma$. In Fig.10.2(a), $\xi(\gamma)$ increases slowly with respect to $\gamma$, thus a choice of $\gamma$ is not critical. In Fig.10.2(b), it is obvious that for values of $\gamma$ exceeding some threshold, $\xi(\gamma)$ attains values close to 1.0: thus it suffices to take a value a little higher than this threshold. In Fig.10.2(c) there is a situation in which significant changes of the fuzzy data are required in order to get a feasible value of the solvability index. For further details, see [4].

## 10.4. Hierarchy in the Data Set

The previous results have a direct impact on the proposed performance index tied to the elements of the data set.

However, here we do not use the solvability index, but another concept in order to search for a hierarchy in the data set ($A_h$, $B_h$) which leads to a solution of the system (10.2).

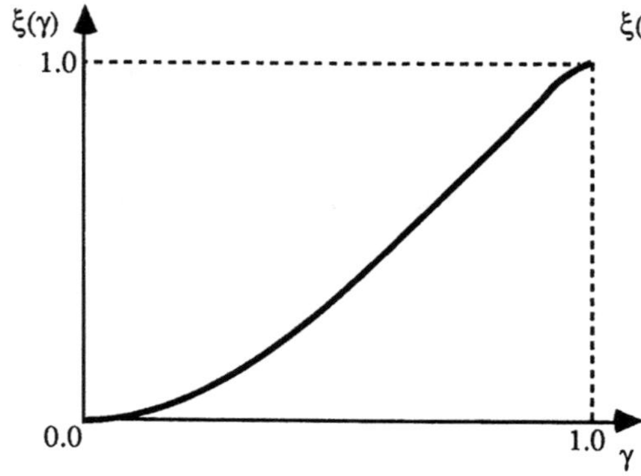


Fig. 10.2 (a)

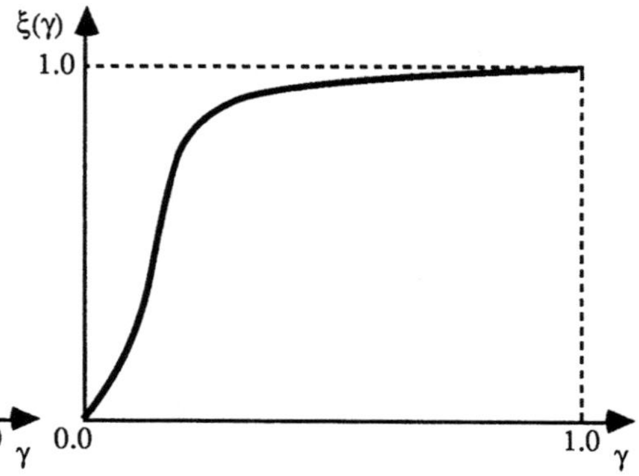


Fig. 10.2 (b)

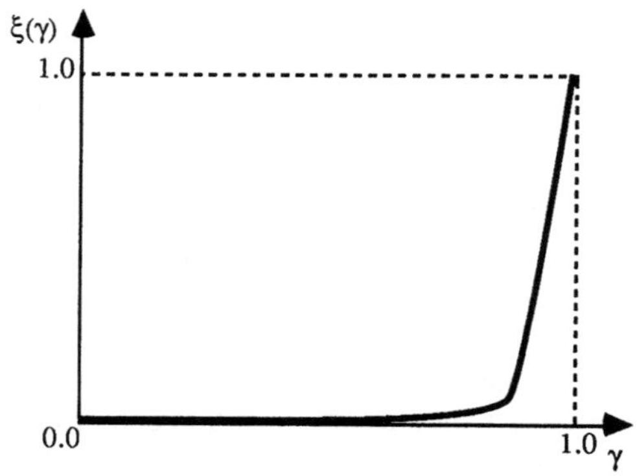


Fig. 10.2 (c)

Plots of $\xi(\gamma)$ vs. $\gamma$

### 10.4.1. PRELIMINARIES. DETERMINATION OF THE HIERARCHY IN THE FAMILIES OF FUZZY SETS

In Sec.10.3, we have seen that inconsistencies occur if $B_h$ and $B_j$ differ significantly for the same $A_h$ and $A_j$, i.e. $A_h=A_j$. Then the inconsistency can be measured comparing the equality indices of $A_h$, $A_j$ and $B_h$, $B_j$. The higher the equality index of $A_h$, $A_j$ and the lower the equality index of $B_h$, $B_j$ , the more inconsistent become the fuzzy data analyzed and the system of equations:

$$\begin{cases} R \text{ t } A_h = B_h, \\ R \text{ t } A_j = B_j, \end{cases} \tag{10.13}$$

is more difficult to solve.

Utilizing the equality index (10.1), we specify t as minimum and $\varphi$ as the "$\alpha$" operator and we put:

$$a_{hj} = [A_h \equiv A_j] \qquad \text{and} \qquad b_{hj} = [B_h \equiv B_j].$$

From now on, we also assume t=min in the systems (10.2) and (10.13).

Thus, it is natural that if $a_{hj} \leq b_{hj}$, we may expect that the pairs $(A_h,B_h)$ and $(A_j,B_j)$ are consistent, while if $a_{hj} > b_{hj}$ we can hardly satisfy the system (10.13).

Introducing the following index:

$$\delta_{hj} = a_{hj} \,\alpha\, b_{hj}\,, \tag{10.14}$$

we express a degree of *feasibility* of the system (10.13). Further, we shall use a modified version of the index (10.14) using the operator "&" introduced in Sec.9.1, defining it as follows:

$$\delta_{hj}^{\sim} = [A_h \,\&\, A_j] \,\alpha\, b_{hj}\,. \tag{10.15}$$

It gives an *optimistic value* of the degree of equality of the two specified fuzzy sets. The degree of feasibility (10.14) takes into account only one pair of data $A_h$ and $A_j$.

To obtain a global view on the whole structure of the data set relatively to the system (10.2), we use well known hierarchical clustering algorithms. Note that $\delta_{hj}$ can be viewed as a *measure of similarity* of the pairs $(A_h,B_h)$ and $(A_j,B_j)$ and consequently we can consider a *similarity matrix* $\mathfrak{D}$ defined as follows:

$$\mathfrak{D} = \begin{Vmatrix} \delta_{11} & \delta_{12} & \dots & \delta_{1m} \\ \delta_{21} & \delta_{22} & \dots & \delta_{2m} \\ \dots & \dots & \dots & \dots \\ \delta_{m1} & \delta_{m2} & \dots & \delta_{mm} \end{Vmatrix}, \tag{10.16}$$

where of course $\delta_{hj} = \delta_{jh}$, $\delta_{hh} = 1.0$ for all $h,j \in \{1,2,\dots,m\}$.

We start by considering m clusters, each constituted by a single pair $(A_h,B_h)$. Successively, we apply an *agglomerative procedure* so that the clusters that are the most similar to each other are merged. The number of clusters diminishes progressively by 1 until one cluster is obtained.

We evaluate the distance between two clusters $\mathfrak{H}$ and $\mathfrak{H}'$ by using the formula:

$$d(\mathfrak{H}, \mathfrak{H}') = \max_{\mathfrak{H},\mathfrak{H}'} \rho_{hj}\,,$$

where $\rho_{hj}=1-\delta_{hj}$ and the maximum is taken over all the pairs $(A_h,B_h)\in\mathfrak{H}$ and $(A_j,B_j)\in\mathfrak{H}'$.

Here $\rho_{hj}$ is viewed as the degree of difficulty of solving the system (10.13).

The hierarchy generated by the distance specified above is also known in clustering technique as complete linkage method. The clustering methods are advantageous in this study because of their suitability to represent the structure of the data set. The example below illustrates the procedure. The data set is that used in [7], where the results were achieved in the setting of the probabilistic sets [5] using iterative clustering techniques (ISODATA [1] and FUZZY C-MEANS [2]).

EXAMPLE 10.1. The data set consists of m=6 pairs of fuzzy sets $A_h,B_h$, h=1,...,6 defined in the spaces $X=\{x_1,x_2,x_3,x_4\}$ and $Y=\{y_1,y_2,y_3,y_4\}$ and the membership functions are specified as follows:

$$A_1 = \| 1.0 \quad 0.6 \quad 0.8 \quad 0.5 \|, \qquad B_1 = \| 0.0 \quad 0.0 \quad 0.3 \quad 0.6 \|,$$

$$A_2 = \| 0.7 \quad 1.0 \quad 0.5 \quad 0.2 \|, \qquad B_2 = \| 1.0 \quad 0.5 \quad 0.4 \quad 0.3 \|,$$

$$A_3 = \| 0.8 \quad 0.9 \quad 1.0 \quad 0.6 \|, \qquad B_3 = \| 0.2 \quad 0.3 \quad 0.5 \quad 1.0 \|,$$

$$A_4 = \| 1.0 \quad 0.5 \quad 0.2 \quad 0.0 \|, \qquad B_4 = \| 0.6 \quad 1.0 \quad 0.3 \quad 0.0 \|,$$

$$A_5 = \| 0.0 \quad 0.0 \quad 0.0 \quad 1.0 \|, \qquad B_5 = \| 0.3 \quad 0.6 \quad 1.0 \quad 1.0 \|,$$

$$A_6 = \| 0.4 \quad 0.8 \quad 1.0 \quad 0.7 \|, \qquad B_6 = \| 1.0 \quad 1.0 \quad 0.6 \quad 0.3 \|.$$

Each equation, viewed separately, has solutions. The resulting fuzzy relations $R_h^{\frown}=(A_h \,\alpha\, B_h)$ are equal to

$$R_1^{\frown}=\begin{Vmatrix} 0.0 & 0.0 & 0.3 & 0.6 \\ 0.0 & 0.0 & 0.3 & 1.0 \\ 0.0 & 0.0 & 0.3 & 0.6 \\ 0.0 & 0.0 & 0.3 & 1.0 \end{Vmatrix}, \quad R_2^{\frown}=\begin{Vmatrix} 1.0 & 0.5 & 0.4 & 0.3 \\ 1.0 & 0.5 & 0.4 & 0.3 \\ 1.0 & 1.0 & 0.4 & 0.3 \\ 1.0 & 1.0 & 1.0 & 1.0 \end{Vmatrix}, \quad R_3^{\frown}=\begin{Vmatrix} 0.2 & 0.3 & 0.5 & 1.0 \\ 0.2 & 0.3 & 0.5 & 1.0 \\ 0.2 & 0.3 & 0.5 & 1.0 \\ 0.2 & 0.3 & 0.5 & 1.0 \end{Vmatrix},$$

$$R_4^{\frown}=\begin{Vmatrix} 0.6 & 1.0 & 0.3 & 0.0 \\ 1.0 & 1.0 & 0.3 & 0.0 \\ 1.0 & 1.0 & 1.0 & 0.0 \\ 1.0 & 1.0 & 1.0 & 1.0 \end{Vmatrix}, \quad R_5^{\frown}=\begin{Vmatrix} 1.0 & 1.0 & 1.0 & 1.0 \\ 1.0 & 1.0 & 1.0 & 1.0 \\ 1.0 & 1.0 & 1.0 & 1.0 \\ 0.3 & 0.6 & 1.0 & 1.0 \end{Vmatrix}, \quad R_6^{\frown}=\begin{Vmatrix} 1.0 & 1.0 & 1.0 & 0.3 \\ 1.0 & 1.0 & 0.6 & 0.3 \\ 1.0 & 1.0 & 0.6 & 0.3 \\ 1.0 & 1.0 & 0.6 & 0.3 \end{Vmatrix}.$$

The matrix (10.16) takes the following entries:

$$
\mathfrak{D} = \begin{Vmatrix} 1.0 & 0.0 & 0.0 & 0.0 & 0.0 & 0.0 \\ 0.0 & 1.0 & 1.0 & 1.0 & 1.0 & 1.0 \\ 0.0 & 1.0 & 1.0 & 1.0 & 1.0 & 0.2 \\ 0.0 & 1.0 & 1.0 & 1.0 & 1.0 & 1.0 \\ 0.0 & 1.0 & 1.0 & 1.0 & 1.0 & 1.0 \\ 0.0 & 1.0 & 0.2 & 1.0 & 1.0 & 1.0 \end{Vmatrix}.
$$

Performing hierarchical clustering procedure, the dendrogram obtained is displayed in Fig.10.3.

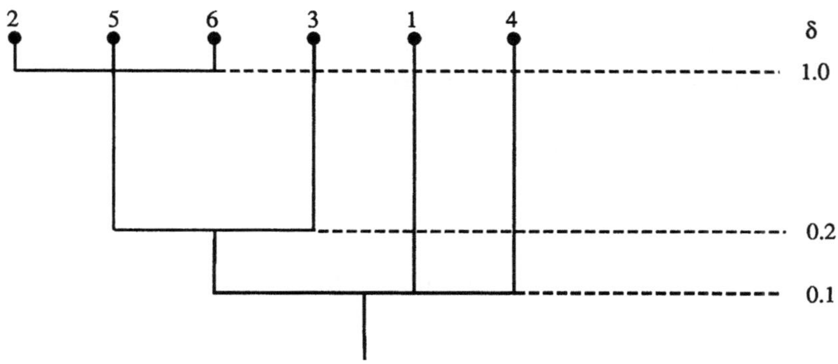


Fig. 10.3

Dendrogram visualizing a hierarchy in the data set

In Fig.10.3, $\delta$ is calculated as the minimum of all $\delta_{hj}$'s involved at that level. At $\delta$=1.0, four clusters may be seen: $\{(A_2,B_2), (A_5,B_5), (A_6,B_6)\}$, $\{(A_3,B_3)\}$, $\{(A_1,B_1)\}$, $\{(A_4,B_4)\}$, any pair is marked by an integer h=1,...,m. If the number of the clusters decreases, $\delta$ tends to zero, i.e. the system (10.2) is more difficult to solve.

In fact, the fuzzy relation $R^{\cap} \in \mathbf{F}(\mathbf{X}\times\mathbf{Y})$, being the finite intersection of the fuzzy relations $(A_h \, \alpha \, B_h) \in \mathbf{F}(\mathbf{X}\times\mathbf{Y})$ and reported below:

$$
R^{\cap} = \begin{Vmatrix} 0.0 & 0.0 & 0.3 & 0.0 \\ 0.0 & 0.0 & 0.3 & 0.0 \\ 0.0 & 0.0 & 0.3 & 0.0 \\ 0.0 & 0.0 & 0.3 & 0.3 \end{Vmatrix},
$$

yields a very poor performance that is visible even without specifying the solvability index. The fuzzy sets $R^{\cap} \odot A_h$ do not have the values of the membership function greater that 0.3

and they are visualized in Figs.10.4.

For instance, taking the fuzzy relation $R_{256}^{\cap}$, equal to the intersection of ($A_2$ α $B_2$), ($A_5$ α $B_5$) and ($A_6$ α $B_6$) and that is based on the data forming a cluster in the dendrogram of Fig.10.3, one gets:

$$R_{256}^{\cap} = \begin{Vmatrix} 1.0 & 0.5 & 0.4 & 0.3 \\ 1.0 & 0.5 & 0.4 & 0.3 \\ 1.0 & 1.0 & 0.4 & 0.3 \\ 0.3 & 0.6 & 0.6 & 0.3 \end{Vmatrix}.$$

For comparison, the resulting fuzzy sets $R_{256}^{\cap} \odot A_h$ are represented in Fig.10.4. Note that now the solvability index of the system (10.2) is higher.

Taking the index $\delta_{hj}^{\sim}$ defined by (10.15), we obtain the related similarity matrix given by

$$\begin{Vmatrix} 1.0 & 0.0 & 0.0 & 0.0 & 0.0 & 0.0 \\ 0.0 & 1.0 & 0.2 & 0.0 & 1.0 & 0.4 \\ 0.0 & 0.2 & 1.0 & 0.0 & 0.2 & 0.2 \\ 0.0 & 0.0 & 0.0 & 1.0 & 1.0 & 0.0 \\ 0.0 & 1.0 & 0.2 & 1.0 & 1.0 & 0.3 \\ 0.0 & 0.4 & 0.2 & 0.0 & 0.3 & 1.0 \end{Vmatrix}.$$

The hierarchical clustering method applied to this matrix has produced a dendrogram presented in Fig.10.5.

A certain difference with respect to the dendrogram of Fig.10.4 can be observed: only two pairs, i.e. ($A_2$,$B_2$) and ($A_5$,$B_5$), have been detected. Indeed, we get $R^{\cap}=(A_2 \alpha B_2) \wedge (A_5 \alpha B_5)$ given by

$$R^{\cap} = \begin{Vmatrix} 1.0 & 0.5 & 0.4 & 0.3 \\ 1.0 & 0.5 & 0.4 & 0.3 \\ 1.0 & 1.0 & 0.4 & 0.3 \\ 0.3 & 0.6 & 1.0 & 1.0 \end{Vmatrix}$$

and $R^{\cap} \odot A_2 = B_2$ and $R^{\cap} \odot A_3 = B_3$.

Let us focus our attention on the previous relation $R_{256}^{\cap}$. The fuzzy sets $A_2$,$A_5$,$A_6$, do not *cover* the entire space **X**. Consequently, the fuzzy relation $R_{256}^{\cap}$ does not yield a good complete map of the entire data set. However, adding another pair, for instance ($A_3$,$B_3$) (see Figs.10.4) the fuzzy relation $R_{2563}^{\cap}=R_{256}^{\cap} \wedge (A_3 \alpha B_3)$ leads to quite a low

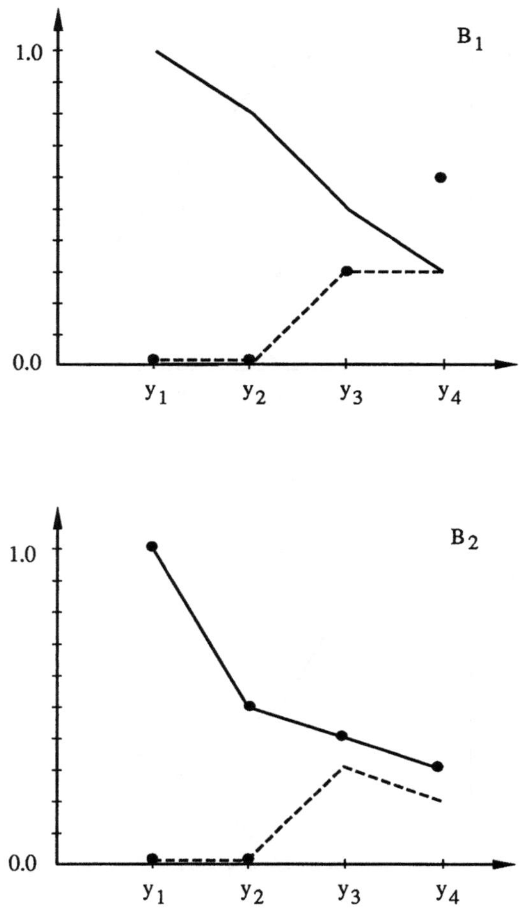


Fig 10.4 (a)

Membership functions of the fuzzy sets $R^{\cap} \odot A_h$(---) and $R_{256}^{\cap} \odot A_h$(—), h=1,2

• denotes the original fuzzy sets $B_1$ and $B_2$

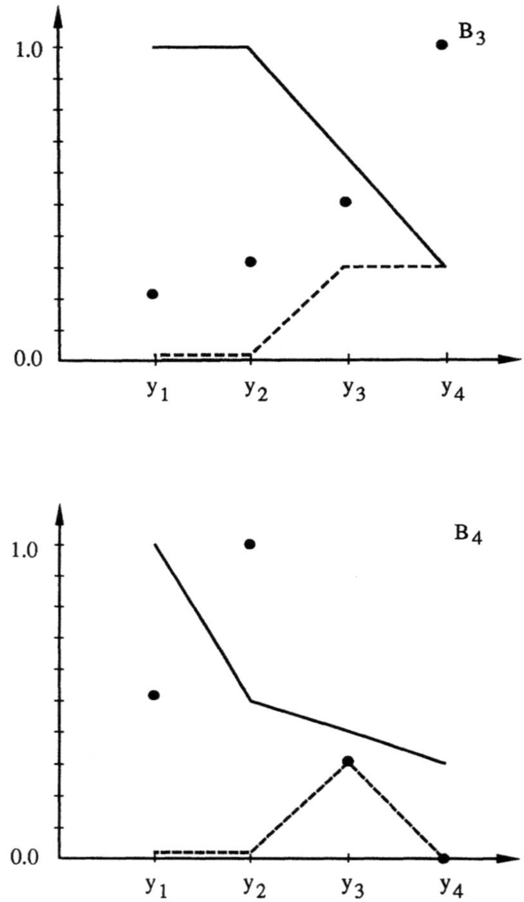


Fig 10.4 (b)

Membership functions of the fuzzy sets $R^{\cap} \odot A_h$(---) and $R_{256}^{\cap} \odot A_h$(——), h=3,4

• denotes the original fuzzy sets $B_3$ and $B_4$

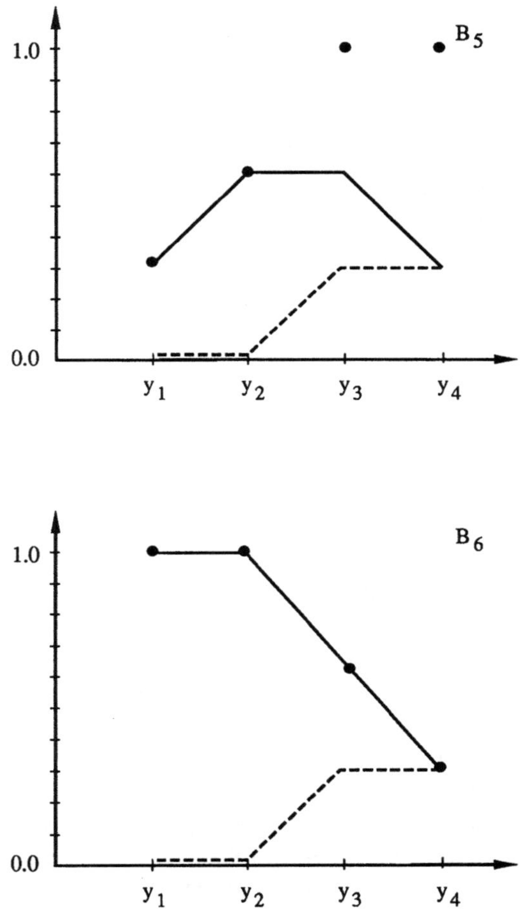


Fig 10.4 (c)

Membership functions of the fuzzy sets $R^{\cap} \odot A_h$(---) and $R_{256}^{\cap} \odot A_h$(—), h=5,6

• denotes the original fuzzy sets $B_5$ and $B_6$

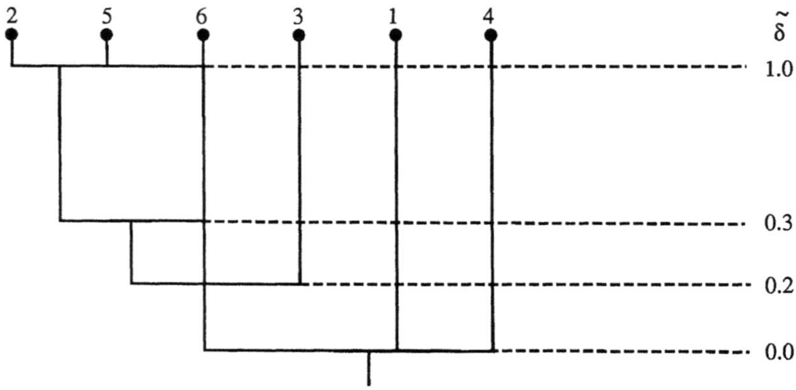


Fig. 10.5

Dendrogram with results of hierarchical clustering for levels of $\tilde{\delta} = \min \tilde{\delta}_{hj}$

degree $\delta$=0.2 and $R_{2563}^{\cap}$ is equal to

$$\begin{Vmatrix} 0.2 & 0.3 & 0.4 & 0.3 \\ 0.2 & 0.3 & 0.4 & 0.3 \\ 0.2 & 0.3 & 0.4 & 0.3 \\ 0.2 & 0.3 & 0.5 & 0.3 \end{Vmatrix}.$$

Performing this analysis, a natural question arises: which number of the pairs of fuzzy data should be taken into account to compute the fuzzy relation. In other words, which is the value of the threshold level in the dendrogram which indicates the fuzzy sets that have to be considered to create a core of the most reliable and consistent subset of the entire data set. The choice is not a trivial task. On the other hand, too many fuzzy data participating in the computation of the fuzzy relation can generate meaningless results, i.e. all entries of the fuzzy relation are equal to zero. Moreover, a too small amount of the fuzzy data cannot enable us to discover the structure of the entire relation. An important question concerns the evaluation of a *representation power* of the fuzzy data $A_1, A_2, \ldots, A_m$ that contribute to the fuzzy relation. This problem is discussed in the next Sec.

## 10.4.2. ON MEASURING THE REPRESENTATION POWER OF THE FUZZY DATA

A concept that directly leads to the method of measuring the representative power of the selected data set is based on a simple idea. Considering Eq.(3.1), it is obvious that only the elements of the fuzzy set A for which $A(x)>B(y)$ contribute to the process of the determination of the fuzzy relation $(A\alpha B)\in \mathbf{F}(\mathbf{X}\times\mathbf{Y})$ (cfr. Ch.3), otherwise the corresponding element $(A\alpha B)(x,y)$ is equal to 1. For simplicity, we put $\mathbf{X}=\{x_1,x_2,\ldots,x_n\}$ and introduce an auxiliary vector $\mathbf{v}=(v_1,v_2,\ldots,v_n)$ defined in $\mathbf{X}$ as follows:

$$v_i=\text{card}\{y\in \mathbf{Y}: A(x_i) \geq B(y)\}$$

for $i=1,2,\ldots,n$. Each component $v_i$ of $\mathbf{v}$ counts the number of the elements $y\in \mathbf{Y}$ where the given value of the membership function of A at $x_i$ exceeds or is equal to $B(y)$. Afterwards, we replace $\mathbf{v}$ by a suitable vector $\mathbf{p}=(p_1,p_2,\ldots,p_n)$ that results from a simple normalization of $\mathbf{v}$, i.e.

$$p_i = v_i / \sum_{i=1}^{n} v_i , \qquad i=1,2,\ldots,n.$$

If the sum of the $v_i$'s is equal to zero, we put $p_i=0$. The vector $\mathbf{p}$ measures to what extent the fuzzy relation R is well-determined and which of its rows are best estimated. The higher the value of the $p_i$'s is, the better the i-th row of R is determined. For instance, if A is a degenerated fuzzy set, i.e. $A(x_i)=0$ and $A(x_s)=1$ for some $s\in\{1,2,\ldots,n\}$, the corresponding vector is equal to

$$(0.0 \quad 0.0 \quad \ldots \quad 1.0 \quad \ldots \quad 0.0 \quad \ldots \quad 0.0),$$

with 1.0 at the s-th component. It is obvious that the s-th row of R is well estimated. Instead, taking A such that all $p_i$'s are equal to zero (i.e. $\mathbf{p}=\mathbf{0}$) we deal with a situation in which the fuzzy relation cannot be estimated.

By taking an intermediate situation where A is sought as *unknown*, i.e. its membership function is equal to 1.0 over $\mathbf{X}$, we have $v_i=\text{card}\mathbf{Y}$ and then $p_i=1/n$ for $i=1,2,\ldots,n$, so that all the rows of the fuzzy relation are estimated at the same degree equal to $1/n$.

For the system (10.2), we can assign to each pair of data $(A_h,B_h)$ a vector $\mathbf{p}_h$, $h=1,2,\ldots,m$ and perform a global evaluation defining another vector $\mathbf{u}$ such that its i-th component is the sup of the i-th components of each vector $\mathbf{p}_h$ and we put, in symbols:

$$\mathbf{u} = \bigvee_{h=1}^{m} \mathbf{p}_h .$$

The vectors $\mathbf{p}_h$ allow us to eliminate some data at the preliminary stage of analysis: all the fuzzy data for which the corresponding vector $\mathbf{p}_h$ is equal to $\mathbf{0}$ can be excluded from further analysis. The highest value of $\mathbf{u}$ is attained when the elements of the data set are carefully chosen.

The values of **u** obtained for the data set give information about the fuzzy sets used to determine the fuzzy relation. A whole picture is reached combining this approach with the results of the overall analysis represented by the dendrogram, e.g., of Fig.10.3. It sheds light on the structure of the data set from two different, competitive, points of view, namely:

(i) *feasibility to solve the system (10.2),*

(ii) *ability to represent the fuzzy relation.*

They are contradictory by nature. Indeed, moving from the top of the dendrogram of Fig.10.3 to its bottom, it is more difficult to satisfy the first requirement, but in the opposite direction, the same holds for the second requirement.

EXAMPLE 10.2. Returning to Ex.10.1, we have

$$\mathbf{p}_1 = \|0.33 \ \ 0.33 \ \ 0.33 \ \ 0.07\|, \qquad \mathbf{p}_2 = \|0.25 \ \ 0.33 \ \ 0.25 \ \ 0.00\|,$$

$$\mathbf{p}_3 = \|0.23 \ \ 0.23 \ \ 0.31 \ \ 0.23\|, \qquad \mathbf{p}_4 = \|0.50 \ \ 0.25 \ \ 0.12 \ \ 0.12\|,$$

$$\mathbf{p}_5 = \|0.00 \ \ 0.00 \ \ 0.00 \ \ 1.00\|, \qquad \mathbf{p}_6 = \|0.11 \ \ 0.22 \ \ 0.44 \ \ 0.22\|.$$

If we restrict ourselves to the data forming the cluster constituted by the pairs $(A_2,B_2)$, $(A_5,B_5)$, $(A_6,B_6)$, the vector $\mathbf{u}'=\mathbf{p}_2\vee\mathbf{p}_5\vee\mathbf{p}_6$ possesses the entries,

$$\mathbf{u}' = \|0.25 \ \ 0.33 \ \ 0.44 \ \ 1.00\|.$$

The addition of the pair $(A_3,B_3)$, that results from the dendrogram of Fig.10.3, does not change the vector **u**'. The ability of the whole data set to estimate the fuzzy relation is expressed by the vector

$$\mathbf{u} = \|0.50 \ \ 0.33 \ \ 0.44 \ \ 1.00\|,$$

which in fact indicates that a restriction to the first subset of the data set containing only three pairs already indicated does not decrease significantly the ability to determine the fuzzy relation. A change in the values of **u**' with respect to **u** is observed for only one element of **X**.

The way presented provides the user with a tool suitable for analysis of the structure in the fuzzy data collected for the purposes of determination of the fuzzy relation. Moreover the index characterizing the ability of the certain subset of data to yield an estimate of the fuzzy relation has been introduced. Nevertheless the choice of the suitable threshold level in the dendrogram is left open and the user must make his own decision on the basis of the indices studied here.

In any situation, a certain compromise should be achieved.

## 10.5. On Solving a System of Max-min Fuzzy Relation Equations by Gathering Statistics of Partial Solutions

Here we deal with the system (10.2) assuming again t=min, i.e. "$\circ$" stands for the max-min composition, the pairs of fuzzy sets $(A_h,B_h)$, $h=1,\ldots,m$, are provided and R is unknown.

The methods proposed in this Sec. have a straightforward idea behind them. In essence they rely on the observation that the values of the membership function of the fuzzy relation should be affected by some collected statistics of the partial results (solutions of the respective equations of the system), rather than simply picking up extremal values of the membership function of R resulting from $\alpha$-composition of the values of the membership functions of $A_h$ and $B_h$.

Following this idea, we investigate two ways to construct the fuzzy relation of the system (10.2) and indicate in which way the results obtained should be interpreted. The main idea of the first method is to gather statistics concerning the distribution of the results of the $\alpha$-composition of the pairs of fuzzy sets $(A_h,B_h)$ and look for a suitable representation of them. The second method makes use of results of the $\alpha$-compositions, transforms them into the unified form of a probabilistic set [5] and furthermore extends the max-min composition allowing us to handle probability and fuzziness. Finally, a careful interpretation of the results is accomplished.

In the sequel, we put $\mathbf{X}=\{x_1,x_2,\ldots,x_n\}$ and $\mathbf{Y}=\{y_1,y_2,\ldots,y_q\}$.

**Method 1**: In this method, for each pair of elements of **X** and **Y**, say $(x_i,y_j)$, one collects the results of the $\alpha$-composition of $A_h(x_i)$ and $B_h(y_j)$ for $h=1,2,\ldots,m$, where $i\in\{1,\ldots,n\}$ and $j\in\{1,\ldots,q\}$. They are arranged in the form of a probability function $P(x_i,y_j,w)$, $w\in[0,1]$, which describes the probability attached to the event that the value of the $\alpha$-composition is equal to w, i.e. $[A_h(x_i)\ \alpha\ B_h(y_j)]=w$. Some illustrative shapes of the probability functions are displayed in Fig.10.6.

It is noticeable that a minimal value $w_0$ of the argument for which the probability function attains nonzero value corresponds precisely to the result obtained in the original approach, i.e.

$$\min_{1\le h\le m}\ [A_h(x_i)\ \alpha\ B_h(y_j)] = w_0\,.$$

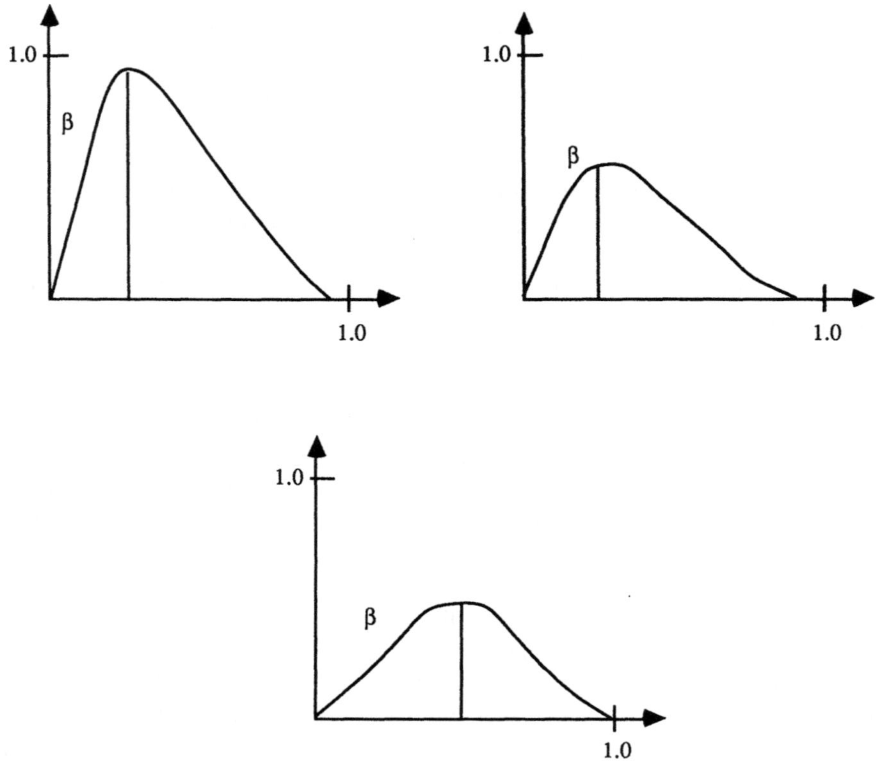


Fig. 10.6

Shapes of probability functions $P(x_i, y_j, w)$

This results from a relationship:

$$\int_0^{w_0} P(x_i, y_j, v)\, dv > 0.$$

More precisely, $w_0$ is the minimal value for which the above inequality holds true.

It is obvious that in many cases the total mass of probability is distributed very far from this value. Thus the idea is to replace this value by another one not so extremal in character. Loosely speaking, we are looking for the value w' with some higher probability standing behind it.

In other words, one accepts the value w' as the (i,j)-th entry of the fuzzy relation for which the cumulative probability (probability distribution function) is not lower than a certain threshold $\beta \in [0,1]$. Then w' results from the following inequality:

$$\int_0^{w'} P(x_i, y_j, v)\, dv \geq \beta .$$

Bearing in mind the diversity of shapes of the probability function, w' could significantly differ from the value $w_0$. Nevertheless, its choice should be made with respect to a sum D of Hamming distances between $B_i$ and the corresponding fuzzy sets obtained from the max-min composition $R \circ A_i$, where R is formed on the basis of the particular threshold level. So R and $D = D(\beta)$ are functions of $\beta$ and therefore $\beta$ could be adjusted in such a way that it minimizes $D(\beta)$. Hence a sequence of steps by which the fuzzy relation is determined forms an iteration loop starting from the lowest nonzero value of the threshold $\beta$ and incrementing it, observing at the same time the value of $D(\beta)$. The value of $\beta$ corresponding to the minimum of D determines the fuzzy relation R of the system (10.2). The higher the values of R obtained (in comparison to those resulting by taking the minimums of all the partial results stemming from individual equations), the higher inconsistency level is in the system reported.

**Method 2**: now we take into account the whole probability distribution functions (not optimizing with respect to any threshold level $\beta$) and perform the respective composition of $A_i$ and R (represented by a family of probability distribution functions for each of its entries). Afterwards the result of the composition is analyzed and its probability function conveys information concerning a search of the particular value of the membership function.

To obtain a closed formula, we denote by

$$F_{R(x_i,y_j)}(w), \qquad w \in [0,1],$$

a probability distribution function of the grades of membership of the fuzzy relation at the specified pair $(x_i,y_j)$. It has been originated as a result of aggregation of the $\alpha$-compositions of $A_h(x_i)$ and $B_h(y_j)$ for all the equations of the system (10.2). In fact, R may also be conveniently viewed as a probabilistic set. Now by using some basic formula of probability calculus and dealing, in general, with fuzzy sets $A \in \mathbf{F}(\mathbf{X})$ and $B \in \mathbf{F}(\mathbf{Y})$, let us express the distribution function of B treating A as a probabilistic set (described by its distribution function). Referring to maximum and minimum of random variables and assuming additionally that A and R are independent (which is rather a critical point in these considerations), we get:

$$F_{B(y_j)}(w) = \prod_{i=1}^{n} [F_{A(x_i)}(w) + F_{R(x_i,y_j)}(w) - F_{A(x_i)}(w) \cdot F_{R(x_i,y_j)}(w)] ,$$

where the distribution function of A is defined as

$$F_{A(x_i)}(w) = 0 \quad \text{if } w<A(x_i),$$

$$= 1 \quad \text{if } w \geq A(x_i).$$

Hence, if $w<\min\{A(x_i): i=1,\ldots,n\}$, we have:

$$F_{B(y_j)}(w) = \prod_{i=1}^{n} F_{R(x_i,y_j)}(w) , \qquad j = 1,2,\ldots,q,$$

i.e. the distribution function of B is precisely the product of the distribution functions of the j-th column of the fuzzy relation R. Then, applying a certain threshold $\beta$ to such a distribution function (as in the first method), the grades of the membership function of B are determined. This approach is clearly more elaborate: it needs to process the distribution functions of the fuzzy relation and A. Additionally, the assumption of independency is rather artificial and difficult to fulfil, but no constructive calculations and comparisons are accessible without such an assumption. This approach is of significant interest in the cases in which A differs from $A_h$, h=1,...,m, and it is a genuine probabilistic set when one has at one's disposal a valuable tool for sensitivity analysis. We illustrate the performance of the two methods with the aid of the following numerical example:

EXAMPLE 10.3. Considering the fuzzy sets $A_h, B_h$, h=1,...,6 of Ex.10.1, we take different values of probability (level of $\beta$) and deal with discrete values in [0,1]. Several ranges of $\beta$ are distinguished and listed below with the relevant fuzzy relations constructed by means of the first method.

$\beta \in [1/6, 2/6]$ ,

$$R = \begin{Vmatrix} 0.0 & 0.0 & 0.3 & 0.0 \\ 0.0 & 0.0 & 0.3 & 0.0 \\ 0.0 & 0.0 & 0.3 & 0.0 \\ 0.0 & 0.0 & 0.3 & 0.3 \end{Vmatrix},$$

$\beta \in [2/6, 4/6]$ ,

$$R = \begin{Vmatrix} 0.2 & 0.3 & 0.3 & 0.3 \\ 0.2 & 0.3 & 0.3 & 0.3 \\ 0.2 & 0.3 & 0.4 & 0.3 \\ 0.2 & 0.3 & 0.5 & 1.0 \end{Vmatrix},$$

$\beta \in [3/6, 1]$ ,
$$R = \begin{Vmatrix} 0.6 & 0.3 & 0.4 & 0.3 \\ 1.0 & 0.5 & 0.4 & 0.3 \\ 1.0 & 1.0 & 0.5 & 0.3 \\ 0.3 & 0.6 & 0.6 & 1.0 \end{Vmatrix},$$

$\beta \in [4/6, 1]$ ,
$$R = \begin{Vmatrix} 1.0 & 1.0 & 0.5 & 0.6 \\ 1.0 & 1.0 & 0.5 & 1.0 \\ 1.0 & 1.0 & 0.6 & 0.6 \\ 1.0 & 1.0 & 1.0 & 1.0 \end{Vmatrix},$$

$\beta = 1$ ,
$$R = \begin{Vmatrix} 1.0 & 1.0 & 1.0 & 1.0 \\ 1.0 & 1.0 & 1.0 & 1.0 \\ 1.0 & 1.0 & 1.0 & 1.0 \\ 1.0 & 1.0 & 1.0 & 1.0 \end{Vmatrix}.$$

In order to choose a particular range of $\beta$, we calculate the sum of the Hamming distances between $R \odot A_h$ and $B_h$ (we emphasize that R has been built with respect to the particular threshold level $\beta$).

| Range of $\beta$ | D |
|---|---|
| [1/6, 2/6] | 9.6 |
| [2/6, 4/6] | 6.8 |
| [3/6, 1] | 7.1 |
| [4/6, 1] | 8.1 |
| 1 | 11.2 , |

where

$$D = \sum_{h=1}^{m} \sum_{j=1}^{q} |B_h(y_j) - (R \odot A_h)(y_j)| .$$

Hence the minimal value of D has been achieved for a relatively low value of $\beta$. For the lowest nonzero value of $\beta$ (which in fact is precisely the minimum of the results of the $\alpha$-compositions), the fuzzy relation is close to the null one, while R consists entirely of 1's if $\beta$=1.

We summarize in Table 10.1 the number of the elements for which $B_h(y_j)$ has the same value, lower or higher than $(R\odot A_h)(y_j)$.

Table 10.1

| Range of β | number of elements with grade of membership lower | equal | higher |
|---|---|---|---|
| [1/6, 2/6] | 18 | 6 | 0 |
| [2/6, 4/6] | 14 | 4 | 6 |
| [3/6, 1] | 12 | 4 | 8 |
| [4/6, 1] | 1 | 7 | 16 |
| 1 | 0 | 7 | 17 |

They give an interesting insight into the properties of the fuzzy relations generated. For low values of β, some inconsistency exists (there is no exact solution), in most cases the resulting grades of membership are lower than the original ones (sometimes identically equal to 0). For an optimal β there is a certain balance, while for too high values of β there is a significant bias towards higher values of the grades of membership. This is a consequence of the structure of the fuzzy relation determined: it is mostly filled with 1's.

Using the second method in which the distribution functions of the fuzzy relation are utilized, the results D=D(β) are given in Fig.10.7.

The minimum value of D is achieved for a relatively low value of β (β=0.35 and D=6.9 being somewhat higher than the value obtained with the aid of the first method).

## 10.6. On Solving a Max-min Fuzzy Relation Equation

In this Sec. we consider Eq.(3.1), i.e. $R\odot A=B$ with A unknown and B and R provided. By the basic Thm.2.4 (applied to Eq.(3.1)), we see that if Eq.(3.1) has solutions, the fuzzy set $A^{\cap}=(R\alpha B)\in \mathbf{F}(\mathbf{X})$ is the greatest solution, where $A^{\cap}$ is pointwise defined as

$$A^{\cap}(x) = \bigwedge_{y\in \mathbf{Y}} [R(x,y)\,\alpha\, B(y)] \tag{10.17}$$

for any $x\in X$. We recall that **X** and **Y** are finite sets and in order to apply the Method 1 of Sec.10.5, we assume $\mathbf{X}=\{x_1,\ldots,x_n\}$ and $\mathbf{Y}=\{y_1,y_2,\ldots,y_m\}$. The original Eq.(3.1) can be seen as a system of equations indexed by $j\in\{1,2,\ldots,m\}$:

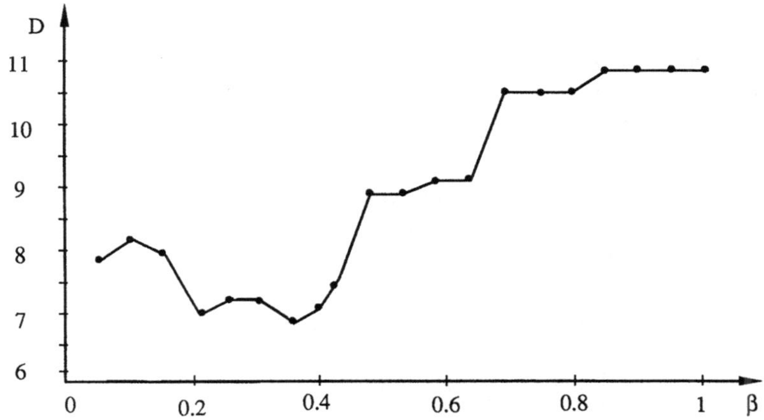


Fig. 10.7

D versus β

$$A \odot R_j = B_j\ , \tag{10.18}$$

where $B_j = B(y_j)$ and $R_j \in \mathbf{F}(\{y_j\} \times \mathbf{X})$ is pointwise defined as $R_j(y_j, x_i) = R(x_i, y_j)$ for $j=1,\ldots,m$. So we get a system of type (10.2) of m max-min fuzzy relation equations. Then for each Eq.(10.18), the corresponding solution is equal to

$$A^{\frown}(x_i, y_j) = R(x_i, y_j)\ \alpha\ B(y_j)\ ,$$

where $i=1,\ldots,n$, $j=1,\ldots,m$ and $A^{\frown} \in \mathbf{F}(\mathbf{X} \times \{y_j\})$ (the second argument in $A^{\frown}$ has been added to underline the origin of this solution: it should be regarded here as a parameter). For each Eq.(10.18), it is a straightforward task to test if there exists a set of solutions: it suffices to satisfy (cfr. Thms.2.8 and 3.3) the condition:

$$\max_{1 \le i \le n} R(x_i, y_j) \ge B(y_j) \tag{10.19}$$

for $j=1,\ldots,m$. Note that $A^{\frown}$, expressed by (10.17) is nothing but the intersection (i.e. the minimum) of the partial results of each Eq.(10.18). Therefore, possessing the problem expressed as a set of equations, the method 1 of Sec.10.5 is now utilized in the following numerical example:

EXAMPLE 10.4. Let $n=m=4$, $R \in \mathbf{F}(\mathbf{X} \times \mathbf{Y})$ and $B \in \mathbf{F}(\mathbf{Y})$ given by

$$R = \begin{Vmatrix} 1.0 & 0.7 & 0.2 & 0.4 \\ 0.4 & 0.5 & 1.0 & 0.5 \\ 0.5 & 0.3 & 1.0 & 0.5 \\ 0.7 & 0.5 & 0.6 & 0.8 \end{Vmatrix}, \qquad B = \| 0.4 \quad 1.0 \quad 0.2 \quad 0.0 \|.$$

Treating Eq.(3.1) as a system of equations, we have:

$$\| 1.0 \quad 0.4 \quad 0.5 \quad 0.7 \| \odot A = 0.4,$$

$$\| 0.7 \quad 0.5 \quad 0.3 \quad 0.5 \| \odot A = 1.0,$$

$$\| 0.2 \quad 1.0 \quad 1.0 \quad 0.6 \| \odot A = 0.2,$$

$$\| 0.4 \quad 0.5 \quad 0.5 \quad 0.8 \| \odot A = 0.0.$$

Then testing whether each equation of the above system has a solution, the second equation has to be eliminated since the condition (10.19) is not satisfied. We build a solution making use of the Method 1 of Sec.10.5. The solutions of each equation are equal to

$$\| 0.4 \quad 1.0 \quad 0.4 \quad 0.4 \|,$$

$$\| 1.0 \quad 0.2 \quad 0.2 \quad 0.2 \|,$$

$$\| 0.0 \quad 0.0 \quad 0.0 \quad 0.0 \|.$$

By calculating the probability functions, we get the results reported in Table 10.2.

Table 10.2

| w / $x_i$ | 0.0 | 0.2 | 0.4 | 1.0 |
|---|---|---|---|---|
| $x_1$ | 1/3 | 0 | 1/3 | 1/3 |
| $x_2$ | 1/3 | 1/3 | 0 | 1/3 |
| $x_3$ | 1/3 | 1/3 | 1/3 | 0 |
| $x_4$ | 1/3 | 1/3 | 1/3 | 0 |

Different levels of the threshold level $\beta$ could be distinguished (observe also the above probability functions); they yield the following results:

$$\beta \le 1/3, \qquad A = \|0.0 \quad 0.0 \quad 0.0 \quad 0.0\|, \qquad D = 1.6,$$

$$1/3 < \beta < 2/3, \qquad A = \|0.4 \quad 0.2 \quad 0.2 \quad 0.2\|, \qquad D = 1.6,$$

$$\beta = 1, \qquad A = \|1.0 \quad 1.0 \quad 0.4 \quad 0.4\|, \qquad D = 2.3.$$

Hence, for the second range of the threshold level, a minimum value of the performance index D (=the Hamming distance between R⊙A and B) has been obtained.

The same approach can be useful for solving α-fuzzy relation equations defined in Ch.6, as in the system:

$$R \,\alpha\, A_h = B_h, \qquad h=1,2,\ldots,m$$

with $R \in \mathbf{F}(\mathbf{X} \times \mathbf{Y})$ unknown and as in a single Eq.(6.1) with $A \in \mathbf{F}(\mathbf{X})$ unknown. The only difference is that β is adjusted starting from high values of w and moving down; hence β and w are tied together via the relationship:

$$\int_w^1 P(x_i, y_j, v)\, dv \ge \beta .$$

The results of Sec.10.5 and 10.6 can be found in Pedrycz [10].

## 10.7. Concluding Remarks

There is no doubt that the search for suitable ways of providing approximate solutions of fuzzy relation equations constitutes a significant topic that ties theoretical results to different applications.

We have discussed some approaches. Two general ways can be distinguished: firstly, an omission of a subset of equations (fuzzy sets) is perhaps the simplest one. A critical point is to indicate which subset should be skipped, secondly, the fuzzy sets are modified in order to get links between them which are consistent with the fuzzy relation.

The use of these methods (or their combination) depends on the nature of the specific problem under discussion.

A last remark concerns reference to the so-called *brute force method*: it consists in the direct use of the method of resolution without justification of the validity of the assumption that the set of the solutions of a fuzzy equation is nonempty. The results produced in this way are usually very poor. They may form a suitable reference point for the methods proposed in this Ch., but a comparison of the quality of the results obtained is always necessary.

## References

[1] G. Ball and A. Hall, A clustering technique for summarizing multivariate data, *Behav. Sci.* 12 (1967), 153-155.

[2] J.C. Bezdek, *Pattern Recognition with Fuzzy Objective Function Algorithms*, Plenum Press, New York, 1981.

[3] A. Di Nola, W. Pedrycz and S. Sessa, A study on approximate reasoning mechanisms via fuzzy relation equations, *Internat. J. Approx. Theory*, to appear.

[4] S. Gottwald and W. Pedrycz, Solvability of fuzzy relational equations and manipulation of fuzzy data, *Fuzzy Sets and Systems* 18 (1986), 45-65.

[5] K. Hirota, Concept of probabilistic sets, *Fuzzy Sets and Systems* 5 (1981), 31-46.

[6] K. Hirota and W. Pedrycz, Fuzzy system identification via probabilistic sets, *Inform. Sciences* 28 (1982), 21-43.

[7] K. Hirota and W. Pedrycz, Analysis and synthesis of fuzzy systems by the use of probabilistic sets, *Fuzzy Sets and Systems* 10 (1983), 1-13.

[8] W. Pedrycz, Structured fuzzy models, *Cybernetics and Systems* 16 (1985), 103-117.

[9] W. Pedrycz, Approximate solutions of fuzzy relation equations, *Fuzzy Sets and Systems* 28 (1988), 183-202.

[10] W. Pedrycz, Algorithms for solving fuzzy relational equations in a probabilistic setting, *submitted.*

# CHAPTER 11

## HANDLING FUZZINESS IN KNOWLEDGE-BASED SYSTEMS

11.1. Handling of Factors of Uncertainty in Knowledge-Based Systems: Approaches and Requirements.
11.2. Selected Aspects of Reasoning with Fuzziness.
11.3. Principle of Consistency of Aggregation and Inference Mechanisms.
11.4. Information Granularity and Fuzzy Sets.
11.5. Conclusions.

*In this Ch., as well as in the following, we will study a unified approach for handling and processing sources of uncertainty in knowledge-based systems. This goal is achieved in the framework of fuzzy relation equations. We point out how the mechanisms of the theory developed in the previous Chs. of this book can be treated as a convenient platform for construction of knowledge-based systems. More precisely, it will be indicated how fuzzy equations contribute to each of the conceptual levels recognized in the construction of these systems (viz. knowledge representation, meta-knowledge, inference techniques, etc.) as well as how they are directly used in formation of the particular elements of the problem-oriented expert systems. It is assumed that the reader has a certain background concerning Knowledge Engineering and Artificial Intelligence, at least on fundamentals of architecture of knowledge-based systems. It is also expected that he is familiar with some of the well-known expert systems, especially those broadly documented in literature (e.g. PROSPECTOR, MYCIN) and mechanisms involved there which are capable of coping with uncertainty, no matter how it has been introduced. This Ch. must be viewed as a concise prerequisite for the successive Chs. and it indicates many problems occurring in knowledge engineering when factors of uncertainty have to be processed.*

*Here we deal with the principal constituents in the structure of any knowledge-based system, namely knowledge base, inference mechanism and system-user interface concerning processes of data acquisition and explanation mechanisms. For each of these elements, the mechanism of handling fuzziness will be discussed in detail. Now we will start with a concise introduction to different methods of coping with uncertainty found in expert systems. We will focus our attention mainly on rule-based systems that are in common use. Moreover they usually cover a broad spectrum of applicational situations; for an extensive list of applications, the reader is referred to existing literature (for instance, cfr. Watermann [16]).*

## 11.1. Handling of Factors of Uncertainty in Knowledge-Based Systems: Approaches and Requirements

In a number of situations we are faced with information simultaneously incomplete and uncertain. Since the very beginning of development of expert systems, it became evident that these factors could not be neglected since they are strongly related to the way in which the problem is handled by a human being. Two aspects of collected data are discussed simultaneously, namely:

- *incompleteness of information,*

- *uncertainty of information.*

The first aspect refers to information that is incomplete, however precise. Here one can encounter some well-known theories and techniques such as non-monotonic logics developed by McDermott [12], truth maintenance systems [11] and reason maintenance systems [4].

The second aspect has been covered by some techniques rooted in probability, subjective probability, evidence theory, fuzzy sets and possibility theory. Also there are known approaches not supported by any consistent theoretical background but working successfully in some particular situations.

The reader is referred to [5], [7], [8], [9] and [19] for a careful analysis of impacts of these techniques on the performance of knowledge-based systems. It would be also of significance to distinguish between uncertainty, being of probabilistic character, and fuzziness. For example, one may deal with the following statement: *within the next two years, the inflation rate will be 5%*. All the concepts in this statement are well defined (two years, 5%) and the uncertainty refers to the fact which is to occur. Another example is just an expression which is certain but contains a linguistic label: *it is cold today*.

Statements containing both types of uncertainty may arise: *within the next few years, the inflation rate will be about 5%*. Now both the time and the value of the inflation rate are not so well defined. Obvioulsy the latter form of statement is often found in real life situations and they cannot be neglected in the construction of a knowledge-based system, if it is to maintain some links to the existing environment.

There is also another aspect of application of the fuzzy sets in the context of system-user communication. As underlined in [1], a necessary condition for good performance of the system is to achieve a good level of communication.

Poor communication could easily lead to lack of robustness of the system itself and hence to extremely unexpected situations which the system itself is unable to cope with. Particular attention has to be paid to the fact that too precise quantification of the collected data is unnecessary. The attempt to perceive a complex situation may fail if an inappropriate degree of precision is applied.

Therefore it is important to have a compensation between qualitative expressions and a certain loss of precision (cfr. [6], [15], [20]). Recent studies recommend strongly a qualitative expression of concepts, mainly in terms of fuzzy sets [21].

It becomes clear now that more fundamental research is needed to formulate some general guidelines concerning the methodological framework of dealing with uncertainty. We badly need a set of requirements to be fulfilled in order to have a plausible method of coping with uncertainty. A list of desiderata, which also underlines shortcomings in the Bayesian approach [14], is formulated accordingly:

(i) *an inference should not depend on any assumptions about the probability distributions of the propositions,*

(ii) *it should be possible to assert common relationships between propositions when the relationships are indeed known,*

(iii) *it should be possible to posit information about any set of propositions and observe the consequences for the whole system,*

(iv) *if the information provided to the system is inconsistent, this fact should be made obvious along with some notion of alternative ways that the information could be made consistent.*

The list of requirements formulated above has been extended [2] and arranged into three groups bearing in mind the distinct layers of the system, namely representation, inference and control. Main requirements tackle the following facts:

– the inference mechanisms should be logically tied to mechanisms utilized before for knowledge acquisition. Thus if the knowledge base is consistent and preserves some properties within the framework of a specified formalism, the same formalism should form a basis for inference layer. In other words, when the knowledge base is created in a probabilistic setting, then the same formalism (Bayesian inference) should be incorporated in the inference layer. In the field of fuzzy sets, it will be indicated that a particular composition operator used for construction of the knowledge base (specified by a family of fuzzy relations, for instance) implies the use of the adjoint one at the inference layer;

– a performance of the knowledge base (in the sense of its consistency and completeness) should be taken into account by any inference procedure. The procedure should return not only a result of inference, viz. a fuzzy conclusion, but also indicate the degree of its precision. In other words, the knowledge base should be evaluated at the stage of knowledge acquisition in such a manner that the results of this evaluation are able to control a propagation of uncertainty which has its roots in uncertainty and inconsistency of the knowledge base itself, fuzziness of information provided by the user in the consultation process and a lack of complete match of the data and the chunks of knowledge contained in the knowledge base (for short, KB).

The methods based on mechanisms of fuzzy equations will be investigated and the satisfaction of these two requirements will also be clarified. In the sequel, selected aspects of reasoning under uncertainty will be considered.

## 11.2. Selected Aspects of Reasoning with Fuzziness

Before referring to the main blocks of the knowledge-based system (see Fig.11.1), we recall the main sources of uncertainty of information processed and some ways in which it is handled.

Main sources of uncertainty with which one is faced in knowledge-based systems arise from the expert side as well as from results in system-user dialogue. It should be borne in mind that almost all the rules are essentially a generalization of particular pieces of

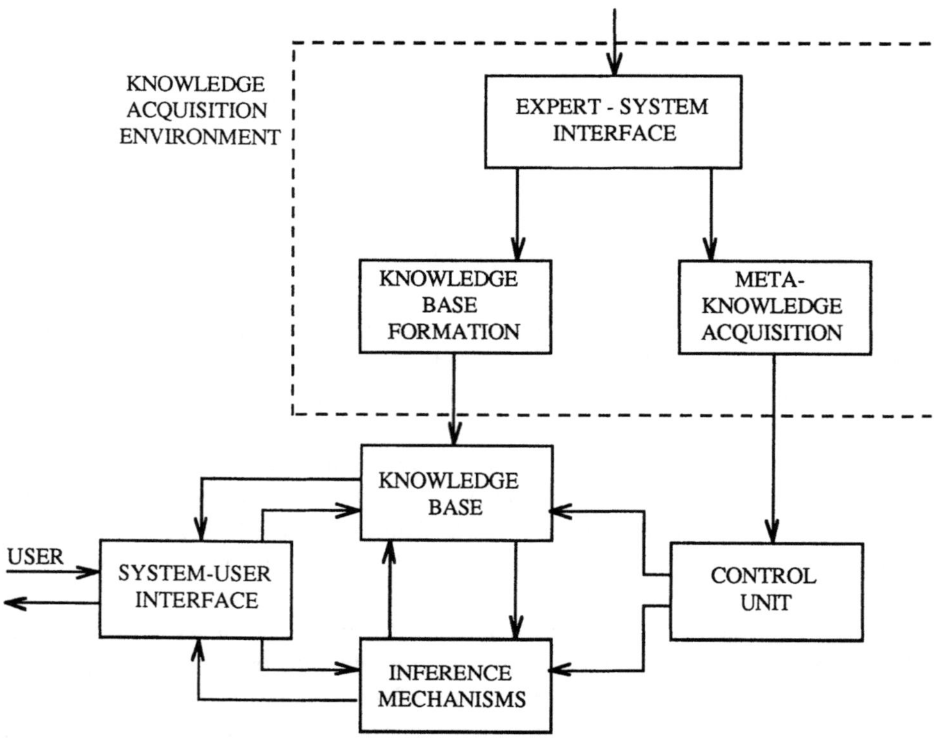


Fig. 11.1

A general structure of knowledge-based systems

knowledge and since any generalization is of a fuzzy character, then the rule is characterized by fuzziness.

Hence in complex tasks, which are essential for these systems, a source of fuzziness is closely related to the stage of formation of the KB. It refers to the uncertainty (or fuzziness) of the particular rule in such a sense that there is no clear notion of the strength of relationship between the set of antecedents and consequents (in other words, between the condition and action parts of the rule). In the simplest case, as for instance in MYCIN, this is modelled by a factor visualizing a strength of relationship between them, say $A \xrightarrow{\omega} B$, where $\omega$ denotes this factor. However one is usually faced with more composite rules. They could be formulated in the following manner:

"*usually* (if $A_h$ then this *strongly* implies $B_h$)",

"*often* (if $A_j$ then this *moderately* implies $B_j$)",

etc. Notice that despite the fact that $A_h$ (resp. $A_j$) and $B_h$ (resp.$B_j$) are fuzzy sets or relations, we also have modifiers referring to

(i) *a strength of relationship (A implies B) between antecedents and consequents,*

(ii) *a frequency of cases in which this rule holds true if fired.*

Description of the strength and frequency is genuinely fuzzy in nature.

## 11.3. Principle of Consistency of Aggregation and Inference Mechanisms

Considering the basic elements of the structure of a knowledge-based system, it is worthwhile to formulate a fundamental statement concerning the role of relation equations in its construction. In the formation of expert systems, the knowledge available at hand is coded into conditional statements "if $A_h$ then $B_h$", where $A_h$ forms a condition part while $B_h$ is treated as an action (or conclusion) part of the h-th rule. Usually at the present stage of development, the process of knowledge acquisition and knowledge refinement, in most of such expert systems is very time consuming, error prone and could produce a large amount of irrelevant information. Simultaneously, we have to take into account a level of consistency which subsequently forms an important characteristic useful for introducing limitations on the length of inference chaining generated by inference mechanisms.

Returning to the condition parts and action (conclusion) parts, we consider $A_h$ and $B_h$ as containing fuzzy quantities, namely $A_h$ is a fuzzy relation defined in the Cartesian product of subspaces, say $A_h:\mathbf{X}=\mathbf{X}_1 \times \ldots \times \mathbf{X}_p \rightarrow [0,1]$ and $B_h:\mathbf{Y} \rightarrow [0,1]$. Since the condition part consists of several subconditions $A_{h1}$, $A_{h2}$, ..., $A_{hp}$, each defined in the respective subspaces $\mathbf{X}_{h1}, \mathbf{X}_{h2}, \ldots, \mathbf{X}_{hp}$, the membership function of A is calculated by utilization of one among conjunction operators existing in fuzzy set theory. Then the KB can be modelled by a fuzzy relation $R \in \mathbf{F}(\mathbf{X} \times \mathbf{Y})$. The values of the membership function of R are computed by means of the following procedure: at the simplest mode, we formulate a hypothesis that all actions available in the rules can be derived by making an inference from the KB with input information equal to $A_h$. In other words, we can state that

$$f(A_h, R)=B_h, \qquad h=1, 2, \ldots, m, \tag{11.1}$$

where "f" forms an operator utilized in the inference procedure.

Moreover the fuzzy relation R standing in (11.1), is formed on the basis of the entire collection of the rules. We write it as

$$R=g(A_1, B_1, A_2, B_2, \ldots, A_m, B_m), \tag{11.2}$$

where "g" can be viewed as an operator of aggregation of all the information covered by the rules. Inserting (11.2) into (11.1) one gets:

$$f(A_h, g(A_1, B_1, A_2, B_2, \ldots, A_m, B_m))=B_h, \qquad h=1, 2, \ldots, m.$$

The above relationship could be viewed as a realization of a principle of *consistency of aggregation and inference mechanisms* involved in the expert system. Therefore one observes that both operations "f" and "g" should coincide in such a sense that the use of one of them predetermines occurrence of the other. In terms of fuzzy equations, "f" and "g" have a very straightforward interpretation. The above problem can be reformulated in the language of these equations accordingly: *given a family of pairs* $(A_h, B_h)$, $h=1, \ldots, m$, *determine a fuzzy relation* R *such that the following set of conditions*:

$$R \Diamond A_h = B_h, \qquad h=1, 2, \ldots, m, \tag{11.3}$$

*is satisfied*, where the symbol "$\Diamond$" replaces any composition operator known in the theory developed in the previous Chs. Such a formulation immediately implies that the fuzzy relation is computed with the aid of the corresponding inverse operator denoted by "$\Diamond^{-1}$", i.e.

$$R=\Diamond^{-1} (A_1, B_1, \ldots, A_m, B_m) \tag{11.4}$$

Thus it is obvious that the composition operator "$\Diamond$" refers to any scheme of forward (data-driven) inference, while "$\Diamond^{-1}$" corresponds to the process of knowledge aggregation. The fuzzy relation R calculated via (11.4) covers all the information conveyed by the set of rules. The formulas (11.3) and (11.4) are useful for realization of a list of interesting problems concerning tests and optimization of the KB:

a) *validation,*

b) *improvement and optimization of the KB,*

c) *choice of a particular fuzzy implication,*

d) *generation of new rules consistent with those already available.*

All these topics will be studied in the sequel. To get a closer look at processing of fuzzy quantities, it is worth considering the concept of granularity of information and its influence on the performance of the expert system.

## 11.4. Information Granularity and Fuzzy Sets

Introduced by Zadeh [18], information granularity becomes one of the crucial points in knowledge representation and significantly enhances a level of effectiveness of system-user communication.

Referring to Szolovitz and Pauker [15], these authors underlined a need for linguistic rather than numerical quantification of frequency estimates (cfr. also [13], [20]). Let us briefly formulate the essence of information granularity. Usually for humans, a level of cognition and information processing is situated at a point where a set of linguistic labels is recognized but individual numerical values are not considered. For instance, consider the concept of temperature. For diagnosis purposes, it is enough to use only a few linguistic labels such as normal, very high, which strongly pertain to the essence of this variable (temperature). Therefore the use of single numerical quantities is not useful at all. The set of linguistic labels (fuzzy sets) constitutes a set of information granulae defined in the space of temperature. One is faced also with a similar phenomenon in a fuzzy controller where inputs of the controller are specified by means of several linguistic labels (namely, error and change of error).

An example of linguistic labels is contained in Figs.11.2 and 11.3. Here we deal with three information granulae defined over a referential set **X**.

Observe that, in the fuzzy partition (Fig.11.2) $A_1$, $A_2$ and $A_3$ overlap, and that for any $x \in \mathbf{X}$, there exists at least one fuzzy set $A_h$ for which the corresponding value of the membership function is greater than zero. Notice that, since the $A_h$'s are fuzzy sets, any information available afterwards, denoted by A, could be easily compared with each of these information granulae. There is however one important point which should be clearly underlined. The result of comparison of A with a fuzzy quantity could lie anywhere in the range between total equality (A and $A_h$ completely match each other) and a point at which A and $A_h$ do not match at all.

There are also many intermediate situations corresponding to only a partial matching of two quantities. For instance, suppose A is a crisp numerical value defined in

**X**, i.e. $A(x_0)=1$ and $A(x)=0$ if $x \neq x_0$, $x, x_0 \in \mathbf{X}$. Thus, making use of a possibility measure Poss $(A/A_h)$ [17] as expressing a measure of matching of fuzzy quantities, its corresponding value is equal to the value of the membership function of $A_h$ at the point $x_0$, i.e. $A_h(x_0)$.

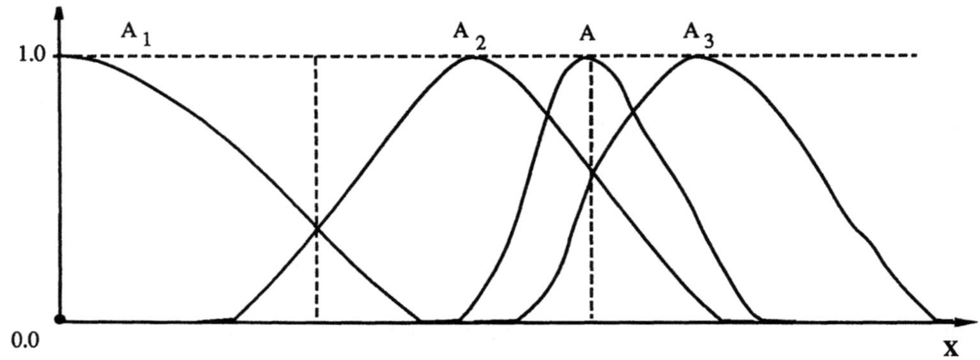


Fig. 11.2

An example of fuzzy information granulae (fuzzy partition)

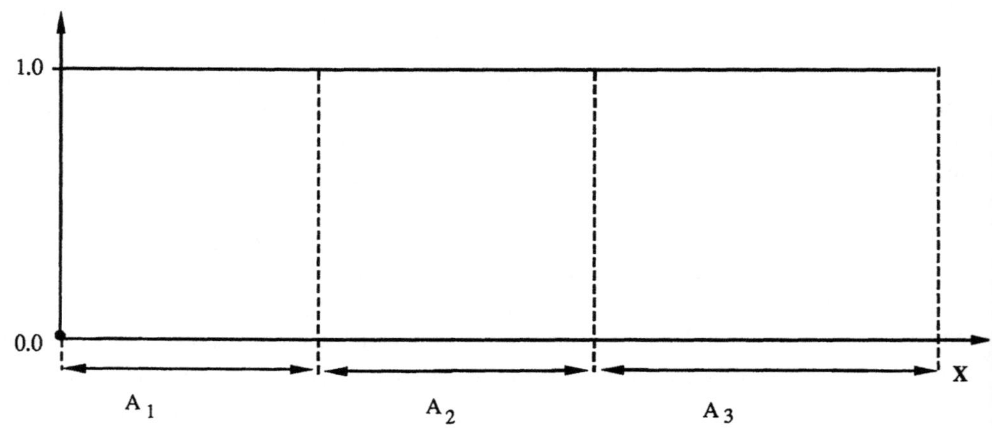


Fig. 11.3

An example of crisp information granulae (hard partition)

The linguistic labels have a straightforward advantage with respect to crisp labels (cfr. Fig.11.3, where a hard partition induced by $A_1$, $A_2$, $A_3$ is provided). They prevent us from an incorrect interpretation of input data. One has to bear in mind that information granularity is established by an expert. Thus if there were differences between the terminology of the expert (designer) and that of an essential user (which is not a rare case), then conclusions would be completely misleading. Let us discuss the case of crisp labels (cfr. again Fig.11.3). The limit values for defining $A_1$, $A_2$, $A_3$ are specified subjectively by the expert or they might come from additional sources. Assigning three verbal notions corresponding to $A_1$, $A_2$, $A_3$, call them *low*, *medium*, *high*, respectively, let us follow the user of the system. He is asked to characterize this variable by simply selecting one of these three alternatives. Simultaneously, he does not know what are the borderline points, in other words, the pattern of a certain notion specified by the expert is not identically equal to the one the user has in mind. Then even a small difference (in this example, it concerns the borderline point) could lead to a completely different inference path. This phenomenon should not be neglected in designing interface schemes in expert systems. Contrary to the undesired effect produced by crisp labels, fuzzy notions are much more robust. The intermediate situation is even self-protecting since the data specified by the user (fuzzy or nonfuzzy) matches simultaneously two linguistic labels but at a relatively low level (see the fuzzy set A in Fig. 11.2).

The above arguments strongly support the use of fuzzy sets at the level of knowledge acquisition as well as utilization of the system.

Yet another phenomenon should be reported here. Since the fuzzy labels do not cover the entire space **X**, i.e. $A_1(x) \vee A_2(x) \vee A_3(x)=1$ for any $x \in \mathbf{X}$, a certain level of robustness is achieved with simultaneous diminution of the degree of matching of the particular linguistic label which, in consequence, leads to a lower value of confidence in the final conclusion. This aspect will be discussed in detail in relation to schemes of fuzzy reasoning. If the fuzzy labels were replaced by crisp sets as already indicated above, precision of inference results would increase significantly, but simultaneously the property of robustness would be lost.

Robustness and precision of derived conclusions are mutually exclusive characteristics. A reasonable threshold should result from an analysis of the performance of the system, especially concerning specificity of input data. A related discussion in terms of trading-off precision and complexity can also be found in [3].

## 11.5. Conclusions

This Ch. forms an introduction to a list of cornerstone aspects of knowledge engineering, designating topics necessary to cope with uncertainty and incompleteness in knowledge-based systems. We should bear in mind that fuzzy sets used in an appropriate context could improve and simplify some procedures. Nevertheless it should be also stressed that they cannot be introduced superficially. This would mean an ad hoc imposition with no reasonable or convincing foundation. There is also another indication which is worth stating.

Most techniques of Artificial Intelligence are symbolic in nature. Fuzzy sets harm this picture in the sense that they are numerically-oriented. Hence we should carefully validate each stage where fuzziness is introduced; in particular one has to look for a tradeoff between benefits and expressiveness of fuzzy sets and an additional computational effort of the numerical character. Despite their numerical form, fuzzy sets should be viewed as a tool for computations performed on a much higher level. Fuzzy sets (or fuzzy labels) deal with aggregates in the universe of discourse rather than single numerical quantities.

If **L**=[0, 1] is not assumed and **L**-fuzzy sets (cfr. [5, Ch.2], [10]) are involved as more abstract concepts, then their calculus is much closer to the nature of symbolic computations.

The scope of fuzzy relation equations focuses the entire discussion on rule-based systems neglecting other schemes of knowledge-based systems that could find an application in these systems. Nevertheless this requires additional research to be concentrated on formulation of problems in sole terms of fuzzy relation equations. For rule-based systems, this language is obvious and, as will be shown in the next Chs., its flexibility and power leads to significant benefits.

## References

[1] R. Beyth-Marom, How probable is probable? Numerical translation of verbal probability expressions, *J. of Forecasting* 1 (1982), 257-269.

[2] P.P. Bonissone, Summarizing and propagating uncertain information with triangular norms, *J. Approx. Reasoning* 1 (1987), 71-101.

[3] P.P. Bonissone and K.S. Decker, Selecting uncertainty calculi and granularity, in: *Uncertainty in Artificial Intelligence* (L. Kanal and J. Lemmer, Eds.), North-Holland, New York (1986), pp.2217-2247.

[4] A.L. Brown, Modal propositional semantics for reason maintenance system, *Proceedings of the 9th Internat. Conference on Artificial Intelligence*, Los Angeles, California, 1985.

[5] P.R. Cohen and M.R. Grinberg, A framework for heuristics reasoning about uncertainty, *Proceedings of the 8th Internat. Joint Conference on Artificial Intelligence*, Karlsruhe, West Germany (1983), 355-357.

[6] J. Fox, D.C. Barber and K.D. Bardhar, Alternative to Bayes? A quantitative comparison with rule-based diagnostic inference, *Method of Information in Medicine* 19 (1980), 210-215.

[7] M.R. Genesereth, An Overview of MRS for AI Experts, *Stanford Heuristic Programming Project*, Report n. HPP-82-27, Dept. of Computer Science, Stanford Univ., 1982.

[8] M.L. Ginsberg, Non-monotonic reasoning using Dempster's rule, *Proceedings of the*

*National Conference on Artificial Intelligence*, Austin, Texas, 1984, 126-129.

[9] M.L. Ginsberg, Implementing Probabilistic Reasoning, *Stanford Heuristic Programming Project*, Report n. HPP-84-31, Dept. of Computer Science, Stanford Univ., 1984.

[10] J.A. Goguen, The logic of inexact concepts, *Synthese* 19 (1986), 325-379.

[11] D.A. McAllester, An Outlook on Truth Maintenance, *MIT Artificial Intelligence Laboratory*, Cambridge, Mass., 1980.

[12] D. McDermott, Non-monotonic logic II: Non-monotonic modal theories, *J. Assn. Comp. Machinery* 29 (1982), 33-57.

[13] L. Philips and W. Edwards, Conservatism in a simple probability inference task, *J. Exp. Psychology* 72 (1966), 346-354.

[14] J.R. Quinlan, INFERNO: A cautious approach to uncertain inference, *Computer J.* 26 (1983), 255-269.

[15] P. Szolovitz and S.G. Pauker, Categorical and probabilistic reasoning in medical diagnosis, *Artificial Intelligence* 11 (1978), 115-144.

[16] D.A. Waterman, *A guide to Expert Systems*, Addison-Wesley, Reading, Mass., 1986.

[17] L.A. Zadeh, Fuzzy sets as a basis for a theory of possibility, *Fuzzy Sets and System* 1 (1978), 3-28.

[18] L.A. Zadeh, Fuzzy sets and information granularity, in: *Advances in Fuzzy Set Theory and Applications* (M.M. Gupta, R.K. Ragade and R.R. Yager, Eds.), North-Holland, Amsterdam (1979), pp.3-18.

[19] L.A. Zadeh, The role of fuzzy logic in the management of uncertainty in expert systems, *Fuzzy Sets and Systems* 11 (1983), 199-227.

[20] A.C. Zimmer, The estimation of subjective probabilities via categorical judgements of uncertainty, in: *Uncertainty in Artificial Intelligence* (L. Kanal and J. Lemmer, Eds.), North-Holland, New York (1986), pp.249-258.

[21] R. Zwick, E. Carlstein and D.V. Budescu, Measures of similarity among fuzzy concepts: a comparative analysis, *Internat. J. Approx. Reasoning* 1 (1987), 221-242.

# CHAPTER 12

## CONSTRUCTION OF KNOWLEDGE BASE, ITS VALIDATION AND OPTIMIZATION



*This Ch. is devoted to central issues arising in any knowledge-based system. Concisely speaking, having already a specified scheme of knowledge representation, we are interested in getting the knowledge concerning the area of interest and, with the aid of the format dictated by the knowledge representation, coding it and indicating a way of effective utilization.*

### 12.1. Preliminaries

Despite tremendous progress in development of more flexible, powerful and more general schemes of knowledge representation, there is an acute lack of methodology and algorithms capable of dealing with the information available and of translating it into a convenient form.

Nowadays there exists a list of different approaches to machine learning such as learning by advice, learning from examples, learning by analogy, etc. ([1], [2], [5]). Still, a lot of systems are developed "manually", and the knowledge, usually covered by rules, is coded by the knowledge engineer (with the cooperation of an expert). There is no doubt that this process is very time-consuming and error prone. Moreover we have to take into account that a level of consistency of the collected *if - then* rules strongly predetermines a length of reasoning scheme to a limit beyond which the conclusions reached are almost meaningless. Moreover, as already underlined, the logical mechanisms imposed at the stage of the construction of the KB should be consistent with the inference mechanisms employed afterwards. All these links are conveniently specified in Fig. 12.1. The inputs of this scheme are formed by rules while the output results are extra messages controlling the inference scheme. In particular, they convey information concerning the composition operators used for creation of the KB (fuzzy relations being its elements) and the measure of consistency achieved for the KB formed, where the set of rules $A_1 \to B_1$, $A_2 \to B_2$, ..., $A_m \to B_m$ is considered.

In fact, the construction of the KB is an iterative process of validation of the set of production rules, formation of the KB and the laborious process of its optimization. Then one must have some well-developed tools by which to follow each of these stages. Here most of the results of Ch.10 are of significant interest. They can be directly formulated in a suitable way to support the solution of the stages of construction of the KB. Each of the algorithms contained there is translated into a language of relevant procedures to detect existing inconsistencies in the rules collected. The first problem is the validation of a family of "if-then" statements.

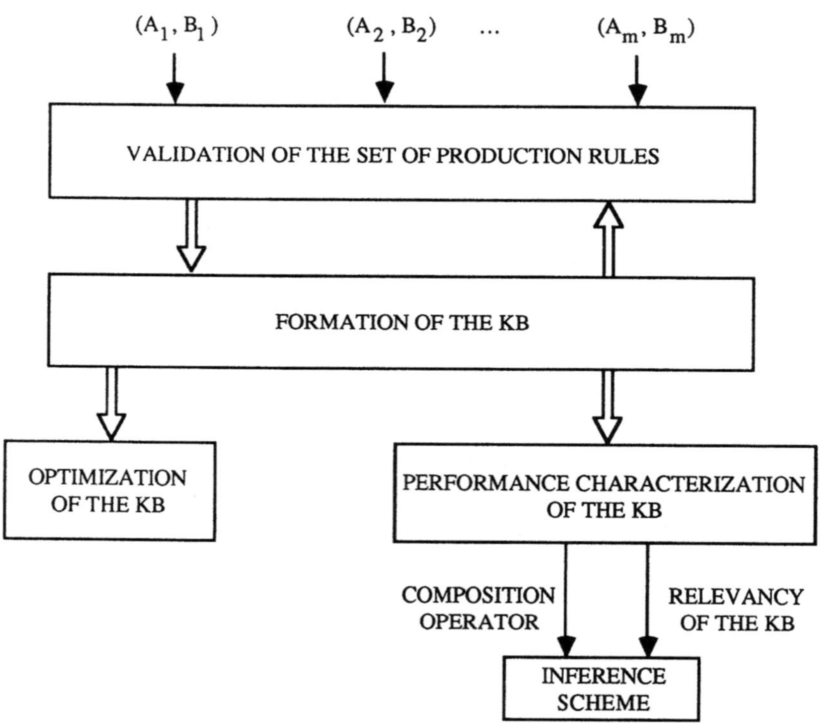


Fig. 12.1

Interrelations between main stages of construction of the KB and their links with the inference scheme

## 12.2. Validation of Production Rules

It may happen that the rules collected from experts are contradictory or incomplete. It is obvious that the inclusion in the KB of two rules possessing the same (or almost the same) condition parts and leading to completely diverse conclusions (actions) could form a misleading inference chain.

By discussing two rules "if $A_h$ then $B_h$" and "if $A_j$ then $B_j$", we can distinguish two cases which indicate drawbacks ([1], [6]): firstly, $A_h$ and $A_j$ are equal to each other or similar in the sense of any similarity measure involved (cfr. Sec.10.4.1) and $B_h$ and $B_j$ are different, i.e. the conclusion for each rule is very different. In this case, the rules are said to be contradictory.

Secondly, $A_h$ and $A_j$ are different from each other but $B_h$ and $B_j$ are close, namely the conclusions derived are almost the same for different conditions. This involves a certain redundancy of information conveyed by two rules which can be omitted to speed up inference procedure.

Both of these phenomena, existing in the rules, are discussed in [6]. A global consistency of the rules is evaluated as follows. We perform a construction of the relation using formula (11.1), denoting by $B_h'$ the fuzzy set of action (consequence) resulting from the composition of $A_h$ and R. Notice that only in very rare cases $B_h'$ has the same membership function as $B_h$. Usually some deviation occurs: it may be caused either by an internal inconsistency stemming from the set of rules (i.e. due to phenomena already described) or by an external inconsistency imposed by implementation, i.e. by the use of a particular implication (combination) operator and the form of the equation involved. Of course, this inconsistency exists to a certain degree depending on the data in the condition and action part.

The first case already discussed concerns the situation in which the rules are contradictory, the second expresses a situation in which some redundancy has been introduced. The degree of contradiction and the degree of redundancy are opposed in nature, however both should be avoided. In the terminology of fuzzy equations, the phenomenon of contradiction of the rules has a very straightforward consequence by diminishing the solvability index of the system of fuzzy equations. Both of these drawbacks can be eliminated by means of the method proposed for visualization of the data structure which pertains to the collected rules. An example of two rules which require particularization is the following:

- if the road is *slippery*, then drive *slowly*,

- if the road is *slippery*, then drive *fast*.

They are obviously contradictory: for the same condition part the resulting action is totally different. These two rules are candidates to be particularized, introducing an additional sub condition which allows a distinction. For instance:

- if the road is *slippery* and *your car is not equipped with snow tyres*, then drive *slowly*,

- if the road is *slippery* and *your car is equipped with snow tyres*, then drive *fast.*

The processes of modification of the rules are called *rule particularization* (specialization) and *rule generalization*, respectively. The relevant procedures make use of graphical means by which the structure of the set of rules can be detected. With the aid of hierarchical clustering techniques, two graphs visualizing candidates for generalization and particularization are derived. An example is given in Fig. 12.2.

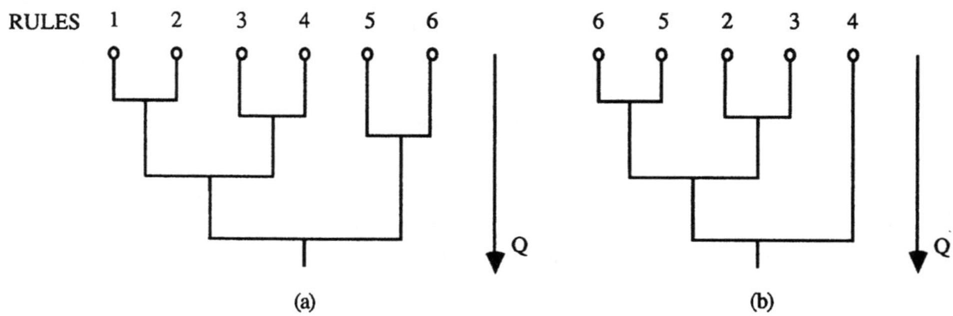


Fig. 12.2

Dendrograms indicating rules for particularization (a) and generalization (b)

The clustering is performed with regard to the same criterion as specified in Sec.10.4.1 when the structure of the data set for fuzzy equations is investigated. Moreover the performance index Q refers to the consistency of the rule derived, viz. a degree to which the fuzzy antecedent forming the cluster and composed by the fuzzy relation fits the consequence formed by the same cluster.

From Fig. 12.2(a), it is observable that rules 1 and 2 as well as rules 3 and 4 are immediate candidates for generalization due to the existence of strong inconsistencies between them. A converse situation holds in the rule generalization, see Fig.12.2(b). Here the rules 5 and 6 are likely to be generalized, then the rules 2 and 3 could be generalized, but to a lesser extent. For more detailed discussion as well as relevant algorithms, the reader is referred to Pedrycz [6].

Having completed the process of validation of the production rules, the KB is formed. It relies on the determination of a solution of the system (10.2), i.e.

$$R \, t \, A_h = B_h, \quad h=1, 2, \ldots, m.$$

Of course $A_h$ and $B_h$ are given, while R is looked for. Then the fuzzy relation R is derived, making use of one of methods provided in Ch.10. To be more precise $A_h$ and $B_h$ should be viewed as modified antecedents and consequents, i.e. they appear after generalization or particularization of the rules. Also m has to be modified with respect to the character of the clusters formed. As indicated in Fig. 12.2, there are feedbacks between these two stages. In fact, we go from one to the other since the process of generalization or particularization is guided by the values of the performance index describing a fit of these rules to the fuzzy equation. It should be stressed that the numbers of clusters strongly depends on the form of the composition operator used to combine antecedents and the fuzzy relation.

Hence, if the performance of the KB is not acceptable, then another form of fuzzy equation (either in the sense of different composition or simply different t-norm) should be tried and this may yield slightly different candidate rules for realizing generalization or particularization in the set of rules.

Referring to the performance of the KB, it could be convenient to evaluate it taking into account only the degree of difference between the consequent derived from the KB when a query was constituted by the sole antecedent of one of the rules, and the consequent of the same rule. Thus, in fact, the construction of the KB is strongly related to the inference mechanism. Hence the composition operator used for the KB uniquely predetermines the form of the inference mechanism. Returning to the fuzzy consequents, an overall performance of the KB is worked out in terms of comparison of $B_h$ and $B_h^{\sim}$, where $B_h^{\sim}$ stands for the consequent obtained by means of the inference process (i.e. by performing a suitable composition of $A_h$ and R) and $h=1,2,\ldots,m$. The comparison of $B_h$ and $B_h^{\sim}$ is done pointwise for each element of $\mathbf{Y}$ and for any h.

For certain h and j, $y_j \in \mathbf{Y}=\{y_1,y_2,\ldots,y_q\}$, one has:

$$[B_h(y_j) \equiv B_h^{\sim}(y_j)] = [B_h(y_j) \, \varphi \, B_h^{\sim}(y_j)] \; t \; [B_h^{\sim}(y_j) \, \varphi \, B_h(y_j)] \, ,$$

specifying to which degree these two grades of membership are equal each other, cfr. Sec.10.2. Then a certain aggregation is suitable by which we get a plausible interpretation of the performance of these equality indices for a certain element of discourse. For a fixed $y_j$, one counts a portion of rules while the equality index attains a value not lower than a threshold $\gamma$, $\gamma \in [0,1]$. This definition refers to an empirical probability function:

$$p(y_j,\gamma) = \frac{\text{card}\,\{h=1, 2, \ldots, m : [B_h(y_j) \equiv B_h^{\sim}(y_j)] \geq \gamma\}}{m}. \qquad (12.1)$$

Noticeable are some properties of $p(y_j,\gamma)$:

- *$p(y_j,\gamma)$ is a nonincreasing function of $\gamma$,*

- $p(y_j,0) = 1$.

Two plots of $p(y_j,\gamma)$ for completely different performances of the KB are shown in Fig. 12.3.

In Fig. 12.3(a), one observes that $p(y_j,\gamma)$ does not depend on $\gamma$ and, even for high values of the threshold $\gamma$, this probability attains high values. This indicates that the KB is highly consistent and the mechanism which has been applied in its construction is a suitable one. On the other hand, in Fig.12.3(b), we get a situation in which the KB is characterized by strong inconsistencies which, unfortunately, have been built into it. This implies that the precision of any conclusion derived could also be significantly lowered.

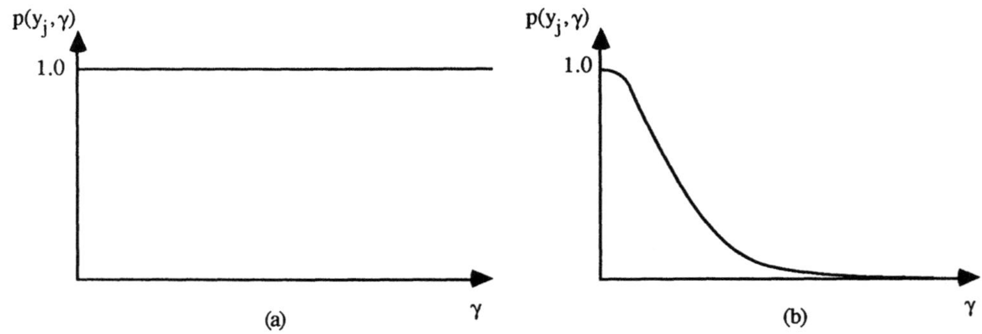


Fig. 12.3

Plots of $p(y_j,\gamma)$ versus $\gamma$

Of course, in applications one can be faced with intermediate situations in which this probability goes down gradually when $\gamma$ attains 1.0. Observe that these probabilities depend on the element of the universe of discourse **Y**, so that within the same KB one might notice elements of **Y** which, with other words, are well-represented, i.e. for a given $\gamma$ the probability $p(y_j,\gamma)$ is high.

For concise treatment, collect the values of probability for all the elements of **Y** into a vector form and denote it by $\mathbf{p}(\gamma)$. The index $\gamma$ is still left to underline the fact that different values of $\gamma$ imply different probability levels. The KB equipped with $\mathbf{p}(\gamma)$ is useful in situations where one is performing an inference process, especially dealing with multiple-step inference procedure. We leave this problem for the next Ch., but we now explain some underlying ideas. Suppose that a fixed value of $\gamma$ is given. Then for any fuzzy set B forming the result of inference, we could establish its bounds which specify the precision of the inferred result. Simply look for all fuzzy sets $B'$:$\mathbf{Y}\rightarrow[0,1]$ which satisfy the inequality:

$$|B'(y_j) - B(y_j)| \geq \gamma$$

for any $y_j \in \mathbf{Y}$. Loosely speaking, we are searching for all fuzzy sets of consequence which are similar to B at the degree not lower than $\gamma$. This process enables us to have a closer look at the result of inference and indicates clearly how precise results are derived. The results of the above inequality are conveniently represented as a so-called interval-valued fuzzy set ($\Phi$-fuzzy sets [7]). We should bear in mind that the precision of the conclusion varies from one element of **Y** to another. Therefore a width of the interval of the $\Phi$-fuzzy set depends on the particular region of **Y**.

One aspect requires additional attention. The choice of $\gamma$ has not been discussed, however it might form a crucial point of the successive steps, i.e. inference procedure. Generally speaking, one could distinguish an infinite family of $\Phi$-fuzzy sets generated by different levels of probability. Therefore it would be interesting to speak about a certain level of probability attached to this equality. If one accepts a high level of probability of events (the equality index is not lower than $\gamma$), then the value of $\gamma$ is relatively low. The higher the probability $p(y_j,\gamma)$, the lower the value of $\gamma$. Referring to the relationship (12.1), notice that for a required value of probability, say $\eta$, the level of $\gamma$ is immediately determined from the inequality $p(y_j,\gamma) \geq \eta$. Unfortunately, contrarily to standards found in statistical analysis, no standard values of $\eta$ are available.

Generally speaking, the process of validation gives two important pieces of knowledge: firstly, a complete mechanism of inference is established that in the sequel will be used in reasoning procedures; secondly one is equipped with a quantitative expression of relevancy of the mechanism of inference involved. This second issue is very important from the point of view of implementation.

## 12.3. Distributed Knowledge Bases

Recalling a phenomenon of conflicting rules in the entire set of rules available as well as noticing that different rules may possess different numbers of antecedents, it would be reasonable to work with the concept of so-called *distributed knowledge bases*. As will be explained, this notion is related to the method of solving a system of fuzzy equations by splitting it into several subsystems sharing a significantly higher level of the solvability property.

Thus a natural partition of the set of production rules arises in the context of a number of antecedents. The rules with antecedents defined in the same universe of discourse are considered to be included (potentially) in the same KB. This procedure implies the formation of several "local" knowledge bases. An illustration of the distributed KB is given in Fig. 12.4.

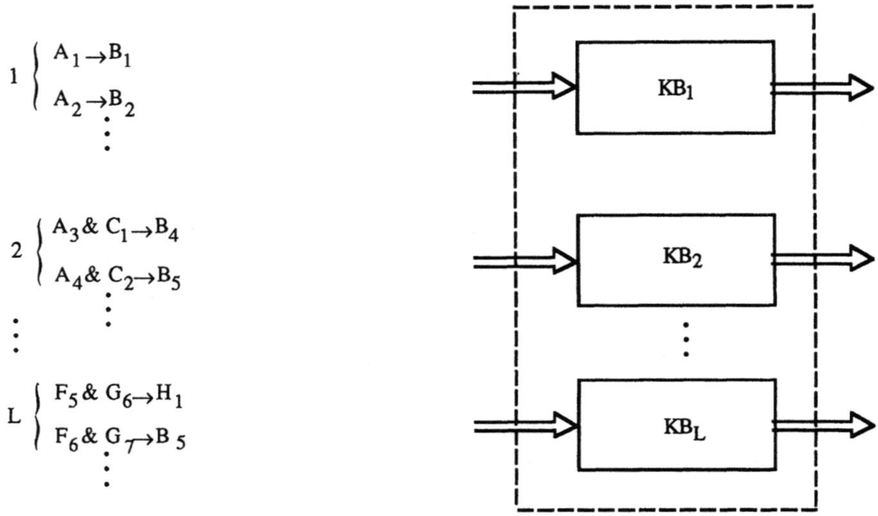


Fig. 12.4

An illustration of the distributed KB

To coordinate each of the components of the KB organized as in Fig. 12.4 and to enable their effective use, it is necessary to give information on each of them. This should contain specification of the universe of discourse of the antecedents and a measure of the precision (relevancy) of any conclusion derived (for each of these knowledge bases, the vector $\mathbf{p}(\gamma)$ is provided).

In the sequel we will discuss an aspect of optimization of the knowledge bases, by making them more efficient (by solving a reduction problem) and by establishing a suitable order of query arrangement.

## 12.4. Reduction Problem in Knowledge Bases

Another aspect of the use of fuzzy equations refers to the optimization of the KB. It will be shown how the production rules can be made more efficient and this is done by a suitable arrangement of subcondition elements in the condition part of the rules available. This problem refers also to a so-called *reduction task*. It often happens, while working with complex problems, that the number of subconditions in the condition part is quite high. This means in practice that the reasoning procedure would (potentially) need all the values of attributes of the object standing in the subcondition part (fuzzy or nonfuzzy) in order to generate a hypothesis concerning the action worked out. It could happen however that some subconditions are difficult to evaluate by the user and/or the reliability of the answer is relatively low. Then it is reasonable to reduce a dimension of the condition space retaining only those subspaces as are most significant for reasoning purposes. Such a reduction procedure may lead to a slight modification of the corresponding conclusion. A main point, while performing reduction of some subcondition parts, is to keep track of a certain balance between imposed changes of the conclusion and the range of the reduction achieved. Assume the condition part of the rule is specified by a Cartesian product $A_h = A_{h1} \times A_{h2} \times \ldots \times A_{hp}$, i.e.

$$A_h \colon \mathbf{X}_1 \times \mathbf{X}_2 \times \ldots \times \mathbf{X}_p \to [0,1],$$

where $X_s$ are suitable universes of discourse and $A_{hs} \colon \mathbf{X}_s \to [0,1]$, s=1,2,...,p. Then, for each fuzzy set $A_{hs}$, we get the following expression:

$$G_{hs} \mathrm{t} A_{hs} = B_{hs}.$$

Here $B_{hs} \colon \mathbf{Y} \to [0,1]$ is the fuzzy set of action resulting from the above reduced form of the fuzzy equation and $G_{hs} \colon \mathbf{X}_s \times \mathbf{Y} \to [0,1]$ is a fuzzy relation viewed as unknown. Thus the reduced KB can be utilized in the sequel for performing inference. Observing an increasing number of subconditions in the production rules, it is convenient to rank them in such a way that the most significant ones are put in first position in the condition part. Another procedure can be applied in order to work out an order of the subconditions of the rules. This enables us to find out the most significant subconditions having the strongest influence on the action part. Then, in the matching stage one starts asking the user for the most significant data instead of looking for relatively insignificant pieces of knowledge. Two basic procedures are discussed here, referring to the s-th coordinate.

(i) *With a negative picture: here we replace all fuzzy sets defined in all the rules by a fuzzy set treated as unknown, i.e. with membership function identically equal to 1. Calculate the fuzzy set of action parts and compare with the corresponding fuzzy sets from the rules. A measure of similarity between them forms also a suitable index for expressing a strength of influence of the s-th subcondition on the action part. The higher the value of this measure, the weaker the influence reported.*

(ii) *With a positive picture: here we replace all fuzzy sets except the s-th subcondition by fuzzy sets with membership function identically equal to 1. Only the fuzzy set forming the s-th subcondition is preserved. Compute now the value of the similarity measure as before. Then if the value of the measure is high, we can conclude that the influence of the s-th condition is the significant one and the remaining subconditions have much less influence and can be viewed as candidates to be reduced.*

The subconditions are arranged according to increasing values of the similarity measure. The name of these approaches given above refers to the essence of the construction involved. In the first one, we neglect information dealt with the s-th subcondition (negative picture) while in the latter a converse situation holds (positive picture). Then the use of fuzzy equations gives a framework for optimization of rules in the sense that most significant subconditions are put at the front of the condition part of the rules. Then one can shorten the time of matching, accepting performance of only those subconditions put as the first ones in the condition part. In parallel mode of firing the rules, as soon as the level of matching of a certain rule is high enough, then there is no need to perform further matching.

To go into detail, assume the universe of discourse for each subcondition to be finite, namely:

$$\mathbf{X}_1 = \{x_{11}, x_{12}, \ldots, x_{1n(1)}\},$$

$$\mathbf{X}_2 = \{x_{21}, x_{22}, \ldots, x_{2n(2)}\},$$

$$\vdots$$

$$\mathbf{X}_p = \{x_{p1}, x_{p2}, \ldots, x_{pn(p)}\}.$$

For simplicity, we assume t=min and h=m=1. Since no misunderstanding can arise, we omit the index h in the successive symbols. Consider the relationship between antecedents $A_1, A_2, \ldots, A_p$ and the consequent B:

$$B = R \odot A_1 \odot A_2 \odot \ldots \odot A_p, \tag{12.2}$$

that in terms of membership functions reads accordingly:

$$B(y_j) = \bigvee_{i(1)=1}^{n(1)} \bigvee_{i(2)=1}^{n(2)} \ldots \bigvee_{i(p)=1}^{n(p)} [A_1(x_{1i(1)}) \wedge \ldots \wedge A_p(x_{pi(p)}) \wedge R(x_{1i(1)}, \ldots, x_{pi(p)}, y_j)]$$

for any $j=1,2,\ldots,q$, where $y_j \in \mathbf{Y}$ and R is a fuzzy relation defined in the Cartesian product $\mathbf{X}_1 \times \ldots \times \mathbf{X}_p \times \mathbf{Y}$.

A reduced form of the KB now takes only the s-th antecedent:

$$B_s = G_s \odot A_s, \tag{12.3}$$

where $G_s$ is a fuzzy relation defined in $\mathbf{X}_s \times \mathbf{Y}$ (see Fig.12.5).

Here we denote the conclusion by $\overline{B}_s$ instead of B underlining that the reduced form of the KB may not, and usually does not, generate the same result, namely B. Of course, we intend to determine the fuzzy relation $G_s$ which allows us to obtain the closest results to B. In an ideal situation, we can seek to satisfy an equality $B=\overline{B}_s$ whatever values of inputs are considered.

In more formal fashion, choose $G_s$ such that the equality:

$$G_s \odot A_s = R \odot A_1 \odot A_2 \odot \ldots \odot A_s \odot \ldots \odot A_p \tag{12.4}$$

is satisfied.

Unfortunately, the above requirement is too restrictive and the equality (12.4) is extremely difficult to verify. Therefore it is reasonable to speak about equivalence of the knowledge bases represented by (12.2) and (12.3) in the sense of an assigned class of inputs.

More clearly, we say that the fuzzy KB (12.2) and KB (12.3) are equivalent with respect to the class of inputs $\mathcal{X}$ if the equality (12.4) holds for any fuzzy set $A_s$, $s=1,2,\ldots,p$.

For simplicity, from now on we put $x_{si(s)}=x_{i(s)}$. Two classes $\mathcal{X}$ of inputs are of significant interest:

(a) *n(s) inputs in the universe $X_s$ are considered as singletons while the inputs defined in the remaining universes are modelled as unknown, i.e. their membership functions are identically equal to 1.0 in the entire universe of discourse,*

(b) *similarly as before, n(s) inputs in the universe $X_s$ are viewed as singletons and the inputs in the remaining universes are also singletons.*

Of course, in both cases (12.2) is viewed as a system of n(s) fuzzy equations, whose solution set is assumed nonempty.

Denoting by **1** the fuzzy set with membership function identically equal to 1 and by $\overline{G}_s$ and $\underline{G}_s$ the fuzzy relations resulting in (a) and (b), the formulae for their membership functions are contained in the following theorems, respectively:

THEOREM 12.1. *If the inputs of Eq.* (12.2) *are specified as follows:*

- *if* $A_k = 1$ *for any* $k\neq s$, k, $s=1,2,\ldots,p$,

- *if* n(s) *input fuzzy sets* $A_s^t$, $t=1,2,\ldots,n(s)$, *defined in* $\mathbf{X}_s$ *are singletons and such that they are disjoint and they completely cover* $\mathbf{X}_s$ *(i.e. their fuzzy union has membership function identically equal to* 1.0*),*

*then the fuzzy relation* $\overline{G}_s$ *of the reduced KB* (12.3), *defined in* $\mathbf{X}_s \times \mathbf{Y}$, *has membership function given by*

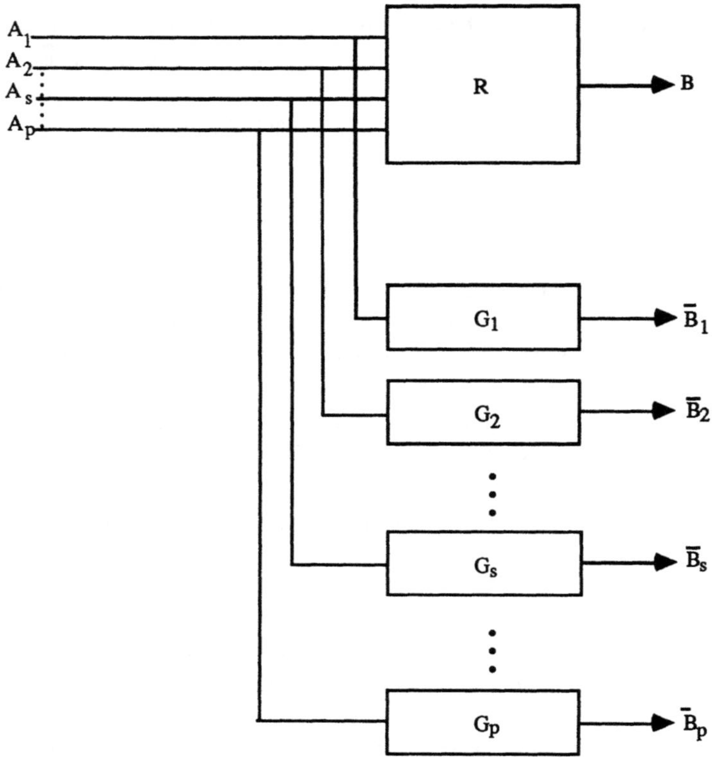


Fig.12.5

The entire KB (R) and the reduced version ($G_1$, $G_2$, ..., $G_p$)

$$\bar{G}_s(x_{i(s)}, y_j) = \bigvee_{i(1)=1}^{n(1)} \dots \bigvee_{i(s-1)=1}^{n(s-1)} \bigvee_{i(s+1)=1}^{n(s+1)} \dots \bigvee_{i(p)=1}^{n(p)} R(x_{i(1)}, \dots, x_{i(s)}, \dots, x_{i(p)}, y_j)$$

*for all* $x_{i(s)} \in \mathbf{X}_s$, $y_j \in \mathbf{Y}$, $i(s)=1,2,\dots,n(s)$ *and* $j=1,2,\dots,q$.

PROOF. In accordance with the assumption made in the statement, we have the following membership function for the output:

$$B_t(y_j) = \bigvee_{i(1)=1}^{n(1)} \dots \bigvee_{i(p)=1}^{n(p)} \{A_s^t(x_{i(s)}) \wedge [A_1(x_{i(1)}) \wedge \dots \wedge A_{s-1}(x_{i(s-1)})$$

$$\wedge A_{s+1}(x_{i(s+1)}) \wedge \ldots \wedge A_p(x_{i(p)})$$

$$\wedge R(x_{i(1)}, \ldots, x_{i(p)}, y_j)]\}$$

$$= \bigvee_{i(s)=1}^{n(s)} \{A_s^t(x_{i(s)})$$

$$\wedge [\bigvee_{i(1)=1}^{n(1)} \ldots \bigvee_{i(s-1)=1}^{n(s-1)} \bigvee_{i(s+1)=1}^{n(s+1)} \ldots \bigvee_{i(p)=1}^{n(p)} R(x_{i(1)}, \ldots, x_{i(s)}, \ldots, x_{i(p)}, y_j)]\},$$

where $t=1,2,\ldots,n(s)$. Since $A^t(x_{i(s)})=1$ if $i(s)=t$ and $A^t(x_{i(s)})=0$ otherwise, then the membership function of $B_t$ simplifies into

$$B_t(b_k) = \bigvee_{i(1)=1}^{n(1)} \ldots \bigvee_{i(s-1)=1}^{n(s-1)} \bigvee_{i(s+1)=1}^{n(s+1)} \ldots \bigvee_{i(p)=1}^{n(p)} R(x_{i(1)},\ldots,x_{i(s)},\ldots,x_{i(p)},y_j) \quad \text{if } i(s)=t,$$

$$= 0 \quad \text{otherwise,}$$

where $t=1,2,\ldots,n(s)$ and $j=1,2,\ldots,q$. Then the solution of the system of Eqs. (12.2) for the antecedents specified above is equal to

$$\overline{G}_s(x_{i(s)},y_j) = \bigwedge_{t=1}^{n(s)} [A_s^t(x_{i(s)} \alpha B_t(y_j)] = B_{i(s)}(y_j)$$

for all $x_{i(s)} \in \mathbf{X}_s$, $y_j \in \mathbf{Y}$, $i(s)=1,2,\ldots,n(s)$ and $j=1,2,\ldots,q$. ■

Similarly it is proved that

THEOREM 12.2. *If the* $n(s)$ *input fuzzy sets* $A_s^t$, $t=1,2,\ldots,n(s)$ *are singletons and such that in every universe of discourse* $\mathbf{X}_s$, $s=1,2,\ldots,p$, *they are pairwise disjoint and they completely cover* $\mathbf{X}_s$ *(as specified in* Thm. 12.1*), then the fuzzy relation* $\underline{G}_s$ *of the reduced KB* (12.3), *defined in* $\mathbf{X}_s \times \mathbf{Y}$, *has membership function given by*

$$\underline{G}_s(x_{i(s)},y_j) = \bigwedge_{i(1)=1}^{n(1)} \ldots \bigwedge_{i(s-1)=1}^{n(s-1)} \bigwedge_{i(s+1)=1}^{n(s+1)} \ldots \bigwedge_{i(p)=1}^{n(p)} R(x_{i(1)}, \ldots, x_{i(s)}, \ldots, x_{i(p)}, y_j)$$

*for all* $x_{i(s)} \in \mathbf{X}_s$, $y_j \in \mathbf{Y}$, $i(s) = 1, 2, \ldots, n(s)$ *and* $j = 1, 2, \ldots, q$.

### 12.5. Reconstruction Problem of Knowledge Bases

We first observe that $\underline{G}_s(x_{i(s)},y_j) \le \overline{G}_s(x_{i(s)},y_j)$ for any $s=1,2,\ldots,p$, i.e. $\underline{G}_s \le \overline{G}_s$ for any $s=1,2,\ldots,p$.

Having at our disposal a collection of reduced models determined with the aid of Thms. 12.1 and 12.2, we may formulate the following problem: assume that a subset **S** of indices of the set $\{1,2,\ldots,p\}$ and related inputs are given, therefore the following system is known:

$$\begin{cases} \overline{G}_s \odot A_s = \overline{B}_s \\ \underline{G}_s \odot A_s = \underline{B}_s, \end{cases}$$

where $s \in \mathbf{S}$. Note that $\underline{B}_s \le \overline{B}_s$ for any $s \in \mathbf{S}$ and we do not deal with single membership functions $B_j$ but with an interval-valued fuzzy set ($\Phi$-fuzzy set). Instead of the above equations, we can introduce a shorthand notation:

$$[\underline{G}_s, \overline{G}_s] \odot A_s = [\underline{B}_s, \overline{B}_s], \qquad s \in \mathbf{S}.$$

It is clear that the reduced KB generates the interval-valued fuzzy set of conclusions. Then, if several fuzzy antecedents are available, the relevant outputs forming $\Phi$-fuzzy sets should be combined to get an overall result.

A reasonable concept to aggregate the results of the reduced models from the subset **S** would work as follows: for any $y_j \in \mathbf{Y}$, calculate the number of $\Phi$-fuzzy sets contributing to the previously fixed values ranging between 0 and card **S**.

By a simple normalization, we obtain a fuzzy set B of the second order; this means that for every $y_j \in \mathbf{Y}$ a fuzzy set defined in [0,1] has been established. More precisely, the value of the membership function B (before normalization) is equal to

$$B(y_j,w) = \frac{\text{card}\,\{s \in \mathbf{S} : \underline{B}_s(y_j) \le w \le \overline{B}_s(y_j)\}}{\text{card}\,\mathbf{S}}$$

for all $w \in [0,1]$, $y_j \in \mathbf{Y}$.

Remembering the main idea behind the concept of reduction of the KB, which relies on diverse grades of difficulty and precision while determining the fuzzy sets of antecedents, the fuzzy set B may be modified to take this into account. We express the level of difficulty to get reliable data from the corresponding input in $\mathbf{X}_s$ by $w_s \in [0,1]$, $s \in \mathbf{S}$. The higher the value of $w_s$, the more severe the difficulties found in dealing with input determination. The results obtained are less reliable and in consequently should not significantly contribute to the final result. The weights $w_S$ can be easily obtained, e.g., by Saaty's well known priority method.

This weight is useful to control the width of the interval-valued fuzzy set $[\underline{B}_S, B_S]$. Now, instead of weights, we have:

$$[\underline{B}_S{}^{w_s}, B_S{}^{w_s}],$$

where by $B_S{}^{w_s}$ we mean the membership function:

$$B_S{}^{w_s}(y_j) = [B_S(y_j)]^{w_s},$$

for any $y_j \in \mathbf{Y}$. We illustrate the above considerations by a numerical example:

EXAMPLE 12.1. We consider a KB with statements consisting of two antecedents and one consequent space, namely $R \odot A_1 \odot A_2 = B$ with $n(1)=n(2)=m=3$ and described by the following fuzzy relation:

$$R = \begin{matrix} x_{11} \\ x_{12} \\ x_{13} \end{matrix} \underset{y_1}{\overset{\begin{matrix} x_{21} & x_{22} & x_{23} \end{matrix}}{\begin{Vmatrix} 0.2 & 0.7 & 0.8 \\ 0.5 & 0.6 & 0.4 \\ 0.9 & 0.0 & 0.7 \end{Vmatrix}}} \underset{y_2}{\overset{\begin{matrix} x_{21} & x_{22} & x_{23} \end{matrix}}{\begin{Vmatrix} 0.4 & 0.3 & 0.9 \\ 0.3 & 1.0 & 0.5 \\ 0.7 & 1.0 & 0.6 \end{Vmatrix}}} \underset{y_3}{\overset{\begin{matrix} x_{21} & x_{22} & x_{23} \end{matrix}}{\begin{Vmatrix} 1.0 & 0.3 & 0.7 \\ 0.4 & 1.0 & 0.2 \\ 0.5 & 0.3 & 0.9 \end{Vmatrix}}}$$

Let us construct two reduced knowledge bases making use of the families of the fuzzy sets introduced earlier. Let $A_1$ be singleton and $A_2$ be "unknown". Then we have for the first and second KB, respectively:

$$G_1 = \begin{matrix} x_{11} \\ x_{12} \\ x_{13} \end{matrix} \overset{\begin{matrix} y_1 & y_2 & y_3 \end{matrix}}{\begin{Vmatrix} 0.8 & 0.9 & 1.0 \\ 0.6 & 1.0 & 1.0 \\ 0.9 & 1.0 & 0.9 \end{Vmatrix}}, \qquad \underline{G}_1 = \begin{matrix} x_{11} \\ x_{12} \\ x_{13} \end{matrix} \overset{\begin{matrix} y_1 & y_2 & y_3 \end{matrix}}{\begin{Vmatrix} 0.2 & 0.3 & 0.3 \\ 0.3 & 0.6 & 0.2 \\ 0.5 & 0.0 & 0.6 \end{Vmatrix}},$$

$$G_2 = \begin{matrix} x_{21} \\ x_{22} \\ x_{23} \end{matrix} \overset{\begin{matrix} y_1 & y_2 & y_3 \end{matrix}}{\begin{Vmatrix} 0.9 & 0.7 & 1.0 \\ 0.7 & 1.0 & 1.0 \\ 0.8 & 0.9 & 0.9 \end{Vmatrix}}, \qquad \underline{G}_2 = \begin{matrix} x_{21} \\ x_{22} \\ x_{23} \end{matrix} \overset{\begin{matrix} y_1 & y_2 & y_3 \end{matrix}}{\begin{Vmatrix} 0.2 & 0.3 & 0.5 \\ 0.3 & 0.6 & 0.0 \\ 0.7 & 0.2 & 0.6 \end{Vmatrix}}.$$

To visualize how the reduced knowledge bases work, we consider two input data $A_1'$ and $A_2'$ such that

$A_1' = \| 0.7 \quad 0.5 \quad 0.2 \|$ , $A_2' = \| 0.3 \quad 1.0 \quad 0.4 \|$.

The first reduced KB gives an interval fuzzy set:

$$\{[0.3\ ,\ 0.5]\ ,\ [0.3\ ,\ 0.7]\ ,\ [0.7\ ,\ 0.7]\}$$

while the second yields:

$$\{[0.3\ ,\ 0.6]\ ,\ [0.4\ ,\ 0.7]\ ,\ [1.0\ ,\ 1.0]\}.$$

Combining them together by the method already given and setting $w_s=1$, $s=1,2$, we get the fuzzy set $B^{\sim}$ of the second order with membership function given by

$$\begin{aligned} B^{\sim}(y_1,w) &= 1 && \text{if } w \in [0.3\ ,\ 0.7], \\ &= 0 && \text{otherwise,} \end{aligned}$$

$$\begin{aligned} B^{\sim}(y_2,w) &= 0.5 && \text{if } w \in [0.5\ ,\ 0.6], \\ &= 1 && \text{if } w \in [0.6\ ,\ 0.7], \\ &= 0 && \text{otherwise,} \end{aligned}$$

$$\begin{aligned} B^{\sim}(y_3,w) &= 0 && \text{if } w \in [0,0.3] \cup [0.7,1], \\ &= 0.5 && \text{if } w \in [0.3,0.4], \\ &= 1 && \text{if } w \in [0.4,0.7]. \end{aligned}$$

For comparison, the fuzzy set resulting from the complete KB is equal to [0.7 0.5 0.5] and is covered by the fuzzy set $B^{\sim}$.

## 12.6. Concluding Remarks

The results of this Ch. can be found in [3] and [4]. The discussion has been focussed on different facets of construction of the knowledge bases. Nowadays it is becoming more and more obvious that when we are faced with a mass of useful not-yet-well-organized information, then time-consuming "manual" methods of knowledge acquisition cannot be accepted. The fuzzy equations naturally fit requirements for automatic (or self-automatic) knowledge acquisition. The constraints imposed within the entire procedure of solving a system of fuzzy equations ensure consistency of the KB derived in such a way. As pointed out, the precision or relevancy of the KB has a direct influence on the process of reasoning and therefore it is convenient to articulate it quantitatively. This has been accomplished by suggesting some indices which, having also a strong statistical background, are transformed in Φ-fuzzy sets. The width of these sets has a natural interpretation and directly refers to the precision of possible inference results. Moreover, the concept of distributed knowledge bases stems from some hints useful for solving fuzzy relation equations. Some optimization tasks, which enable us to process the KB and arrange it in a more effective fashion (especially for friendly system-user conversation), have been proposed and discussed in detail.

The contents of this Ch. are centered around some procedures involved with statical modification and improvement of the KB. Notice, for instance, that we have not yet studied dynamical properties of the knowledge-based systems such as occurrence of circular rules. This would require extra effort to get a reliable set of tools which are capable of coping with this undesirable phenomenon.

## References


[1] P.R. Cohen and E. Feigenbaum, *The Handbook of Artificial Intelligence*, vol. III, Chapter : Learning and Inductive Inference, Addison Wesley, Reading, Mass., 1986.

[2] T.G. Dietterich and R.S. Michalski, Inductive learning of structural descriptions: Evaluation criteria and comparative review of selected methods, *Artificial Intelligence* 16 (1981), 257-294.

[3] A. Di Nola, W. Pedrycz and S. Sessa, Fuzzy relation equations and its applications to knowledge engineering, in: *1st. Suppl. Volume to Systems & Control Encyclopedia* (M.G. Singh, Ed.), Pergamon Press Ltd., to appear.

[4] A. Di Nola, W. Pedrycz and S. Sessa, Reduction procedures for rule-based expert systems as a tool for studies of properties of expert's knowledge, *submitted.*

[5] R. Kling, A paradigm for reasoning by analogy, *Artificial Intelligence* 2 (1971), 147-178.

[6] W. Pedrycz, Generalization and particularization of production rules in expert systems, in: *Cybernetics and Systems* '86 (R. Trappl Ed.), D. Reidel Publ. Co., Dordrecht (1986), pp.783-790

[7] R. Sambuc, Fonctions Φ-floues: application à l'aide en diagnostic in pathologie thyroidienne, *Ph.D. Thesis*, Marseille, 1975.

# CHAPTER 13

## INFERENCE ALGORITHMS IN KNOWLEDGE-BASED SYSTEMS



*This Ch. summarizes some common techniques of inference utilized for fuzzy data. Special attention has been paid to the implementation of modus ponens (which realizes a data-driven mode of reasoning) and modus tollens (corresponding to a goal-driven mode of reasoning). The detachment principle (corresponding to a means of expressing a similarity between fuzzy statements) is also investigated. We discuss how different forms of fuzzy relation equations are used to handle each of these modes of inference. Also the question of a direct link between the relevancy of the KB and the length of the inference chain leading to meaningful conclusions is considered. This is of primordial importance; it has to be analyzed to interpret the results of inference and, in particular, to visualize precision. A proper reformulation of the problem in terms of fuzzy equations makes it possible to consider this knowledge transformation in a greater detail.*

### 13.1. Preliminaries

The inference mechanism (cfr. [1], [3], [6], [8], [13], [18]) forms one of the most important functional blocks of knowledge-based systems. Equipped with the KB already constructed, the system is able to infer a useful conclusion with the aid of facts collected from the user, if necessary. Recall that among strategies in rule-based systems, two are in common use. The first one, called *goal-driven* or *backward chaining* is based on backward reasoning. For a formulated goal, the system tries to validate (prove) it making use of knowledge already contained in the KB and, if not accessible, formulating a relevant question to the user. The essence of this method relies on chaining one rule to another in such a way that an antecedent of one rule is matched to consequents of other rules. The information gathered from the user is acquired if the system is unable to infer it itself from already existing rules.

Backward chaining is common in all situations in which the number of goals is small enough while the amount of information required a priori is significant. The user is thus ensured that the number of questions formulated by the system is kept as low as possible; no redundant information is grasped. It is worth underlining that backward chaining is an essential strategy guiding the entire querying process realized in PROLOG. A schematic illustration of a goal-driven scheme is displayed in Figs.13.1(a), (b).

Each node of the net represents the goal (conclusion of the rule) contained in the KB, while the connections refer to the rules contained in the KB. The arrows indicate a direction of reasoning from the final goal situated at the top of the structure to successive subgoals. As already mentioned above, the idea of backward chaining resides in attempting to satisfy a set of goals starting from the very general one.

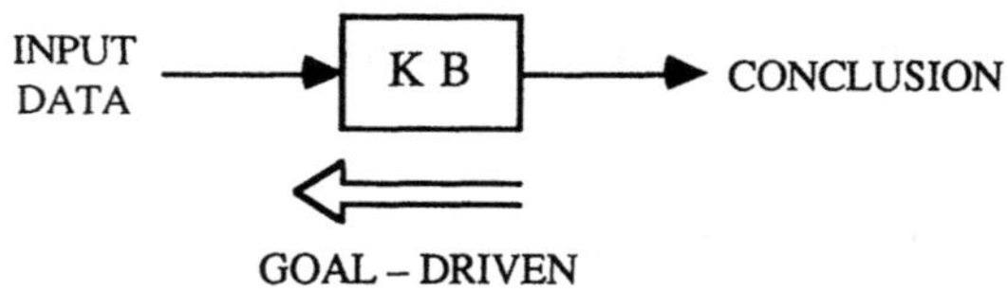


Fig. 13.1 (a)

Generic idea of goal – driven mode of reasoning

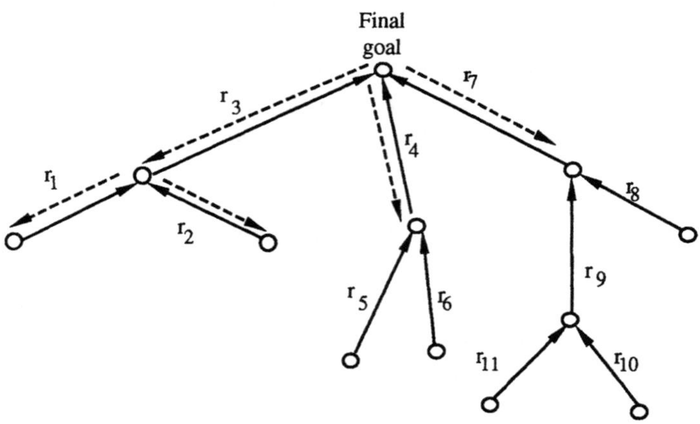


Fig. 13.1 (b)

Generic idea of goal-driven mode of reasoning

The second mode of reasoning is called *data-driven* or *forward chaining*. Within this mode, firstly data are gathered and afterwards the rules are activated. Having one rule activated (i.e. getting a matching between the antecedents of the particular rule and the data gathered), the consequent of this rule is matched to antecedents standing in other rules. Then the remaining rules are chained in sequel. An illustration of data-driven mode of reasoning is displayed in Figs.13.2(a), (b).

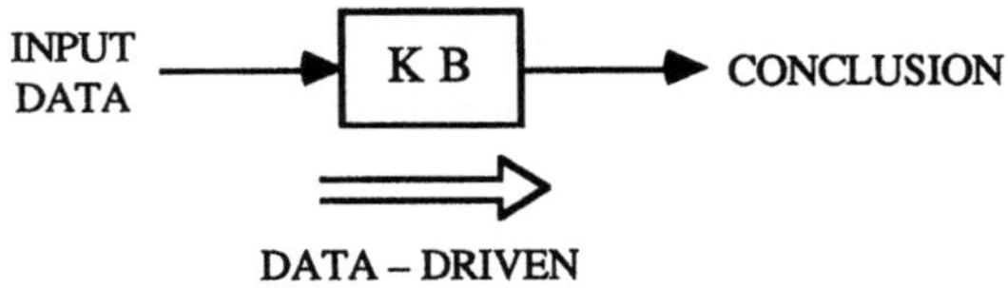


Fig. 13.2 (a)

Generic idea of data – driven mode of reasoning

In Fig.13.2(b), we have a set of rules having the same structure as in Fig.13.1(b). We observe that the method of inference is the same as marked in the rules. This underlines the fact that, firstly, the data are gathered and, secondly, the inference is performed.

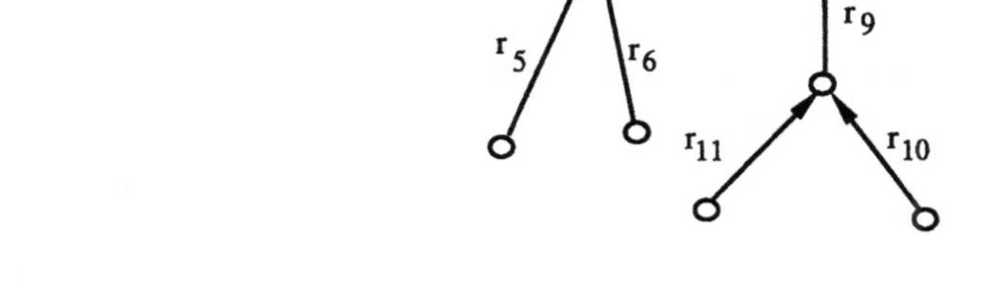


Fig. 13.2 (b)

An illustration of forward chaining

This scheme of reasoning is characteristic for situations in which there is not so much input information and, contrary to the goal-driven reasoning scheme, there is an extensive set of goals. Such cases are often found in planning: only few data are required while the number of possible plans is quite considerable.

In the sequel, we will discuss three patterns of reasoning useful for processing both of the modes discussed above:

- *modus ponens,*

$$\frac{\begin{array}{l}A\\ A\rightarrow B\end{array}}{B}$$

- *modus tollens,*

$$\frac{\begin{array}{l}B\\ A\rightarrow B\end{array}}{A}$$

- *detachment principle,*

$$\frac{\begin{array}{l}A\\ A\leftrightarrow B\end{array}}{B.}$$

The next Secs. will be devoted to a detailed discussion of these schemes.

## 13.2. Representation of Modus Ponens with Fuzzy Premises

The principle of modus ponens relies on the determination of the fuzzy consequence B, given an antecedent A and a conditional statement A→B. At first glance, it is clear that modus ponens implements forward chaining in the expert system.

A significant proportion of research stems from the pioneering works of Zadeh [9, Ch.1], Mamdani [9] and Mamdani and Assilian [10] in the field of fuzzy controllers. We recall briefly the main facts.

The implication A→B is viewed as a fuzzy relation defined in the Cartesian product $\mathbf{X}\times\mathbf{Y}$ (we assume in this Ch. both the universes of discourse $\mathbf{X}$ and $\mathbf{Y}$ to be finite). Then for $A\in\mathbf{F}(\mathbf{X})$ and $R\in\mathbf{F}(\mathbf{X}\times\mathbf{Y})$ given, the fuzzy consequence $B\in\mathbf{F}(\mathbf{X})$ is viewed as a max-min composition of A and R, namely the resulting fuzzy set B is calculated as $B=R\odot A$ and hence this form of composition enables us to use the theory of fuzzy relation equations. In this stream of investigation, the implication "→" has been modelled in a different way, mainly with the use of some well known multivalued implications. In some applications, it was proposed to consider minimum as a model of implication (cfr., e.g., [10]).

However we observe that the minimum is not a genuine implication since it is symmetrical and does not satisfy conditions which hold for Boolean truth values, say 0 and

1. For any implication "$\rightarrow$", one has $(0\rightarrow1)=1$ and $(1\rightarrow0)=0$ while for both of these cases, the minimum gives zero.

The scheme of modus ponens is extended to cover a set of implications:

$$\frac{\begin{array}{c}A\\ A_h\rightarrow B_h\end{array}}{B} \qquad h=1,2,\ldots,m,$$

which is a case often found in applications, i.e. we deal with a set of rules $A_h\rightarrow B_h$ describing our knowledge about the problem under discussion. Following the rule of composition, the implementation of the modus ponens follows accordingly (as usual, A, $A_h$ and B, $B_h$ are defined in the respective spaces $\mathbf{X}$ and $\mathbf{Y}$):

(i) *express each rule* $A_h\rightarrow B_h$ *as*

$$R_h = A_h \times B_h,$$

*which in terms of membership functions is read as*

$$R_h(x,y) = \min\{A_h(x), B_h(y)\}$$

*for all* $x\in\mathbf{X}, y\in\mathbf{Y}$,

(ii) *combine the rules by setting a fuzzy relation* R *which is formed by the union of all the rules:*

$$R = \bigvee_{h=1}^{m} R_h,$$

$$R(x,y) = \bigvee_{h=1}^{m} R_h(x,y),$$

*for all* $x\in\mathbf{X}, y\in\mathbf{Y}$,

(iii) *to infer the conclusion* B *for any* A, *perform max-min composition of* A *and* R:

$$B = R \odot A = [\bigvee_{h=1}^{m} (A_h \times B_h)] \odot A. \qquad (13.1)$$

This form of interpretation of the inference procedure has been extensively studied in relation to fuzzy controllers; the reader interested in some practical experiments is referred to [7], [12] and [19] which treat techniques of fuzzy control.

It is noticeable that (13.1) has an interesting and transparent interpretation which has also been recently implemented by means of hardware [15].

It is easily verified that the fuzzy set B can be computed as the union of fuzzy sets $B_h$ intersected by fuzzy sets $\Lambda_h$ (with a constant membership function) that play a role of coefficients representing a degree of rule firing (degree of their satisfaction). More clearly one has:

$$B(y) = \bigvee_{h=1}^{m} [\Lambda_h(y) \wedge B_h(y)] \tag{13.2}$$

for any $y \in \mathbf{Y}$, where

$$\Lambda_h(y) = \lambda_h \qquad \text{for any } y \in \mathbf{Y}$$

and

$$\lambda_h = \text{Poss}\,(A/A_h) = \sup_{x \in \mathbf{X}} [A(x) \wedge A_h(x)]$$

for any $h=1,2,\ldots,m$. The number $\lambda_h$ describes to what extent the fuzzy set (premise) A matches $A_h$, i.e. the fuzzy premise activates or fires the h-th rule. From an applicational point of view, B(y) could be conveniently computed following a three stage scheme displayed in Fig. 13.3.

The scheme of Fig.13.3 comprises three stages:

(a) *a matching stage: the fuzzy data A is matched against $A_h$ and the value of the possibility measure is obtained. Observe that it returns a number expressing the degree to which the fuzzy quantities A and $A_h$ overlap,*

(b) *an activation stage: the fuzzy consequence (conclusion) $B_h$ is intersected by the number $\lambda_h$, thus the higher $\lambda_h$ the more fireable $B_h$,*

(c) *a combination stage: all the results coming from different rules are put together by means of the union composition.*

This scheme has been realized in VLSI techniques. Despite its implementational simplicity, the way in which modus ponens is performed has some drawbacks or at least shortcomings which have to be reported. Firstly, it is not obvious which composition

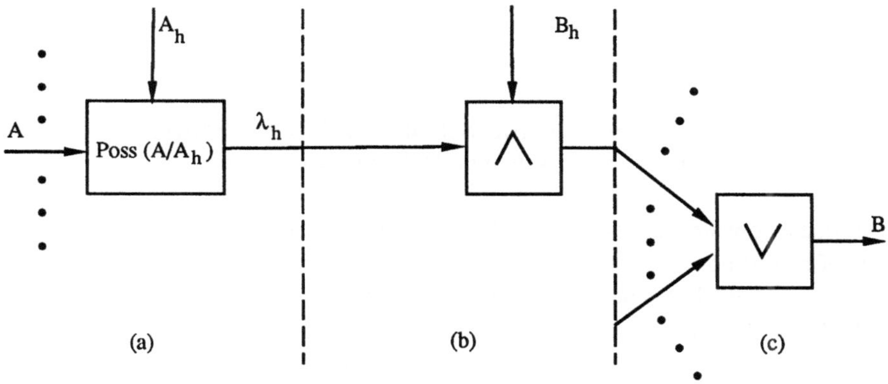


Fig. 13.3
Modus ponens performed in terms of the expression (13.2)

operator, especially at the stage of formation of the fuzzy relation, has to be used. Secondly, and more acutely in many cases, there is a lack of formal mechanism of the rules combination. Consider, for instance, two strongly contradictory rules which, possessing almost the same antecedent, generate totally different conclusions. Putting them together by means of (13.2) and setting A equal to this antecedent, we derive a conclusion B which is a fuzzy set in **Y**. Hence an inconsistency is visualized by a lack of convexity of the resulting fuzzy set. For detailed treatment of this topic, we refer to [2].

We will also report another implementation phenomenon.

When the conclusion B is realized by means of (13.2), we take a union of fuzzy conclusions of each rule. This implies an increasing fuzziness of the fuzzy set B. Take, for instance, $A=A_j$. Originally from the j-th rule, the true conclusion is equal to $B_j$. Now, in virtue of some overlap between $A_h$'s, B resulting from (13.2) usually satisfies the fuzzy inclusion $B_j \leq B$ (we assume all $A_h$'s and $B_h$'s are normal fuzzy sets or relations). In other words, the scheme of calculation (13.2) fuzzifies the result (the fuzzy set of conclusion) in comparison to the original one.

In [5] and [11] some studies have been focused on a suitable choice of implication in the modus ponens expressed by (13.2). The problem is worked out under several intuitively appealing criteria concerning a form of acceptable conclusion when the antecedent is specified as, e.g., "unknown", not A, very A, etc. In the entire scheme, even the implication "$\rightarrow$" has been modified and the sup-min composition for A and R has been left unchanged. This would be somewhat surprising since no relationship of implication and inference mechanism has been imposed. Further studies indicate a rather astonishing result: a very "crisp" form of implication, e.g. the drastic product (cfr. Sec.8.1) satisfies the criteria introduced to the highest extent. Moreover a certain shortcoming stems from the fact that these criteria are formulated for a single rule and do not refer to analysis for all the rules.

## 13.3. Expressing the Scheme of Modus Ponens in Terms of Fuzzy Relation Equations

To investigate how the theory of fuzzy relation equations contributes to the realization of the scheme of modus ponens, it is worth paying attention to different types of these equations since any conclusion B can be easily derived by performance of a suitable composition with the KB (R). Contrarily to the previous approach, the composition operator is now uniquely predetermined at the level of construction of the KB. Remembering that this stage of building of the system also provides a performance index for dealing with relevancy. It will be successively used to determine the length of any inference chain.

Consider two types of fuzzy equations, say with max-t and the min-$\varphi$ composition, with t lower semicontinuous (cfr. Ch.8). Recall that in their basic form, they are written as

$$B = R\, t\, A, \tag{13.3}$$

$$B = R\, \varphi\, A, \tag{13.4}$$

respectively, while the fuzzy relations R are equal to

$$R = \bigwedge_{h=1}^{m} A_h\, \varphi\, B_h,$$

$$R = \bigvee_{h=1}^{m} A_h\, t\, B_h.$$

Let us have a look at the main feature of these compositions. The result of the max-t composition (13.3) is the degree to which the antecedent A matches the KB. The composition can be treated as a generalization of the possibility measure, within which A is matched to the entire KB(R).

The inference results will be specified more formally in terms of the following propositions:

PROPOSITION 13.1. *If for any* $y \in \mathbf{Y}$ *there exists a point* $x \in \mathbf{X}$ *such that* $A_h(x) \leq B_h(y)$ *for any* $h=1,2,\ldots,m$ *and the fuzzy antecedent* A *in* (13.3) *represents a notion "unknown"* (*i.e.* $A(x)=1$ *for any* $x \in \mathbf{X}$), *then* B *is also "unknown", i.e.* $B(y)=1$ *for any* $y \in \mathbf{Y}$.

PROOF. Rewriting (13.3), we obtain that for any $y \in \mathbf{Y}$:

$$B(y) = \max_{x \in \mathbf{X}} \{A(x)\, t\, [\bigwedge_{h=1}^{m} (A_h(x)\, \varphi\, B_h(y))]\} = \max_{x \in \mathbf{X}} \{ \bigwedge_{h=1}^{m} [A_h(x)\, \varphi\, B_h(y)]\} =$$

$$= \{ \max_{x \in \mathbf{X}_y} \bigwedge_{h=1}^{m} [A_h(x)\, \varphi\, B_h(y)]\} \vee \{ \max_{x \notin \mathbf{X}_y} \bigwedge_{h=1}^{m} [A_h(x)\, \varphi\, B_h(y)]\},$$

where $\mathbf{X}_y = \{x \in \mathbf{X} : A_h(x) \leq B_h(y)$ for any $h=1,2,\ldots,N\}$.

In virtue of the assumption, $\mathbf{X}_y$ is nonempty and hence the thesis since the first maximum is equal to 1. ■

Thus the above proposition specifies that under some conditions for a totally unreliable fuzzy antecedent, the resulting conclusion is also meaningless.

It is seen immediately that

PROPOSITION 13.2. *If the fuzzy antecedent* A *in* (13.4) *is viewed as "unknown", the fuzzy conclusion* B *is given by the following membership function:*

$$B(y) = \min_{x \in \mathbf{X}} \{ \bigvee_{h=1}^{m} [A_h(x)\, t\, B_h(y)]\}$$

*for any* $y \in \mathbf{Y}$.

It is worth observing that, in virtue of the minimum on **X**, the conclusion B tends to be a fuzzy set with a low grade of membership function.

The above propositions shed light on the performance of the modus ponens worked out in a certain format of fuzzy equations. Remembering that the fuzzy set B can be equipped with a measure of relevancy which stems from the relevancy of the entire KB, we could control a length of the chain of inference. In other words, we can specify the precision of the conclusion reached at each stage of chaining (see Fig. 13.4), simply by specifying the stage at which the conclusion, expressed as an interval-valued fuzzy set ([7, Ch. 12]), has very different bounds. This in turn would require more precise input information or extra data. Each KB (at each stage of reasoning) is characterized by its own relevancy.

This relevancy is related to the level $\gamma$ at which this inequality is solved, e.g.:

$$|B(y_j) - B^{\frown}(y_j)| \geq \gamma_j$$

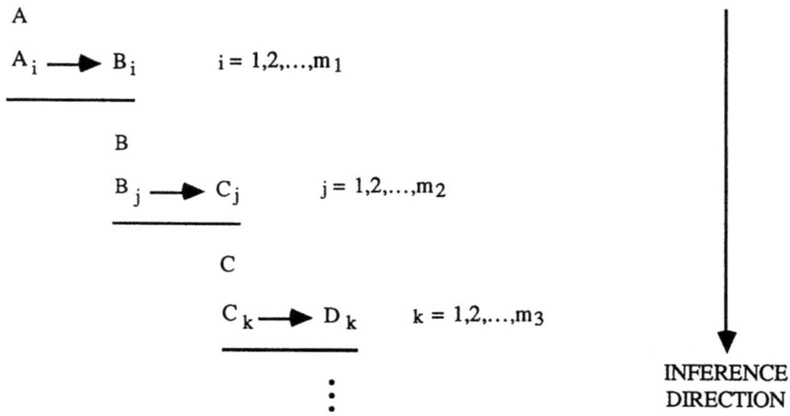


Fig. 13.4

Scheme of forward chaining

where j=1,2,...,q, q=card **Y**, B denotes the fuzzy set calculated with the fuzzy equation considered and $B^{\cap}$ stands for any fuzzy set satisfying the above inequality. Processing the resulting interval-valued fuzzy set and treating its bounds separately at the next inference level, the entire procedure is repeated now taking into account the measure of relevancy of the appropriate KB. A schematic illustration of forward chaining with a simultaneous propagation of fuzziness and imprecision is summarized in Fig. 13.5.

As pointed out in the previous discussion, a set of if-then statements is partitioned into local knowledge bases; each of them is characterized by its own form of implication operator. The resulting fuzzy relations are denoted by $R_1, R_2, \ldots$ Also the relevancy of each of them takes a different level specified by $\gamma_1, \gamma_2, \ldots$ . Because of the fact that not all (and this is an unusual case) the $\gamma_j$'s are equal to 1.0, we get the interval-valued fuzzy set as the result of any inference. This propagates along the next inference step producing a broader range of values of the membership function of the conclusion.

The length of the chain of inference is controlled by a straightforward observation of the width of the resulting interval-valued fuzzy set. When the bounds move too "far", the reasoning should be stopped and it is thus obvious how the relevancy of the KB influences the precision of the inferred results.

## 13.4. Backward Chaining

As mentioned above, backward chaining results in an opposite direction as a data-driven mode of reasoning. Now a certain goal is formulated and subgoals are to be determined.

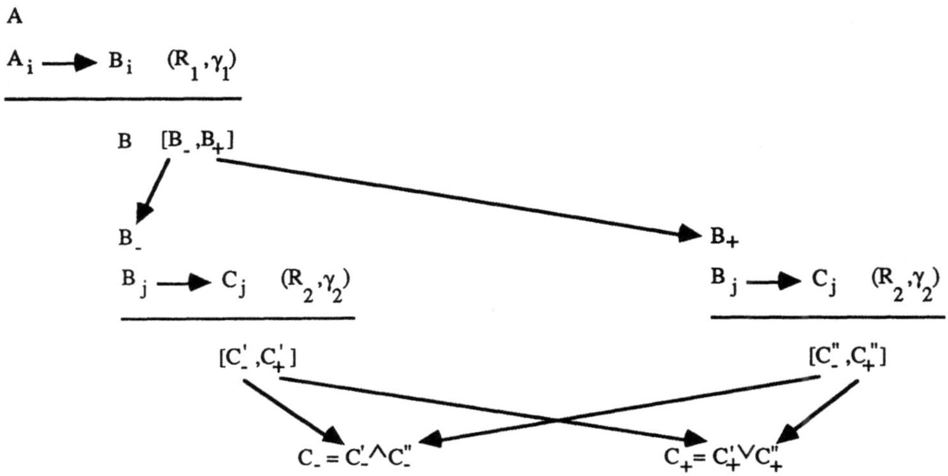


Fig. 13.5

Fuzziness and imprecision in forward chaining of rules

This refers to modus tollens: knowing a conclusion (or its complement) the antecedent is inferred, i.e. the scheme of Fig. 13.4 is reversed. Having formulated D, the fuzzy antecedent C is determined by solving the fuzzy equation:

$$C \lozenge R_3 = D, \tag{13.5}$$

where "◊" is any composition used in fuzzy equations. At the next step, C is used to solve the equation $B \lozenge R_2 = C$, etc. If there is no solution of a certain equation corresponding to the given level of inference (or its solvability index is extremely low), then it means that there is no applicable information in the KB (or in terms of fuzzy sets, any conclusion of the rules from the next inference step matches D very weakly).

The question is as follows: given D, specify which is the possible C related to it. Such questions are of value in knowledge acquisition. Sometimes they are more natural and obvious than those formulated in the opposite direction, i.e. those where the knowledge is acquired backwards. Also, the relevancy of the KB has a direct impact on the formulation of the final goal that one is trying to reach. Since each output is equipped with the vector of levels of equality $\gamma$, the goal should be transformed into an interval-valued fuzzy set with the bounds uniquely defined as $D_-$ and $D_+$ (they stem (cfr. Sec.10.2) from inequality $[D \equiv A] \geq \gamma$, hence A lies in between $D_-$ and $D_+$). Then the problem of solving the fuzzy Eq. (13.5) converts into the problem: determine all C (or bounds of C) which satisfy the two inequalities:

$$D_- \leq C \lozenge R_3 \leq D_+ .$$

For all the calculated bounds, (call them $C_-$ and $C_+$), one is trying to look for the conclusion of any rule from the next inference layer. Then $C_-$ and $C_+$ are modified bearing in mind the relevancy of the respective KB, hence the fuzzy sets are computed accordingly to the formulae:

$$[C_- \equiv A] \geq \gamma \quad \text{and} \quad [C_+ \equiv A] \geq \gamma \tag{13.6}$$

and the antecedents are deduced from the inequalities:

$$C'_- \leq B \lozenge R_2 \leq C_+',$$

where $C'_-$ and $C_+'$ stand for lower bound and upper bound of the solutions of (13.6).

### 13.5. Modus Ponens Performed in the Framework of Fuzzy Logic

It is of interest to discuss another approach to work with modus ponens in the framework of fuzzy logic [20]. This means that the truth value $\tau(v)$ of any statement "v" is the fuzzy set corresponding to a suitable linguistic expression modelling the truth of the statement in question; for instance one can deal with expressions like *very true*, *true*, *more or less true*, etc. An example of some forms of their membership functions is conveyed by Fig. 13.6.

Following the idea of attaching fuzzy truth values to statements in the scheme of modus ponens, the authors of [16] and [17] proposed an extension of multivalued Lukasiewicz logic to the fuzzy case. Recall that the implication studied there takes the form:

$$| r | = \min (1, 1 - | p | + | q |),$$

where $|p|$ and $|q|$ are viewed as truth values of the antecedent "p" and conclusion "q" and $|r|$ denotes the resulting truth value of the implication $r=(p \rightarrow q)$ for p and q known.

In the sequel, for simplicity, the truth value $|p|$ will be simply denoted by p. Now it is assumed that both the antecedent and the conclusion are specified as fuzzy sets P and Q, respectively:

$$P, Q : [0,1] \rightarrow [0,1]$$

and of course, the resulting truth value of the implication is a fuzzy set R as well. Therefore, to calculate the membership function of R, we use the extension principle of Zadeh ([17, Ch. 11]), which gives:

$$Q(q) = \sup_{p \in [0,1]} \{\min [P(p), R(r)] : r = \min (1,1-p+q)\} \tag{13.7}$$

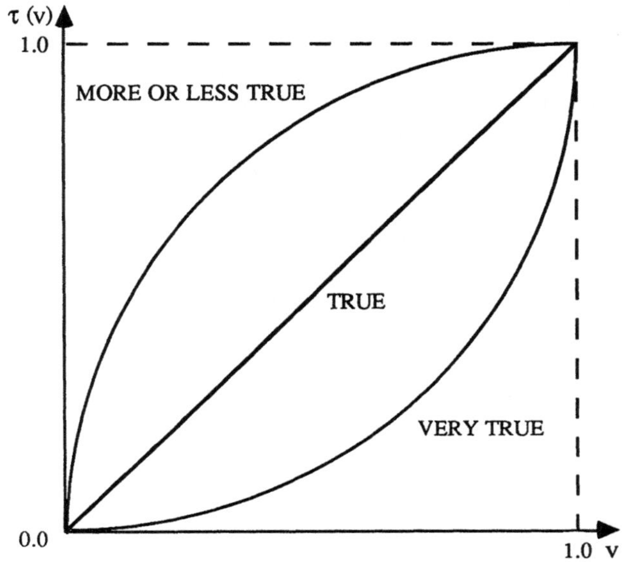


Fig. 13.6

Examples of linguistic truth values

for any $q \in [0,1]$. Under an additional hypothesis [16], P and Q have cuts given in the form of closed subintervals of [0,1], the corresponding cut of Q is developed; this avoids solving a tedious nonlinear optimization task.

Since P, Q and R are defined in [0,1], the entire discussion is translated from the physical spaces in which the original propositions were specified into the unit interval [0,1]. This requires an additional mechanism called *truth qualification*. Loosely speaking, it enables us to determine the fuzzy truth value $P_h$ of the proposition:

"*(antecedent of the rule is* $A_h$*) is* $P_h \equiv$ *antecedent of the rule is* A, h=1,2,...,m."

The truth value $P_h$ for $A_h$ and A provided, specifying a degree of their compatibility, is computed as

$$P_h(v) = A\,(A_h^{-1}(v))$$

if $A_h$ is a one-to-one mapping, or in general case we write:

$$P_h(v) = \sup_{x \in [0,1]} \{A(x) : A_h(x) = v\}$$

otherwise, where $v \in [0,1]$. It is noticeable that for $A=A_h$, $P_h(v)=v$ for any $v \in [0,1]$, i.e. $P_h$ is specified by the identity function (cfr. Fig.13.6). This method of deriving the fuzzy truth value is called *converse truth qualification* [14]. The inference method proceeds as follows: for $R_h$ already known and $P_h$ calculated, $Q_h$ is obtained from (13.7). Next a truth qualification method is applied which transforms the fuzzy set $Q_h$ from [0,1] into the physical space again. Then the conclusion $B_h$, obtained from the h-th rule, is equal to

$$B_h(y) = Q_h(B(y))$$

for any $y \in \mathbf{Y}$. An overall conclusion, stemming from combination of all the rules, is derived as an intersection of $B_h$'s, say:

$$B = \bigwedge_{h=1}^{m} B_h.$$

Observe that now we take intersection of partial results instead of their union (cfr. Sec.13.2). But the mechanism underlying their influence is completely different. It is easy to verify that for a very low level of matching of A and $A_h$, the corresponding truth value $Q_h$ has a membership function almost identically equal to 1.0, hence the resulting fuzzy conclusion has the same property, i.e. the h-th rule generates the notion "unknown". Hence the use of this rule is somewhat meaningless and in this case it does not convey any useful information. The intersection of the results simply eliminates this information. An illustration of the scheme of modus ponens worked out by means of fuzzy logic is contained in Fig.13.7.

Of course, instead of speaking in Fig.13.7 of fuzzified Lukasiewicz logic, one can refer to fuzzification of any of the many-valued logics. Generally, for any implication modelled by φ-operator (cfr. Ch. 8) one gets for any $r \in [0,1]$:

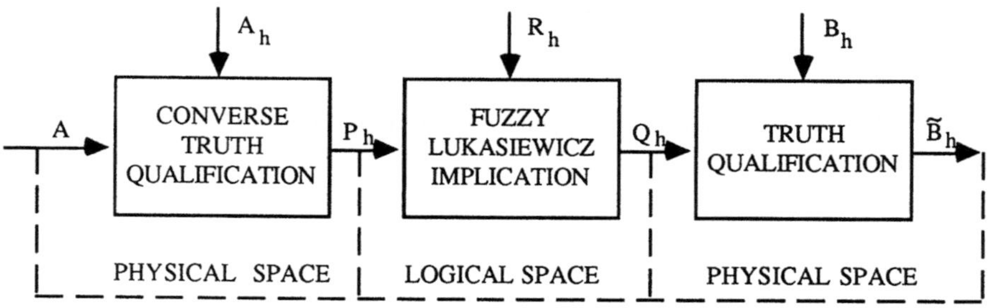


Fig. 13.7

Physical and logical spaces in the realization of modus ponens

$$R(r) = \sup_{p,q \in [0,1]} \{\min \{P(p), Q(q)\} : r = p\varphi q\}.$$

To avoid constraints in the above expression, let us introduce a Boolean relation:

$$B(p,q,r) = 1 \quad \text{if } r=p\varphi q, \qquad = 0 \quad \text{otherwise.}$$

Therefore we have:

$$R(r) = \sup_{p,q \in [0,1]} [B(p,q,r) \wedge P(p) \wedge Q(q)],$$

for any $r \in [0,1]$. This explicitly yields the following fuzzy relation equation:

$$R = B \odot P \odot Q.$$

If Q is unknown and P, B and R given, the conclusion in the logical space, if it exists, is given by

$$Q = (B \odot P) \varphi R$$

(cfr. Sec.8.3). Therefore, repeating calculations for each rule h=1,2,...,N, the corresponding $Q_h$'s are obtained and in sequel they are transformed into the fuzzy sets $B_h$ in the physical space.

## 13.6. Detachment Principle and Its Implementation with the Aid of Fuzzy Equations

The idea of the detachment principle being of importance in reasoning procedures, can be formulated as follows (cfr. [1, Ch. 11]):

*Given is a statement "P" with a certainty value lying in a closed subinterval of [0,1], say [$p_1$,$p_2$], where $0 \leq p_1, p_2 \leq 1$. Provided is also an inference rule "P↔Q" specifying an equality between "P" and "Q", where "Q" is another statement. The rule is equipped with bounds of its sufficiency and necessity, which are two numbers, "suff" and "necc", of [0,1] such that suff≤necc. Our aim is to derive the boundaries of the certainty value of the statement "Q".*

Thus the essence of the detachment principle is to infer the certainty of the conclusion "Q" if we know that both the rule is uncertain and the statement "P" is true to a certain degree. The detachment principle is written accordingly:

$$\begin{array}{ll} P & [p_1,p_2] \\ \underline{P \leftrightarrow Q} & [suff,necc] \\ Q & [x,y], \end{array} \tag{13.8}$$

where x and y denote lower and upper bounds of certainty assigned to "Q". Notice moreover that the equivalence "P↔Q" can be considered as two implications:

$$\begin{array}{ll} P \rightarrow Q & suff \\ Q \rightarrow P & necc. \end{array}$$

Returning to (13.8), it has been pointed out [4] that the bounds of certainty of "Q" are equal to

$$x = suff\ t\ p_1$$

$$y = p_2\ s\ (1 - necc).$$

The bounds for "Q" are simply formed by the combination of the bounds of certainty of "P" and suff and necc of the statement "P↔Q", while the combination is performed with the aid of triangular norms and conorms.

It is instructive to distinguish some characteristic cases. They are summarized in Table 13.1.

The terms "certain" and "unknown" are represented by a single numerical value in [0,1] and by the unit interval, respectively.

The inputs of this table result from computations specified above.

Table 13.1

| P | P↔Q | Q |
|---|---|---|
| unknown | certain | unknown |
| certain | certain | certain |
| unknown | unknown | unknown |
| certain | unknown | unknown |

Let us formulate the above problem in a more general setting incorporating suitable mechanisms of fuzzy relation equations. Denote by $\tau$ a truth value of the statement "P"; now instead of two bounds in [0,1], one takes a fuzzy set $\tau : [0,1] \to [0,1]$. Suppose that "P↔Q" is modelled by a binary relation $R : [0,1]^2 \to [0,1]$ and $\eta : [0,1] \to [0,1]$ stands for the truth value of the statement "Q". Hence (13.8) refers to the transformation in logical rather than in physical space, however the truth values are only pointwise defined. Here we adopt a form of fuzzy relation equation with the equality composition operator (cfr. Ch.9):

$$\eta = R \,\&\, \tau,$$

which in terms of membership functions is read as

$$\eta(u) = \sup_{v \in [0,1]} [\tau(v) \,\&\, R(v,u)]$$

for any $u \in [0,1]$. Thus (13.8) can also be written as follows:

$$\begin{array}{cc} P & \tau \\ \underline{P \leftrightarrow Q} & \underline{R}\,. \\ Q & \eta \end{array}$$

Now, for practical purposes one is interested in determining the fuzzy relation R, knowing that the detachment principle generates some pairs of data, say $(P_h, Q_h)$, $h=1,2,\ldots,m$. Thus we see that their fuzzy truth values are equal to $\tau_h$ and $\eta_h$, respectively. To calculate R, we consider the following system:

$$\eta_h = R \,\&\, \tau_h, \qquad h = 1,2,\ldots,m, \tag{13.9}$$

whose solution is guaranteed by the following theorem:

THEOREM 13.3. *Let* $\mathbf{Z}_h$, $h=1,2,\ldots,m$, *be the set of each Eq.* (13.9) *and assume that the set*

$$\mathbf{Z} = \bigcap_{h=1}^{m} \mathbf{Z}_h$$

*is nonempty. If each* $\eta_h$ *is not normal, i.e.* $\eta_h(u)<1$ *for any* $u \in [0,1]$, *then the fuzzy relation* $R^{\frown}:[0,1]^2 \to [0,1]$ *defined by*

$$R^{\cap}(v,u) = \bigwedge_{h=1}^{m} [\tau_h(v)\ \xi\ \eta_h(u)] \tag{13.10}$$

*for all* $u,v \in [0,1]$, *belongs to* $\mathbf{Z}$ *and* $R^{\cap} \geq R$ *for any* $R \in \mathbf{Z}$ *provided that* $(R^{\cap} \& \tau_h)(u) < 1$ *for any* $u \in [0,1]$ *and* $h=1,2,\ldots,m$.

PROOF. Since $\mathbf{Z} \neq \emptyset$, let $R \in \mathbf{Z}$ and then $R \& \tau_h = \eta_h$, hence $R \leq (\tau_h \xi \eta_h)$ for any $h=1,\ldots,m$ by Thm. 9.4 since each $\eta_h$ is not normal. This means that $R \leq R^{\cap}$ for any $R \in \mathbf{Z}$.

Since $(R \& \tau_h)(u) = \eta_h(u) < 1$ and $(R^{\cap} \& \tau_h)(u) < 1$ for any $u \in [0,1]$, we have:

$$\eta_h(u) = (R\ \&\ \tau_h)(u) \leq (R^{\cap}\ \&\ \tau_h)(u) \leq [(\tau_h\ \xi\ \eta_h)\ \&\ \tau_h]\ (u) = \eta_h(u)$$

for any $u \in [0,1]$ by Lemma 9.3. This implies that $R^{\cap} \& \tau_h = \eta_h$ for any $h=1,\ldots,m$ and therefore $R^{\cap} \in \mathbf{Z}$. ■

For practical use, one cannot assume a priori that the hypotheses of Thm.13.3 are satisfied. However one can utilize (13.10) and afterwards observe to what extent the fuzzy sets obtained from this composition are equal to those standing at the left-side of (13.9). If the differences observed are neglected, then we can treat $R^{\cap}$, given by (13.10), as an acceptable solution of the system (13.9). Further details concerning this method can be found in [2, Ch.9], where Thm. 13.3 was established. Related applications of fuzzy relation equations with difference operator (cfr. Sec.9.5) can be found in [3, Ch.9].

## References


[1] B. Buchanan and E.M. Shortliffe, *Rule-Based Expert Systems*, Addison-Wesley, Reading, Mass., 1984.

[2] E. Czogala and W. Pedrycz, Some problems concerning the construction of algorithms of decision-making in fuzzy systems, *Internat. J. Man-Machine Studies* 15 (1981), 201-211.

[3] A. Di Nola, W. Pedrycz and S. Sessa, Towards handling fuzziness in intelligent systems, in *Fuzzy Computing* (M.M. Gupta and T. Yamakawa, Eds.), Elsevier Science Publishers B.V. (North-Holland), Amsterdam (1988), pp.365-374.

[4] D. Dubois and H. Prade, The principle of minimum specificity as a basis for evidential reasoning, in: *Uncertainty in Knowledge-Based Systems* (B. Bouchon and R.R. Yager, Ed.) Lecture Notes in Computer Science, Vol. 286, Springer-Verlag, Berlin (1987), pp.75-84.

[5] S. Fukami, M. Mizumoto and S. Tanaka, Some considerations on fuzzy conditional inferences, *Fuzzy Sets and Systems* 4 (1980), 243-273.

[6] F. Hayes-Roth, Rule-Based Systems, *Communications of ACM* 28 (1985), 921-932.

[7] W.J.M. Kickert and E.M. Mamdani, Analysis of a fuzzy logic controller, *Fuzzy Sets and Systems* 1 (1978), 29-44.

[8] R. Kowalski, *Logic for Problem Solving*, North-Holland, New York, 1979.

[9] E.M. Mamdani, Advances in the linguistic synthesis of fuzzy controllers, *Internat. J. Man-Machine Studies* 8 (1976), 669-678.

[10] E.M. Mamdani and S. Assilian, An experiment in linguistic synthesis with a fuzzy logic controller, *Internat. J. Man-Machine Studies* 7 (1978), 1-13.

[11] M. Mizumoto and M.J. Zimmermann, Comparison of fuzzy reasoning methods, *Fuzzy Sets and Systems* 8 (1982), 253-283.

[12] W. Pedrycz, *Fuzzy Control and Fuzzy Systems*, J. Wiley (Research Studies Press), 1988, to appear.

[13] H. Prade, A computational approach to approximate and plausible reasoning with applications to expert systems, *IEEE Trans. on Pattern Analysis and Machine Intelligence*, 7 (1985), 260-283.

[14] E. Sanchez, On truth-qualification in natural languages, *Proc. Internat. Conf. Cybernetics and Society*, Tokyo-Kyoto (Japan), 3-7 Nov. 1978, Vol.II, 1233-1236.

[15] M. Togai and H. Watanabe, Expert-Systems on a chip: An engine for real-time approximate reasoning, *IEEE Expert* 1, no.3, (1986), 55-62.

[16] Y. Tsukamoto, *Fuzzy logic based on Lukasiewicz logic and its application to diagnosis and control*, Ph. D. Thesis, Tokyo Inst. of Technology, Tokyo, 1979.

[17] Y. Tsukamoto, T. Takagi and M. Sugeno, Fuzzification of Aleph-1 and its application to control, *Proc. Internat. Conf. Cybernetics and Society*, Tokyo-Kyoto (Japan), 3-7 Nov. 1978, Vol.II, 1217-1221.

[18] S. Weiss and C.A. Kulikowski, *A Practical Guide to Designing Expert Systems*, Rowman & Allanheld, Philadelphia, 1984.

[19] D. Willaeys and N. Malvache, The use of fuzzy sets for the treatment of fuzzy information by computer, *Fuzzy Sets and Systems* 3 (1981), 323-328.

[20] L.A. Zadeh, Syllogistic reasoning in fuzzy logic and its application to usuality and reasoning with dispositions, *IEEE Trans. Syst. Man. Cybern.* SMC-15 (1985), 754-763.

# CHAPTER 14

## A FUZZY CONTROLLER AND ITS REALIZATION



*The schemes of inference, introduced in Sec.2, are used in Sec.3 to discuss the special case of an expert system oriented towards control of an industrial process consisting of a steam engine and a boiler. We consider here only max-min fuzzy equations and $\alpha$-fuzzy equations (cfr. Ch.6) defined on finite sets.*

### 14.1. Theoretical Preliminaries

In this Ch. we discuss a topic that is very close to applications of fuzzy relation calculus in Knowledge Engineering, namely fuzzy controller (fuzzy logic controller). These studies were initiated in the 70's by Mamdani [9, Ch.13] in search of a novel approach to the control of processes difficult to manage by traditional methods. It is worth underlining two aspects which are linked with their utilization:

- *existence of an appropriate, relevant model of the process,*

- *an explicite formulation of goals of control as well as their constraints.*

It often happens, however, that these are not provided, especially in cases where the systems are controlled manually by a human operator. Therefore, to realize computer support, it would be useful to incorporate his knowledge concerning the control strategy. Usually this knowledge is coded in the form of rules, namely conditional statements "if state of the process, then control". Since the reasoning of the operator is not very precise in the sense of states and control actions involved in the process, the fuzzy sets are considered as labels standing in the statements. Simply bearing in mind the structure of the knowledge gathered, the fuzzy controller can be viewed as a rule-based knowledge system. There are, however, two main differences:

*- a reasoning chain is one-step (state→control), no explaination mechanism is inbuilt. In this sense the system is a real-time system,*

*- secondly, the system and the fuzzy controller are put together in a closed loop. This implies that a stage of fuzzification and defuzzification is required.*

In the following discussion, we will consider in depth the logical background of the fuzzy controller. We will not discuss a variety of problems that are related to the nonfuzzy transformation of the fuzzy control into a nonfuzzy value. This has been the object of extensive studies over the last decade.

Let $\mathbf{X}_i$, i=1,2,...,n and $\mathbf{Y}$ be finite referential sets. Denoting by $\mathbf{X}$ the Cartesian product of the n sets $\mathbf{X}_i$, we consider the max-min fuzzy equation (cfr. Sec.3.1):

$$B = R \odot A, \tag{14.1}$$

i.e.

$$B(y) = \bigvee_{\mathbf{x}\in \mathbf{X}} [A(\mathbf{x}) \wedge R(\mathbf{x}, y)],$$

for any $y\in Y$, where $\mathbf{x}=(x_1,x_2,\ldots,x_n)$, $x_i\in \mathbf{X}_i$ for i=1,2,...,n and $B\in \mathbf{F}(\mathbf{Y})$, $A\in F(\mathbf{X})$, $R\in F(\mathbf{X}\times\mathbf{Y})$ are fuzzy sets from the specified spaces into [0,1]. Further, we consider the α-fuzzy equation (cfr. Sec.6.1):

$$B = R \,\alpha\, A, \tag{14.2}$$

i.e.

$$B(y) = \bigwedge_{\mathbf{x}\in \mathbf{X}} [A(\mathbf{x}) \,\alpha\, R(\mathbf{x}, y)],$$

for any $y\in \mathbf{Y}$. As usual, we assume A and B known and R unknown in Eqs. (14.1) and (14.2). Moreover, we have two systems, each of m fuzzy equations:

$$B_h = R \odot A_h, \tag{14.3}$$

$$B_h = R \,\alpha\, A_h, \tag{14.4}$$

where h=1,2,...,m, $A_h\in \mathbf{F}(\mathbf{X})$ and $B_h\in \mathbf{F}(\mathbf{Y})$ are given and $R\in \mathbf{F}(\mathbf{X}\times\mathbf{Y})$ is unknown. If $\mathcal{R}_h=\mathcal{R}_h(A_h,B_h)$ and $\mathcal{R}_h'=\mathcal{R}_h'(A_h,B_h)$ denote the sets of the solutions $R\in \mathbf{F}(\mathbf{X}\times\mathbf{Y})$ satisfying Eqs. (14.3) and (14.4), respectively, the following theorems hold (cfr. Thms.8.6 and Remark 8.2):

THEOREM 14.1. *If* $(\mathcal{R}_1 \cap \ldots \cap \mathcal{R}_m) \neq \emptyset$, *then the fuzzy relation:*

$$R^{\cap} = \bigwedge_{h=1}^{m} R_h^{\cap} \tag{14.5}$$

*with membership function defined as*

$$R^{\cap}(\mathbf{x},y) = \min_{1\le h\le m} R_h^{\cap}(\mathbf{x},y) = \min_{1\le h\le m} [A_h(\mathbf{x})\ \alpha\ B_h(y)]$$

*for all* $\mathbf{x}\in \mathbf{X}$, $y\in \mathbf{Y}$, *is the greatest element of the set* $(\mathcal{R}_1 \cap \ldots \cap \mathcal{R}_m)$.

THEOREM 14.2. If $(\mathcal{R}_1' \cap \ldots \cap \mathcal{R}_m') \neq \emptyset$, *then the fuzzy relation:*

$$R^{\cup} = \bigvee_{h=1}^{m} R_h^{\cup} \tag{14.6}$$

*with membership function defined as*

$$R^{\cup}(\mathbf{x},y) = \max_{1\le h\le m} R_h^{\cup}(\mathbf{x},y) = \max_{1\le h\le m} [A_h(\mathbf{x}) \wedge B_h(y)]$$

*for all* $\mathbf{x}\in \mathbf{X}$, $y\in \mathbf{Y}$, *is the smallest element of the set* $(\mathcal{R}_1' \cap \ldots \cap \mathcal{R}_m')$.

## 14.2. Schemes of Inference with Fuzzy Information

We consider the KB consisting of conditional production rules:

$$KB = \{p_1, p_2, \ldots, p_m\}.$$

The h-th proposition "$p_h$" assumes the following form:

$$p_h : \text{"if } A_h \text{ then } B_h\text{"}, \tag{14.7}$$

where $A_h$ and $B_h$, h=1,2,...,m, express parts of bodies of evidence given by fuzzy quantities. For instance, if n=2 and m=1, the proposition (14.7) may be considered as the rule of a driver: "if the road is *wide* and you are a *good* driver, then you can drive *fast*", where *wide*, *good* and *fast* are fuzzy quantities expressed in the respective spaces $\mathbf{X}_1$, $\mathbf{X}_2$ and $\mathbf{Y}$ (cfr. Sec.12.2).

Returning to our discussion, we are interested in algorithms of inference starting from an unconditional statement "p" containing a fuzzy or non-fuzzy quantity. Formally, we have the following scheme:

$$\begin{array}{ll} p\ : A & \\ \underline{p_h : \text{if } A_h \text{ then } B_h} & h=1,2,\ldots,m, \\ \qquad\qquad B=? & \end{array}$$

where, using modus ponens of Sec.13.2, the antecedents (14.7) are put together in the form of the fuzzy relation $R^{\cup}$ given by (14.6) and B is calculated using (14.1) as compositional rule of inference, i.e. $B=R^{\cup}\odot A$.

We note that (14.1) and (14.6) represent implementation of the inference scheme formulated above and therefore we need further empirical validation of the construction proposed, i.e. we have to see if the combination of the antecedents (14.7) by means of (14.6) is appropriate in every situation (cfr. Sec.12.2). We remember that it may lead to occurrence of interactivity that results from the proposed algorithm and not from the KB handled. As a simple example, we take A to be exactly equal to one of the $A_h$' s, say $A_k$, and combining with $R^{\cup}$ using (14.1), we expect $B_k$ to be obtained but this requires special hypotheses on the $A_h$' s, as is shown in the following result:

THEOREM 14.3. *If the inference scheme is implemented by* (14.1) *and* (14.6) *and*

(i) $A_h \cap A_k = \emptyset$, *i.e.* $A_h(\mathbf{x}) \wedge A_k(\mathbf{x})=0$ *for all* $\mathbf{x}\in \mathbf{X}$, $h\in \{1,2,\ldots,m\}-\{k\}$,

(ii) $A_k$ *is normal, i.e.*

$$\max_{\mathbf{x}\in \mathbf{X}} A_k(\mathbf{x}) = 1,$$

*then* $R^{\cup}\odot A_k=B_k$.

PROOF. Using (14.1) and (14.6), we have for any $y\in \mathbf{Y}$:

$$(R^{\cup}\odot A_k)(y) = \max_{\mathbf{x}\in \mathbf{X}} \{A_k(\mathbf{x}) \wedge \{ \bigvee_{h=1}^{m} [A_h(\mathbf{x}) \wedge B_h(y)]\}$$

$$= \max_{\mathbf{x}\in \mathbf{X}} \{ \bigvee_{h=1}^{m} [A_k(\mathbf{x}) \wedge A_h(\mathbf{x}) \wedge B_h(y)]\}$$

$$= \max_{x \in X} [A_k(x) \wedge B_k(y)] \vee \{ \max_{x \in X} \{ \bigvee_{h \neq k} [A_k(x) \wedge A_h(x) \wedge B_k(y)] \} \}$$

$$= \{B_k(y) \wedge [\max_{x \in X} A_k(x)]\} \vee \{0\} = B_k(y) \wedge 1 = B_k(y) .$$

Then the thesis follows by using assumptions (i) and (ii). ■

The assumptions (i) and (ii) are difficult to verify in any applicable scheme of fuzzy reasoning: there is evidently an overlap between the fuzzy sets.

Some discussions concerning interaction of propositions of the KB in relation to the fuzzy controllers are pointed out in [1] and [2]. However, we propose another inference mechanism combining the antecedent (14.7) by means of the fuzzy relation $R^\cap$ given by (14.5). If the rules of the KB are consistent, i.e. $\mathfrak{R}_h \neq \emptyset$ for any h=1,2,...,m, then this way of calculating the fuzzy relation $R^\cap$ ensures that there is no interaction between parts of the KB. If $\mathfrak{R}_h = \emptyset$ for some h, we apply the same way and calculate additionally the measure of fitness of $B_h^\cap = R^\cap \odot A_h$ and $B_h$ coming from the conditional statements. This may be expressed by values of the Hamming distance between $B_h^\cap$ and $B_h$:

$$d(B_h^\cap, B_h) = \sum_{y \in Y} |B_k^\cap(y) - B_k(y)|.$$

By putting:

$$d_{min} = \min_{1 \le h \le m} d(B_h^\cap, B_h) \quad \text{and} \quad d_{max} = \max_{1 \le h \le m} d(B_h^\cap, B_h),$$

the values of the index

$$1 - \frac{d(B_h^\cap, B_h) - d_{min}}{d_{max} - d_{min}}$$

will be interpreted as a certainty factor $CF_h$ associated with the h-th conditional statement (14.7), which is now read as

"if $A_h$ then $B_h$ ($CF_h$)".

This allows us to extract the most reliable rules of the KB. If the original conditional statements are equipped with certainty factors, they should be included in formulas expressing the fuzzy relation of the KB. Strictly speaking, we propose the following expression for (14.5):

$$R^{\cap} = \bigwedge_{h=1}^{m} \tau R_h^{\cap} \tag{14.8}$$

where $\tau R_h^{\cap} \in \mathbf{F}(\mathbf{X}\times\mathbf{Y})$ has membership function equal to

$$\tau R_h^{\cap}(\mathbf{x},y) = [R_h^{\cap}(\mathbf{x},y)]^{CF_h}$$

for all $\mathbf{x}\in\mathbf{X}$, $y\in\mathbf{Y}$. If $CF_h=1$, then (14.8) becomes (14.5). If $CF_h=0$ (or nearly equal to zero), one excludes this contribution.

We propose the following formula for (14.6):

$$R^{\cup} = \bigvee_{h=1}^{m} \xi R_h^{\cup} \tag{14.9}$$

where $\xi R_h^{\cup} \in \mathbf{F}(\mathbf{X}\times\mathbf{Y})$ has membership function equal to

$$\xi R_h^{\cup}(\mathbf{x},y) = [R_h^{\cup}(\mathbf{x},y)]^{2-CF_h}$$

for all $\mathbf{x}\in\mathbf{X}$, $y\in\mathbf{Y}$. Note that (14.9) becomes (14.6) if $CF_h=1$.

## 14.3. Realization of the Fuzzy Controller and an Applicational Example

In this Sec. we discuss the determination of the KB of the fuzzy controller and its evaluation. For this purpose we present the fuzzy controller of Mamdani and Assilian [10, Ch.13] which was constructed for the control of an industrial process consisting of a steam engine and a boiler. This process is viewed as consisting of two-inputs and two-outputs, as shown in Fig. 14.1.

We observe, as already stated, that in comparison to conventional knowledge-based systems, the reasoning chain is reduced to one step and the explanation mechanisms are not necessary.

The inputs are constituted by heat input (HI) to the boiler and throttle opening (TO) at the engine. The outputs are formed by the steam pressure and speed of the engine. The goal of the control is to achieve the steam pressure and the speed of the engine at certain fixed levels despite disturbances (noises) existing in the system. The control actions (HI, TO) depend on the error of the steam pressure and the speed of the engine (error = actual value of the HI or TO) and their derivatives that reflect trends in changes of the error.

The inputs of the fuzzy controller are formed by pressure error (PE) and its derivative change of error (CPE), speed error (SE) and its derivative change (CSE). The collection of the statements are summarized in Tables 14.1 and 14.2. We have two control algorithms: the first one, consisting of 15 rules, deals with the boiler, while the second one provides the control strategy for the steam engine by means of 9 rules. Each of them contains the fuzzy labels describing the state of the process under control. The shorthands in the tables denote the following fuzzy sets: NB (Negative Big), NM (Negative Medium), NS (Negative Small), NO (Negative Zero), PO (Positive Zero), PS (Positive Small), PM (Positive Medium), PB (Positive Big).

The fuzzy set ANY has membership function equal to 1 in every point of the universe of discourse.

We analyze the control task of the boiler. It is transparent from Table 14.1 that the variables describing the steam engine do not influence the control of heat exchange (SE and CSE are given by the fuzzy set ANY for every fuzzy set expressing pressure error and its change).

The format of the control rules is the following:

$$\text{“if } (PE_h \text{ and } CPE_h) \text{ then } HI_h\text{”},$$

where $h=1,\dots,15$ and $PE_h \in \mathbf{F}(\mathbf{PE})$, $CPE_h \in \mathbf{F}(\mathbf{CPE})$ and $HI_h \in \mathbf{F}(\mathbf{HI})$ are fuzzy sets defined in the specified spaces.

We use directly Thms. 14.1 and 14.2 for the construction of the fuzzy relation of the KB (we assume that the hypotheses of these theorems are satisfied). Precisely, if the inference rule is

$$HI = R^{\cap} \circ (PE \times CPE), \tag{14.10}$$

applying Thm.14.1, we have

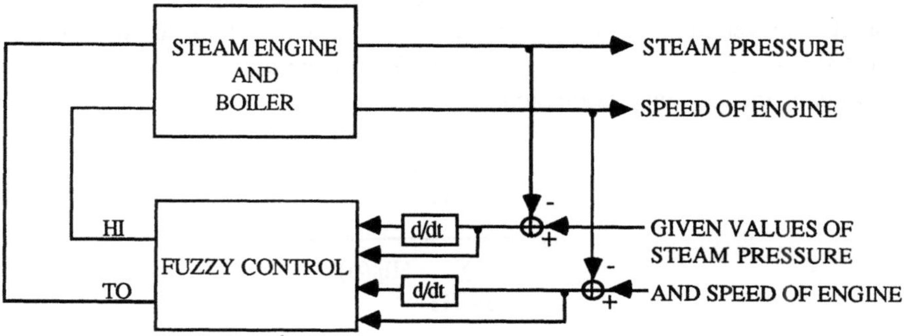


Fig.14.1

Structure of the system under control and the fuzzy controller

$$R^{\cap} = \bigwedge_{h=1}^{15} [(PE_h \times CPE_h)\ \alpha\ HI_h].$$

If the inference rule is

$$HI = R^{\cup}\ \alpha\ (PE \times CPE), \tag{14.11}$$

applying Thm.14.2, we deduce that

$$R^{\cup} = \bigvee_{h=1}^{15} (PE_h \times CPE_h \times HI_h) \tag{14.12}$$

Let us use the performance index given by

$$D = \sum_{h=1}^{15} \sum_{u \in \mathbf{HI}} |HI_h{}^*(u) - HI_h(u)| \tag{14.13}$$

where $HI_h{}^*$ is the fuzzy set resulting from the scheme inference used. This performance index reflects the quality of the mapping realized by the fuzzy relation and the inference mechanism, and it yields the following values: D=10.4 for the inference rule (14.10) and D=1.7 for (14.11).

It is obvious that the second method is preferable. The values of the certainty factors for each of the two methods discussed are shown in Table 14.3.

Usually, for purposes of implementation of the fuzzy controller, the fuzzy set of control HI is converted into a single numerical value. It is chosen in various modes (cfr.[2, Ch.13], [3], [4], [5], [7, Ch.13], [9, Ch.13]). One of them is obtained as an averaged mean by taking the following expression:

$$\bar{u}_h = \frac{\sum_{u \in \mathbf{HI}} u \cdot HI_h(u)}{\sum_{u \in \mathbf{HI}} HI_h(u)},$$

$$\bar{u}_h{}^* = \frac{\sum_{u \in \mathbf{HI}} u \cdot HI_h{}^*(u)}{\sum_{u \in \mathbf{HI}} HI_h{}^*(u)}.$$

Plots of these representatives of fuzzy control $\bar{u}_h$ and $\bar{u}_h{}^*$ are presented in Fig.14.2 and Fig.14.3, respectively.

In Fig.14.2, the points are significantly scattered around the straight line. In Fig.14.3, we have a very good fit of empirical data and no scatter is visible.

Most results of this Ch. are contained in [4, Ch.6].

For comparison, we summarize the control actions resulting from the classic form of the classifier introduced in Mamdani's papers already cited. We recall that the fuzzy relation of the controller is computed by means of (14.12) while the fuzzy control is obtained from max-min composition. They are contained in Table 14.4. This way of implementing the fuzzy controller reveals that a large proportion of the rules is characterized by relatively low values of the certainty factor. Also the performance index (14.13) attains significantly higher values than in the two previous methods.

## 14.4. Concluding Remarks

This Ch. has covered aspects of utilizing fuzzy relation equations in designing of control algorithms. The well known structure of the fuzzy controller arising more than a decade ago is reconsidered by means of methods discussed in the previous Chs. One has to be aware that all these considerations indicate clearly how the pieces of knowledge (rules) are combined to assure the highest possible consistency and reproducibility of the fuzzy sets standing in the control rules.

With respect to the original concept of the fuzzy controller, the mechanism of fuzzy relation equations enable to design it in more sistematic way. First of all, one has at his disposal different types of equations which can be investigated simultaneously to get the best fit to the data conveyed by the rules. Also a straightforward link between the operators of combination of pieces of knowledge (rules) and those utilized for inference purposes is underlined. It is evident that a weak correspondence between the fuzzy sets of control contained in the rules and those from the fuzzy controller has two sources, which unfortunately cannot be clearly identified. The first one stems from the implementation level. This means that the fuzzy relation equation is unable to cope with the real structure of the rules and interrelationships between them. The second one finds its origin in some inconsistencies in the overall set of rules.

They have to be extracted before constructing the fuzzy controller. Both of these deficiencies can be eliminated using methods of getting approximate solutions of equations (cfr. Ch.10).

It should be stressed that the way of working with the fuzzy controller refers to its construction at the level of fuzzy modelling. We do not pay any attention to its application at the level of nonfuzzy input and nonfuzzy control which has to be reached when one realizes a closed loop of any physical system and the fuzzy controller. These considerations are not dealt in this Ch. but for further details, the reader is referred to the existing literature, e.g., [3] and [5].

Table 14.1
A set of control rules concerning the boiler

| CONDITIONAL STATEMENT | PE | CPE | SE | CSE | HI |
|---|---|---|---|---|---|
| 1 | NB | not(NB or NM) | ANY | ANY | PB |
| 2 | NB or NM | NS | ANY | ANY | PM |
| 3 | NS | PS or NO | ANY | ANY | PM |
| 4 | NO | PB or PM | ANY | ANY | PM |
| 5 | NO | NB or NM | ANY | ANY | NM |
| 6 | PO or NO | NO | ANY | ANY | NO |
| 7 | PO | NB or NM | ANY | ANY | PM |
| 8 | PO | PB or PM | ANY | ANY | NM |
| 9 | PS | PS or NO | ANY | ANY | NM |
| 10 | PB or PM | NS | ANY | ANY | NM |
| 11 | PB | NB or NM | ANY | ANY | NB |
| 12 | NO | PS | ANY | ANY | PS |
| 13 | NO | NS | ANY | ANY | NS |
| 14 | PO | NS | ANY | ANY | PS |
| 15 | PO | PS | ANY | ANY | NS |

Table 14.2
A set of control rules concerning the steam engine

| CONDITIONAL STATEMENT | PE | CPE | SE | CSE | TO |
|---|---|---|---|---|---|
| 1 | ANY | ANY | NB | not (NB or NM) | PB |
| 2 | ANY | ANY | NM | PB or PM or PS | PS |
| 3 | ANY | ANY | NS | PB or PM | PS |
| 4 | ANY | ANY | NO | PB | PS |
| 5 | ANY | ANY | PO or NO | PS or NS or NO | NO |
| 6 | ANY | ANY | PO | PB | NS |
| 7 | ANY | ANY | PS | PB or PM | NS |
| 8 | ANY | ANY | PM | PB or PM or PS | NS |
| 9 | ANY | ANY | PB | not (NB or NM) | NB |

Table 14.3

| CONDITIONAL STATEMENT | CSE INFERENCE RULE (14.10) | TO INFERENCE RULE (14.11) |
|---|---|---|
| 1 | 1 | 1 |
| 2 | 1 | 1 |
| 3 | 0.70 | 1 |
| 4 | 1 | 1 |
| 5 | 1 | 1 |
| 6 | 0.39 | 1 |
| 7 | 1 | 1 |
| 8 | 1 | 1 |
| 9 | 0.91 | 1 |
| 10 | 0.96 | 1 |
| 11 | 1 | 1 |
| 12 | 0 | 0 |
| 13 | 0.39 | 1 |
| 14 | 0.04 | 1 |
| 15 | 0.13 | 0.3 |

Table 14.4

| CONDITIONAL STATEMENT | CF |
|---|---|
| 1 | 1 |
| 2 | 0.89 |
| 3 | 0.53 |
| 4 | 0.81 |
| 5 | 0.79 |
| 6 | 0 |
| 7 | 0.72 |
| 8 | 0.89 |
| 9 | 0.60 |
| 10 | 0.79 |
| 11 | 0.91 |
| 12 | 0.72 |
| 13 | 0.43 |
| 14 | 0.36 |
| 15 | 0.74 |

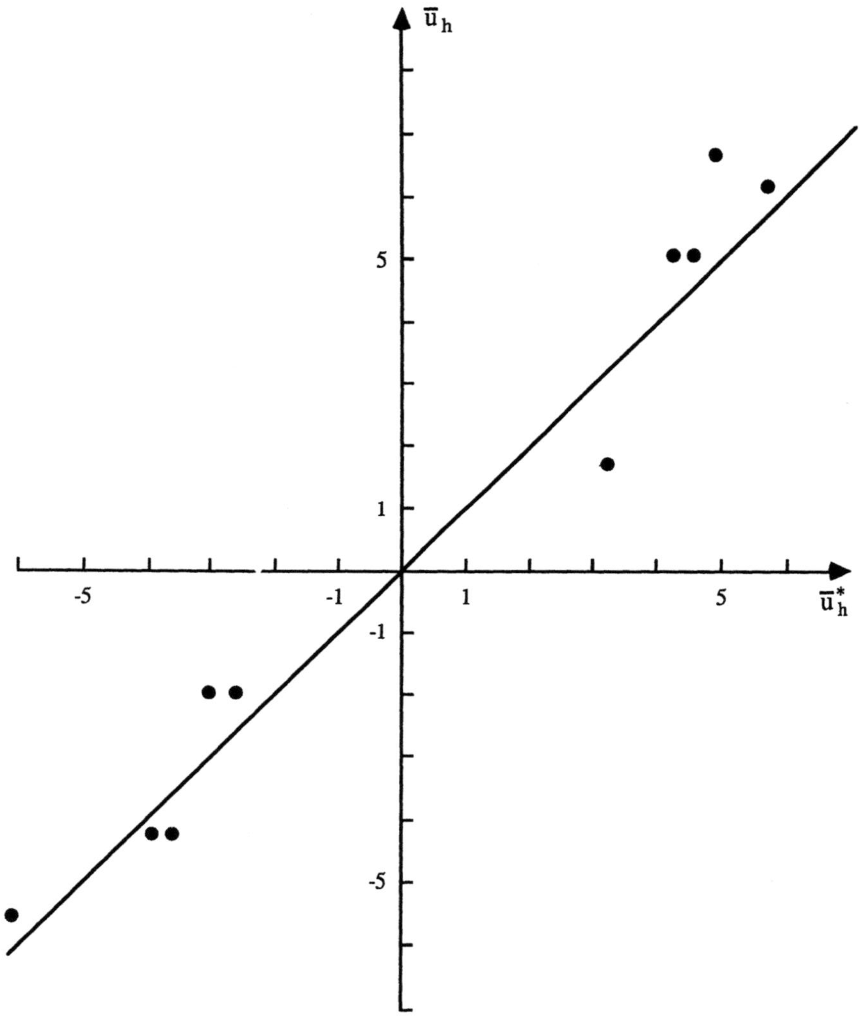


Fig.14.2

Plots of $\overline{u}_h$ vs. $\overline{u}_h^*$ for inference rule (14.10)

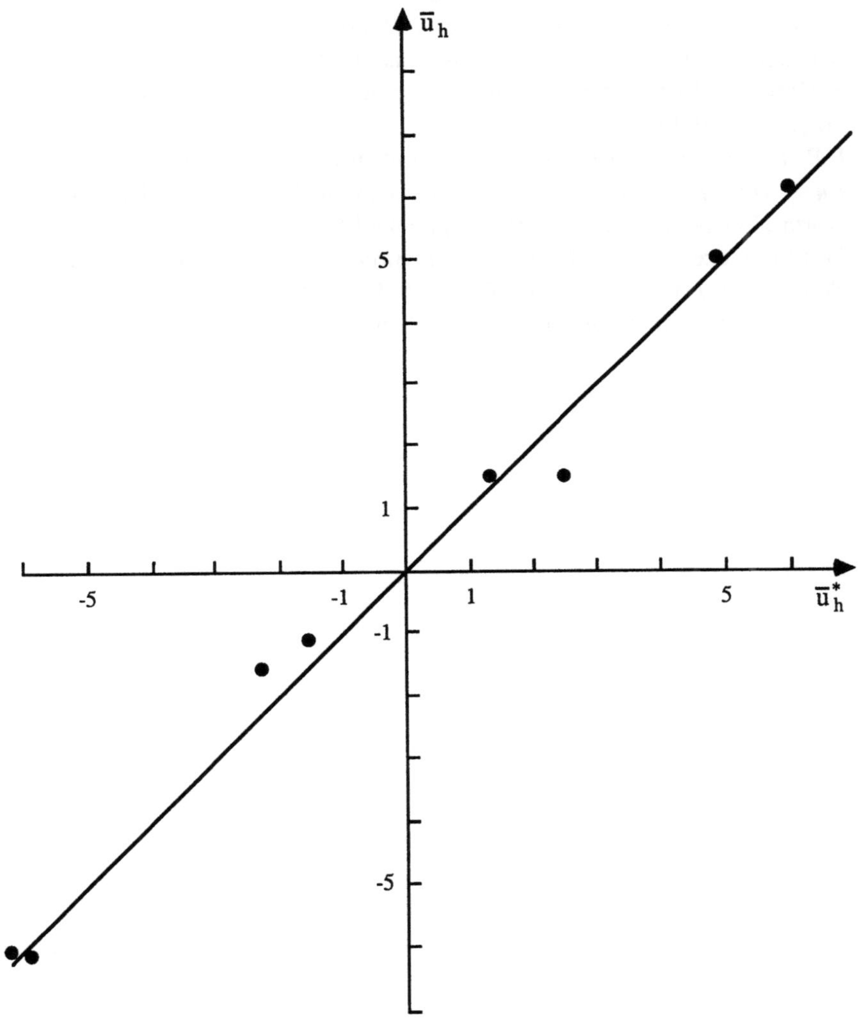


Fig.14.3

Plots of $\overline{u}_h$ vs. $\overline{u}_h^*$ for inference rule (14.11)

## References

[1] E. Czogala, S. Gottwald and W. Pedrycz, Aspects for the evaluation of decision situations, in: *Fuzzy Information and Decision Processes* (M.M. Gupta and E. Sanchez, Eds.), North-Holland, Amsterdam (1982), pp.41-49.

[2] S. Gottwald, Criteria for non-interactivity of fuzzy logic controller rules, *BUSEFAL* 16 (1983), 69-79.

[3] L.P. Holmblad and J.J. Ostergaard, Control of a cement kiln by fuzzy logic, in: *Fuzzy Information and Decision Processes* (M.M. Gupta and E.Sanchez, Eds.), North-Holland, Amsterdam (1982), pp.389-399.

[4] W.J.M. Kickert, *Fuzzy Theories on Decision-Making*, M. Nijhoff, Leiden, 1978.

[5] W.J.M. Kickert and H.R. Van Nauta Lemke, Application of a fuzzy controller in a warm water plant, *Automatica* 12 (1976), 301-308.

# CHAPTER 15

## BIBLIOGRAPHIES

15.1. A List of Papers on Fuzzy Relation Equations and Related Topics.
15.2. A List of Papers on Fuzzy Relations and Related Topics.

*We conclude this book exhibiting an useful list of papers on fuzzy relation equations which, to the best of our knowledge, have been published on several books and international journals. This list (of course to be considered as not exhaustive) also includes papers dealing with applications of fuzzy relation equations and close topics. Thus the reader can have a global point of view on the present literature, which, of course, this book could not entirely cover. We avoid recalling the papers and the books already cited in the references of the previous Chs. An additional list of papers covering several topics close to the theory of fuzzy relations and their applications is also enclosed. For some papers, unfortunately, an updated reference is not complete (we have only related preprints). Due to the abundance of the present literature, some authors could have been involuntarily omitted.*

### 15.1. A List of Papers on Fuzzy Relation Equations and Related Topics


K.P. Adlassing, A survey on medical diagnosis and fuzzy subsets, in: *Approximate Reasoning in Decision Analysis* (M.M. Gupta and E. Sanchez, Eds.), North-Holland Publ. Co. (1982), pp.203-217.

K.P. Adlassing, Fuzzy Set Theory in Medical Diagnosis, *IEEE Trans. Syst. Man Cybern.* 16, n.2 (1986), 260-265.

R. Ambrosio and G. Martini, Resolution of max-min equations with fuzzy symbols, *BUSEFAL* 16 (1983), 61-68.

A. Asse, P. Mangin, D. Wiart and D. Willaeys, Assisted diagnosis using fuzzy information: A comparative analysis of several methods for the research of solutions of the max-min inverse composition, *Proceedings of NAFIP Congress*, 19-21 May 1982, Logan (Utah), USA.

A. Asse, P. Mangin and D. Willaeys, Assisted diagnosis using fuzzy information, *Proceedings of NAFIP Congress*, 29 June - 1 July 1983, Schenectady, New York.

A. Asse, P. Mangin and D. Willaeys, Assisted diagnosis using fuzzy information. Realization of interactivity in a system of assisted industrial diagnosis based on fuzzy information, *Proceedings of 1st Napoli Meeting: Mathematics of Fuzzy Systems*, Napoli, 15-17 June 1984, 13-28.

A. Asse, P. Mangin and D. Willaeys, Assisted diagnosis using fuzzy information: method of inverting equations of fuzzy relations with Φ-fuzzy sets, in: *Analysis of Fuzzy Information, Vol.2: Artificial Intelligence and Decision Systems* (J.C. Bezdek, Ed.), CRC Press, Boca Raton, FL. (1987), pp.153-162.

L. Bour, G. Hirsch and M. Lamotte, Opérateur de minimalisation pour la résolution d'équations de relations floues avec la composition inf-conorm, *BUSEFAL* 28 (1986), 68-77.

L. Bour, G. Hirsch and M. Lamotte, Syllogismes direct et indirect, condition pour que le syllogisme indirect soit parfait, *BUSEFAL* 29 (1987), 48-57.

K. Burdzy and J.B. Kiszka, The reproducibility property of fuzzy control systems, *Fuzzy Sets and Systems* 9 (1983), 161-177.

B.Y. Cao, The theory and practice of solution for fuzzy relative equations in max-product, *BUSEFAL* 35 (1988), 124-131.

Z.G. Cao, Eigenvectors of fuzzy matrices under incline operations, *Fuzzy Math.* 6 (1986), 53-56 (in chinese).

Z.Q. Cao, The eigen fuzzy sets of a fuzzy matrix, in: *Approximate Reasoning in Decision Analysis* (M.M. Gupta and E. Sanchez, Eds.), North-Holland Publ. Co., Amsterdam (1982), pp.61-63.

Z.Q. Cao and J.Q. Lin, Eigenvectors of incline matrices, *preprint.*

R. E. Cavallo and G.J. Klir, Reconstruction of possibilistic behavior systems, *Fuzzy Sets and Systems* 8 (1982), 175-197.

Y.Y. Chen, Solving fuzzy relation equations, *Fuzzy Math.* 3 (1983), 109-124 (in chinese).

Y.Y. Chen, Fuzzy permutation with lower solution of fuzzy relation equation, *BUSEFAL* 21 (1985), 66-76.

L.C. Cheng, The inverse of the union or intersection preserving mapping and its applications in fuzzy relation equation with multi-operators, in: *Proceedings of the Internat.*

*Symposium on Fuzzy Systems and Knowledge Engineering* (X.H. Lin and P.Z. Wang, Eds), Guangzhou and Guiyang, China, July 1987, Guangdong Higher Education Publishing House, pp.672-676.

S.Q. Chien, Analysis for multiple fuzzy regression, *Fuzzy Sets and Systems* 25 (1988), 59-65.

E. Czogala and W. Pedrycz, On identification in fuzzy systems and its applications in control problems, *Fuzzy Sets and Systems* 6 (1981), 73-83.

E. Czogala and W. Pedrycz, Some problems concerning the construction of algorithms of decision-making in fuzzy systems, *Internat. J. Man-Machine Studies* 15 (1981), 201-211.

E. Czogala and W. Pedrycz, Control problems in fuzzy systems, *Fuzzy Sets and Systems* 7 (1982), 257-273.

E. Czogala and W. Pedrycz, Fuzzy rules generation for fuzzy control, *Cybern. and Systems* 13 (1982), 275-293.

E. Czogala and W. Pedrycz, On the concept of fuzzy probabilistic controllers, *Fuzzy Sets and Systems* 10 (1983), 109-121.

E. Czogala and W. Pedrycz, Identification and control problems in fuzzy systems, *TIMS / Studies in the Management Sciences* 20 (1984) 447-466.

E. Czogala and H.J. Zimmermann, Some aspects of synthesis of probabilistic fuzzy controllers, *Fuzzy Sets and Systems* 13 (1984), 169-177.

M. De Glas, A mathematical theory of fuzzy systems, in: *Fuzzy Information and Decision Processes* (M.M. Gupta and E. Sanchez, Eds.), North-Holland Publ. Co., Amsterdam (1982), pp.401-410.

M. De Glas, Theory of fuzzy systems, *Fuzzy Sets and Systems* 10 (1983), 65-77.

M. De Glas, Invariance and stability of fuzzy systems, *J. Math. Anal. Appl.* 99 (1984), 299-319.

A. Di Nola, Fuzzy equations in infinitely distributive lattices, *Preprints of II IFSA Congress*, Tokyo, 20-25 July 1987, 533-534.

A. Di Nola, W. Pedrycz and S. Sessa, Selected topics in theory and applications of fuzzy relational equations in system analysis, in: *Large Scale Systems: Theory and Applications*,

Proceedings of IFAC/IFORS Symposium (A. Straszak, Ed.), Warsaw, 11-15 July 1983, Pergamon Press, pp.247-252.

A. Di Nola, W. Pedrycz and S. Sessa, Coping with uncertainty for knowledge acquisition and inference mechanism, *Kybernetes* 15 (1986), 243-249.

A. Di Nola, W. Pedrycz and S. Sessa, Models of decision making realized via fuzzy relation equations, in: *Proc. NAFIPS '86* (W. Bandler and A. Kandel, Eds.), New Orleans, 2-4 June 1986, pp.92-106.

A. Di Nola, W. Pedrycz and S.Sessa, Processing of fuzzy numbers by fuzzy relation equations, *Kybernetes* 15 (1986), 43-47.

A. Di Nola, W. Pedrycz and S. Sessa, Using fuzzy relation equations in the construction of inference mechanisms in expert systems, *Cybern. and Systems* 18 (1987), 49-56.

A. Di Nola and S. Sessa, Finite fuzzy relation equations with a unique solution in linear lattices, *J. Math. Anal. Appl.* 132 (1988), 39-49.

J. Drewniak, Systems of equations in a linear lattice, *BUSEFAL* 15 (1983), 88-96.

J. Drewniak, Fuzzy relation equations and inequalities, *Fuzzy Sets and Systems* 14 (1984), 237-247.

J. Drewniak and W. Pedrycz, A note on the inverses of fuzzy relations, *BUSEFAL* 17 (1983), 36-37.

D. Dubois and H. Prade, Inverse Operations for fuzzy numbers, in: *Fuzzy Information, Knowledge representation and Decision Analysis* (M.M. Gupta and E. Sanchez, Eds.), IX IFAC Symposium, Marseille, 19-21 July 1983, Pergamon Press, pp.391-396.

D. Dubois and H. Prade, Fuzzy numbers: an overview, in: *Analysis of Fuzzy Information, Vol.I: Mathematics and Logic*, (J.C. Bezdek, Ed.), CRC Press, Boca Raton, FL. (1987), pp.3-39.

R. Goetschel, Jr. and W. Voxman, Eigen fuzzy number sets, *Fuzzy Sets and Systems* 16 (1985), 75-85.

S. Gottwald, Generalization of some results of E. Sanchez, *BUSEFAL* 16 (1983), 54-60.

S. Gottwald, Generalized solvability criteria for fuzzy equations, *Fuzzy Sets and Systems* 17 (1985), 258-296.

S. Gottwald, Generalized solvability criteria for systems of fuzzy equations defined with t-norms, *BUSEFAL* 21 (1985), 77-81.

S. Gottwald, On solvability degrees for fuzzy equations, in: *Proceedings of the Polish Symposium on Interval and Fuzzy Mathematics* (J. Albrycht and H. Wisniewski, Eds.), Poznan, 26-29 Aug. 1983, 1985, pp.97-102.

S. Gottwald, On the suitability of fuzzy models: an evaluation through fuzzy integrals, *Internat. J. Man-Machine Studies* 24 (1986) 141-151.

S. Gottwald, Solvability considerations for generalized set equations through many-valued logic, *Proceedings of 4th Easter Conference on Model Theory*, Seminar Report 86, Humboldt Universität, Sektion Mathematik, Berlin (1986), 56-62.

S. Gottwald and W. Pedrycz, Analysis and synthesis of fuzzy controller, *Problems of Control and Information Theory* 14 (1985), 33-45.

S. Gottwald and W. Pedrycz, On measuring controllability property for systems described by fuzzy relation equations, *BUSEFAL* 22 (1985), 75-79.

S. Gottwald and W. Pedrycz, Problems of the design of fuzzy controllers, in: *Approximate Reasoning in Expert Systems* (M.M. Gupta, A. Kandel, W. Bandler and J.B. Kiszka, Eds.), Elsevier Science Publishers B.V. (North-Holland), Amsterdam (1985), pp. 393-405.

S. Gottwald and W. Pedrycz, On the methodology of solving fuzzy relational equations and its impact on fuzzy modelling, in: *Fuzzy Logic in Knowledge-Based Systems, Decision and Control* (M.M. Gupta and T. Yamakawa, Eds.), Elsevier Science Publishers B.V. (North-Holland), Amsterdam (1988), pp.197-210.

M.M. Gupta, J.B. Kiszka and G.M. Trojan, Controllability of fuzzy control systems, *IEEE Trans. Syst. Man Cybern.* SMC-16 (1986), 576-582.

M.M. Gupta, J.B. Kiszka and G.M. Trojan, Multivariable structure of fuzzy control systems, *IEEE Trans. Syst. Man Cybern.* SMC-16 (5) (1986), 638-657.

A.A. Gvozdik, Solution of fuzzy equations, *Soviet. J. Comput. Systems Sci.* 23, n.1 (1985), 60-67.

S. Heilpern, Fuzzy equations, *Fuzzy Math.* 7 (1987), 77-84.

M. Higashi and G.J. Klir, Identification of fuzzy relation systems, *IEEE Trans. Syst. Man.*

*Cybern.* SMC-14 (1984), 349-355.

M. Higashi and G.J. Klir, Application of finite fuzzy relation equations to systems problems, *Preprints of II IFSA Congress*, Tokyo, 20-25 July 1987, 547-550.

M. Higashi, G.J. Klir and M.A. Pittarelli, Reconstruction families of possibilistic structure systems, *Fuzzy Sets and Systems* 12 (1984), 37-60.

H. Hirota and W. Pedrycz, On identification of fuzzy systems under the existence of vagueness, *Summary of Papers on General Fuzzy Problems*, Tokyo, 6 (1980), 37-40.

K. Hirota and W. Pedrycz, On identification of fuzzy systems under the existence of vagueness II, *Proceedings of IFAC Symposium on Theory and Application of Digital Control*, New Delhi, 1982, 11-15.

K. Izumi, Duality of fuzzy systems based on the intuitionistic logic calculus LJ, *Preprints of II IFSA Congress*, Tokio, 20-25 July 1987, 535-539.

K. Izumi, H. Tanaka and K. Asai, Adjointness of fuzzy systems, *Fuzzy Sets and Systems* 20 (1986), 211-221.

K. Izumi, H. Tanaka and K. Asai, Adjoint fuzzy systems based on L-fuzzy logic and their application, in: *Analysis of Fuzzy Information, Vol.1: Mathematics and Logic* (J. Bezdek, Ed.), CRC Press, Boca Raton, FL. (1987), pp.231-239.

J.B. Kiszka, M.M. Gupta and P.N. Nikiforuk, Some properties of Expert Control Systems, in: *Approximate Reasoning in Expert Systems* (M.M. Gupta, A. Kandel, W. Bandler and J.B. Kiszka, Eds.), Elsevier Science Publishers B.V. (North-Holland), Amsterdam (1985), pp.283-306.

J. Kitowski and M. Bargiec, Diagnostics of faulty states in complex physical systems using fuzzy relational equations, in: *Approximate Reasoning in Intelligent Systems, Decision and Control* (E. Sanchez and L.A. Zadeh, Eds.), Pergamon Press, Oxford (1987), pp.175-193.

G.J. Klir, *Architecture of Systems Problem Solving*, Plenum Press, New York and London, 1985.

G.J. Klir and T.A. Folger, *Fuzzy Sets, Uncertainty and Information,* Prentice Hall, Englewood Cliffs, New York, 1988.

G.J. Klir and E.C. Way, Reconstructability analysis: aims, results, open problems,

*Systems Research* 2, n.2 (1985), 141-163.

A. Lettieri and F. Liguori, Some results on fuzzy relation equations provided with one solution, *Fuzzy Sets and Systems* 17 (1985), 199-209.

B.X. Li, Rudimentary transformations for solving fuzzy relation equations, *Fuzzy Math.* 1 (1981), 53-60 (in chinese).

H. Li, A method for solving fuzzy relation equations by means of the basic element sequence, *Fuzzy Math.* 2 (1982), 67-71 (in chinese).

H.X. Li, The stability of the solutions of fuzzy relation equations, *BUSEFAL* 19 (1984), 105-112.

H.X. Li, Construction of the set of solutions of a general fuzzy relation equations, *Fuzzy Math.* 4 (1984), 59-60 (in chinese).

H.X. Li, Fuzzy similarity matrix equations of varied order, Fuzzy Math. 4 (1984), 15-20 (in chinese).

H.X. Li, Fuzzy perturbation analysis, Part 1: Directional perturbation, *Fuzzy Sets and Systems* 17 (1985) 189-197.

H.X. Li, Fuzzy perturbation analysis, Part 2: undirectional perturbation, *Fuzzy Sets and Systems* 19 (1986), 165-175.

S.H. Li, Fuzzy proximity matrix equations of varied order, *BUSEFAL* 17 (1983), 38-47.

Z.W. Liao, Necessary and sufficient conditions for the existence of solutions of fuzzy equations, *BUSEFAL* 5 (1980), 20-29.

Z.W. Liao, A simple method for finding the general solution of fuzzy relation equations, *Fuzzy Math.* 2 (1982), 127-128 (in chinese).

L.Y. Lu, The minimal solutions of fuzzy relation equations on a finite set, in: *Proceedings of the Internat. Symposium on Fuzzy Systems and Knowledge Engineering* (X.H. Liu and P.Z. Wang, Eds.), Guangzhou and Guiyang, China, July 1987, Guangdong Higher Education Publishing House, pp.717-719.

C.Z. Luo, Fuzzy relation equation on infinite sets, *BUSEFAL* 26 (1986), 57-66.

A.N. Melikhov, L.S. Bernstein and S.Y. Korovkin, Making control decisions in fuzzy systems invariants to a change of external conditions, *Fuzzy Sets and Systems* 22 (1987), 93-105.

M.C. Miglionico, Other solutions of fuzzy equations with extended operation, *BUSEFAL* 25 (1986), 95-106.

M. Miyakoshi and M. Shimbo, Sets of solution-set-invariant coefficient matrices of simple fuzzy relation equations, *Fuzzy Sets and Systems* 21 (1987), 59-83.

T. Murai, M. Miyakoshi and M. Shimbo, A modeling of search oriented thesaurus use based on multivalued logic inference, *Inform. Sciences* 45 (1988), 185-212.

S. Nakanishi and H. Hoshino, Composite type 2 - fuzzy relation equations, *Systems and Control* 28, n.7 (1984), 483-490 (in japanese).

S. Nakanishi and H. Hoshino, Composite fuzzy relational equations expressed by min-bounded sum and max-bounded difference, in: *Analysis of Fuzzy Information, Vol. 1: Mathematics and Logic* (J.C. Bezdek, Ed.), CRC Press, Boca Raton, FL. (1987), pp.41-49.

C.V. Negoita and D.A. Ralescu, *Applications of fuzzy sets to system analysis*, Birkhäuser Verlag, Basel, 1975.

C.V. Negoita, *Fuzzy Systems*, Cybernetics and Systems Series, Abacus Press, Turnbridge Wells, U.K., 1981.

C.V. Negoita, *Expert Systems and Fuzzy Systems*, The Benjamin/Cummings Publ. Co., Inc., Menlo Park, CA, 1985.

V. Novak and W. Pedrycz, Fuzzy sets and t-norms in the light of fuzzy logic, *Internat. J. Man-Machine Studies* 29 (1988), 113-127.

A. Ohsato and T. Sekiguchi, Method of solving the polynomial form of composite fuzzy relation equations and its application to a group decision problem, in: *Approximate Reasoning in Decision Analysis* (M.M. Gupta and E. Sanchez, Eds.), North-Holland Publ. Co., Amsterdam (1982), pp.33-45.

A. Ohsato and T. Sekiguchi, Solution of the convexly combined form of sup-min and inf-max composite fuzzy relation equations, *Trans Soc. Instrum. Control Engin.* 19, n.3 (1983), 212-219 (in japanese).

A. Ohsato and T. Sekiguchi, Convexly combined form of fuzzy relational equations and its application to knowledge representation, *Proceedings of IEEE Internat. Conf. on Systems, Man and Cybernetics*, Bombay and New Delhi, 29 Dec. 1983 - 7 Jan. 1984, 294-299.

A. Ohsato and T. Sekiguchi, Solution of the convexly combined form of composite fuzzy relation equations, *Trans. Soc. Instrum. Control Engin.* 21, n.5 (1985), 423-428 (in japanese).

A. Ohsato and T. Sekiguchi, Convexly combined fuzzy relational equations and several aspects of their application to fuzzy information processing, *Inform. Sciences* 45 (1988), 275-313.

A. Ohsato and M. Sugeno, Identification of ill-defined systems by means of convexly combined fuzzy relational equations, *Proceedings of III Fuzzy System Symposium*, Osaka (1987), 89-94 (in japanese).

C. P. Pappis, Resolution of Cartesian products of fuzzy sets, *Fuzzy Sets and Systems* 26 (1988), 387-391.

K.G. Peeva, Behaviour, reduction and mnimization of finite L-automata, *Fuzzy Sets and Systems* 28 (1988), 171-181.

W. Pedrycz, An approach to the analysis of fuzzy systems, *Internat. J. Control* 34, n.3 (1981), 403-421.

W. Pedrycz, Fuzzy relational equations with triangular norms in modelling of decision-making processes, in: *Fuzzy Information, Knowledge Representation and Decision Analysis* (M.M. Gupta and E. Sanchez, Eds.), IX IFAC Symposium, Marseille, 19-21 July 1983, Pergamon Press, pp.275-279.

W. Pedrycz, Identification in fuzzy systems, *IEEE Trans. Syst. Man Cybern.* SMC-14 (2) (1984), 361-366.

W. Pedrycz, An identification algorithm in fuzzy relational systems, *Fuzzy Sets and Systems* 13 (1984), 153-167.

W. Pedrycz, A model of decision-making in a fuzzy environment, *Kybernetes* 13 (1984), 99-102.

W. Pedrycz, On handling fuzziness and randomness in fuzzy relational models, *BUSEFAL* 20 (1984), 115-119.

W. Pedrycz, A method of classifier design for fuzzy data, *BUSEFAL* 21 (1985), 113-116.

W. Pedrycz, Default production rules, *BUSEFAL* 24 (1985), 100-104.

W. Pedrycz, Classification in a fuzzy environment, *Pattern Recognition Letters* 3 (1985), 303-308.

W. Pedrycz, Design of fuzzy control algorithms with the aid of fuzzy models, in: *Industrial Applications of Fuzzy Control* (M. Sugeno, Ed.), North-Holland, Amsterdam (1985), pp.153-174.

W. Pedrycz, A unified approach for ranking fuzzy sets in the same space, *BUSEFAL* 23 (1985), 67-72.

W. Pedrycz, Ranking multiple aspect alternatives – Fuzzy relational equation approach, *Automatica* 22 (1986), 251-253.

W. Pedrycz, Analysis of fuzzy data for fuzzy models, *BUSEFAL* 25 (1986), 113-117.

W. Pedrycz, An algorithmic approach to studies of relevancy of fuzzy models, *Preprints of II IFSA Congress*, Tokyo, 20-25 July 1987, 74-77.

W. Pedrycz, On solution of fuzzy functional equations, *J. Math. Anal. Appl.* 123 (1987), 589-604.

W. Pedrycz, Fuzzy models and relational equations, *Mathematical Modelling* 6 (1987), 427-434.

W. Pedrycz, An inference scheme based on transformation of equality indices, in: *Proceedings of the Internat. Symposium on Fuzzy Systems and Knowledge Engineering* (X.H. Liu and P.Z. Wang, Eds.), Guangzhou and Guiyang, China, July 1987, Guangdong Higher Education Publishing House, pp.54-60.

W. Pedrycz, E. Czogala and K. Hirota, Some remarks on the identification problem in fuzzy systems, *Fuzzy Sets and Systems* 12 (1984), 185-189.

W. Pedrycz and S. Gottwald, Evaluating fuzzy controllers through possibility and certainty measures, *BUSEFAL* 19 (1984), 119-122.

E. Sanchez, Resolution of eigen fuzzy sets equations, *Fuzzy Sets and Systems* 1 (1978), 69-74.

E. Sanchez, Medical diagnosis and composite fuzzy relations, in: *Advances in Fuzzy Set Theory and Applications* (M.M. Gupta, R.K. Ragade and R.R. Yager, Eds.), North-

Holland, Amsterdam (1979), pp.437-444.

E. Sanchez, Inverses of fuzzy relations. Application to possibility distribution and medical diagnosis, *Fuzzy Sets and Systems* 2 (1979), 75-86.

E. Sanchez, Solution of fuzzy equations with extended operations, *Fuzzy Sets and Systems* 12 (1984), 237-248.

E. Sanchez, Possibility distributions, fuzzy intervals and possibility measure in a linguistic approach to pattern classification in medicine, in: *Knowledge Representation in Medicine and Clinical Behavioural Science* (L.J. Kohout and W. Bandler, Eds.), Abacus Press, Turnbridge Wells (1986), Ch.12.

E. Sanchez, Equations de relations floues: méthodologie et applications, in: *Fuzzy Sets Theory and Applications* (A. Jones, A. Kaufmann and H.J. Zimmermann, Eds.), D. Reidel Publ. Co., Dordrecht (1986), pp.213-229.

S. Sessa, Finite fuzzy relation equations with unique solution in complete Brouwerian lattices, *Fuzzy Sets and Systems*, 29 (1989), 103-113.

E.W. Shi, The hypothesis on the number of lower solutions of a fuzzy relation equation, *BUSEFAL* 31 (1987), 32-41.

G. Soula and E. Sanchez, Soft deduction rules in medical diagnostic processes, in: *Approximate Reasoning in Decision Analysis* (M.M. Gupta and E. Sanchez, Eds.), North-Holland Publ. Co., Amsterdam (1982), pp.77-88.

R.G. Sun, G.X. Chen and J.J. Yan, On iterative formula of the number of lower solutions for fuzzy relation equation on finite set, *BUSEFAL* 32 (1987), 61-63.

T. Takagi and M. Sugeno, Fuzzy identification of systems and its applications to modeling and control, *IEEE Trans. Syst. Man Cybern.* SMC-15 (1) (1985), 116-132.

H. Tanaka and J. Watada, Possibilistic linear systems and their application to the linear regression model, *Fuzzy Sets and Systems* 27 (1988), 275-289.

F. Tang, Fuzzy bilinear equations, *Fuzzy Sets and Systems* 28 (1988), 217-226.

M. Togai, A fuzzy inverse relation based on Gödelian logic and its applications, *Fuzzy Sets ans Systems* 17 (1985), 211-219.

G.M. Trojan, J.B. Kiszka, M.M. Gupta and P.N. Nikiforuk, Solution of multivariable fuzzy equations, *Fuzzy Sets and Systems* 23 (1987), 271-279.

T. Tsukiyama, H. Tanaka and K. Asai, Logical systems identification by fuzzy input-output data, *Systems and Control* 28 (1982), 319-326.

S. Umeyama, The complementary process of fuzzy medical diagnosis and its properties, *Inform. Sciences* 38 (1986), 229-242.

M. Wagenknecht and K. Hartmann, On the construction of fuzzy eigen solutions in given regions, *Fuzzy Sets and Systems* 20 (1986), 55-65.

M. Wagenknecht and K. Hartmann, Fuzzy modelling with tolerances, *Fuzzy Sets and Systems* 20 (1986), 325-332.

M. Wagenknecht and K. Hartmann, On direct and inverse problems for fuzzy equation systems with tolerances, *Fuzzy Sets and Systems* 24 (1987), 93-102.

M. Wagenknecht and K. Hartmann, Applications of fuzzy sets of type 2 to the solution of fuzzy equation systems, *Fuzzy Sets and Systems* 25 (1988), 183-190.

A.M. Wang, Q.G. Wang, W.Z. Yu and H.Q. Zhao, Fuzzy equation solving, *BUSEFAL* 31 (1987), 15-23.

H.F. Wang, An algorithm for solving iterated composite relation equations, *Proceedings of NAFIPS* 88, San Francisco, 8-10 June, 242-249.

H.X. Wang, Fuzzy relational non-deterministic equation and solution of the Schein rank of a fuzzy matrix, *BUSEFAL* 27 (1986), 88-93.

H.X. Wang and G.M. Chui, Indeterminate fuzzy relation equations, *Fuzzy Math.* 4 (1984), 17-26 (in chinese).

J.Y. Wang, An iterative formula finding the number of lower solutions of finite fuzzy relation equations, *preprint.*

J.Y. Wang, A simple and convenient method for solving fuzzy relation equation $X \odot A = B$, *preprint.*

M.X. Wang, Realizability condition for fuzzy matrices and their content, *Fuzzy Math.* 4 (1984), 51-58 (in chinese).

P.Z. Wang, *Fuzzy Set Theory and Practice*, Shangai Science and Technology Publ.

House, 1983 (in chinese).

W.M. Wu, Fuzzy reasoning and fuzzy relational equations, *Fuzzy Sets and Systems* 20 (1986), 67-78.

Z. Xiu and Z. Yang, An approach to solve fuzzy relation equations using iteration, *Preprints of II IFSA Congress*, Tokyo, 20-25 July 1987, 650-652.

C.W. Xu and Y.Z. Lu, Fuzzy model identification and self-learning for dynamic systems, *IEEE Trans. System Man Cybern.* SMC-17 (4) (1987), 683-689.

R.R. Yager, On solving fuzzy mathematical relationships, *Inform. and Control* 49 (1979), 29-55.

R.R. Yager, On the lack of inverses in fuzzy arithmetic, *Fuzzy Sets and Systems* 4 (1980), 73-82.

R.R. Yager, Some properties of fuzzy relationships, *Cybern. and Systems* 12 (1981), 123-140.

B.S. Zhay, A method for solving fuzzy relation equations by defining the symbolic value and generalized inverse of a fuzzy matrix, *Fuzzy Math.* 3 (1983), 7-18 (in chinese).

J. Zhu and R. Cheng, Some properties of fuzzy eigen set, *BUSEFAL* 30 (1987), 51-57.

## 15.2. A List of Papers on Fuzzy Relations and Related Topics

Autori Vari, *Insiemi Sfocati e Decisioni*, Edizioni Scientifiche Italiane, Napoli, 1983.

S.M. Baas and H. Kwakernaak, Rating and ranking of multi-aspect alternatives using fuzzy sets, *Automatica* 13 (1977), 47-58.

J.F. Baldwin and N.C.F. Guild, Comparison of fuzzy sets on the same decision space, *Fuzzy Sets and Systems* 2 (1979), 213-231.

W. Bandler and L.J. Kohout, Special properties, closures and interiors of crisp and fuzzy relations, *Fuzzy Sets and Systems* 26 (1988), 317-331.

J.C. Bezdek and J.D. Harris, Convex decomposition of fuzzy partitions, *J. Math. Anal. Appl.* 67 (1979), 490-512.

J.C. Bezdek, B. Spillman and R. Spillman, Fuzzy measures of preference and consensus in group decision making, *Proc. IEEE Conf. on Decision and Control* (K.S. Fu, Ed.), 1977, 1303-1309.

J.C. Bezdek, B. Spillman and R. Spillman, A fuzzy relation space for group decision theory, *Fuzzy Sets and Systems* 1 (1978), 225-268.

J.C. Bezdek, B. Spillman and R. Spillman, Coalition analysis with fuzzy sets, *Kybernetes* 8 (1979), 203-211.

J.C. Bezdek, B. Spillman and R. Spillman, A fuzzy analysis of consensus in small groups, in: *Fuzzy Set Theory and Applications to Policy Analysis and Information Sciences* (P.P. Wang and S.K. Chang, Eds.) Plenum Press, New York (1980), pp.291-308.

J.C. Bezdek, B. Spillman and R. Spillman, A dynamic perspective on leadership, development of a fuzzy measurement procedure, *Fuzzy Sets and Systems* 7 (1982), 19-33.

N. Blanchard, Cardinal and ordinal theories of fuzzy sets, in: *Fuzzy Information and Decision Processes* (M.M. Gupta and E. Sanchez, Eds.), North-Holland Publ. Co., Amsterdam (1982), pp.149-160.

N. Blanchard, Fuzzy relations on fuzzy sets: fuzzy orderings, fuzzy similarities (revised notions); two applications to ecology, *IEEE Internat. Conf. on Systems, Man and Cybernetics*, Bombay and New Delhi, 30 Dec. 1983 - 7 Jan. 1984, 132-135.

J.M. Blin, Fuzzy relations in group decision theory, *J. Cybern.* 4 (1974), 17-22.

A.N. Borisov and G.V. Merkuryeva, Methods of utility evaluation in decision-making problems under fuzziness and randomness, in: *Fuzzy Information, Knowledge Representation and Decision Analysis* (M.M. Gupta and E. Sanchez, Eds.), IX IFAC Symposium, Marseille, 19-21 July 1983, Pergamon Press, pp.305-310.

J.J. Buckley and W. Siler, Echocardiogram analysis using fuzzy sets and relations, *Fuzzy Sets and Systems* 26 (1988), 39-48.

Z.Q. Cao, K.H. Kim and F.W. Roush, *Incline Algebra and Applications*, John Wiley & Sons, New York, 1985.

C. Cella, A. Fadini and R. Sarno, A sharpening operator for a class of fuzzy preference relations, *BUSEFAL* 6 (1981), 76-86.

U. Cerruti, Completion of L-fuzzy relations, *J. Math. Anal. Appl.* 94 (1983), 312-327.

M.K. Chakraborty and M. Das, On fuzzy equivalence I, *Fuzzy Sets and Systems* 11 (1983), 185-193.

M.K. Chakraborty and M. Das, On fuzzy equivalence II, *Fuzzy Sets and Systems* 11 (1983), 299-307.

M.K. Chakraborty and M. Das, Studies on fuzzy relations over fuzzy subsets, *Fuzzy Sets and Systems* 9 (1983), 79-89.

M.K. Chakraborty, S. Sarkar and M. Das, Some aspects of [0,1]-fuzzy relation and a few suggestions toward its use, in: *Approximate Reasoning in Expert Systems* (M.M. Gupta, A. Kandel, W. Bandler and J.B. Kiszka, Eds.), Elsevier Science Publishers B.V. (North-Holland), Amsterdam (1985), pp.139-156.

Y.Y. Chen and T.Y. Chen, Fuzzy relations, *Fuzzy Math.* 1 (1981), 125-132 (in chinese).

M. Das and M.K. Chakraborty, An introduction to fuzzy tolerance relation, *Proceedings of the Seminar on Fuzzy Systems and Non-Standard Logic*, Calcutta, 14-16 May 1984, 10-32.

M. Delgado, J.L. Verdegay and M.A. Vila, A procedure for ranking fuzzy numbers using fuzzy relations, *Fuzzy Sets and Systems* 26 (1988), 49-62.

O.P. Dias, The R&D project selection problem with fuzzy coefficients, *Fuzzy Sets and Systems* 26 (1988), 299-316.

A. Di Nola and A. Fadini, A hyperspatial representation of a particular set of fuzzy preference relations, *Fuzzy Sets and Systems* 7 (1982), 79-87.

J.P. Doignon, B. Monjardet, M. Roubens and P.Vincke, Biorder families, valued relations and preference modelling, *J. Math. Psychology* 30 (4) (1986), 435-480.

J.C. Dunn, A graph theoretic analysis of pattern classification via Tamura's fuzzy relation, *IEEE Trans. Syst. Man Cybern.* SMC-4 (1974), 310-313.

A. Fadini, A Lettieri and F. Liguori, Classification of some fuzzy preference relations, *BUSEFAL* 5 (1980), 30-41.

A. Fadini and S. Sessa, Measures of fuzziness in a set of preference relations, *BUSEFAL* 6 (1981), 66-75.

M. Fedrizzi, *Gli insiemi sfocati nei modelli di supporto alle decisioni*, Provincia Autonoma di Trento, Studio Monografico, Servizio Statistica, 1988.

M. Fedrizzi and J. Kacprzyk, On measuring consensus in the setting of fuzzy preference relations, *Preprints of II IFSA Congress*, Tokyo, 20-25 July 1987, 625-628.

M. Fedrizzi and J. Kacprzyk, Soft consensus measures for monitoring real consensus reaching processes under fuzzy preferences, *Control and Cybern.* 15 (1987), 309-324.

S.Z. Guo, An algorithm for fuzzy clustering based on fuzzy relations, *BUSEFAL* 23 (1985), 111-119.

H. Hashimoto, Subinverses of fuzzy matrices, *Fuzzy Sets and Systems* 12 (1984), 155-168.

H. Hashimoto, Properties of negatively transitive fuzzy relations, *Preprints of II IFSA Congress*, Tokyo, 20-25 July 1987, 540-543.

S. Higuchi, S. Tamura and K. Tanaka, Pattern classification based on fuzzy relations, *IEEE Trans. Syst. Man Cybern.* SMC-1 (1971), 61-66.

U. Höhle, Quotients with respect to similarity relations, *Fuzzy Sets and Systems* 27 (1988), 31-44.

J. Jacas and L. Valverde, A metric characterization of T-transitive relations, *Proceedings Fall Internat. Seminar on Applied Logic*, Palma de Mallorca (1986), 81-89.

J. Kacprzyk, Group decision making with a fuzzy linguistic majority, *Fuzzy Sets and Systems* 18 (1986), 105-118.

J. Kacprzyk and M. Roubens (Eds.), *Non - Conventional Preference Relations in Decision Making*, Lecture Notes in Economics and Mathematical Systems, Vol. 301, Springer-Verlag, Berlin, 1988.

J. Kacprzyk and R.R. Yager (Eds.), *Management Decision-Support Systems Using Fuzzy Sets and Possibility Theory*, ISR N.83, Verlag TÜV Rheinland, Köln, 1985.

J. Kacprzyk, S. Zadrozny and M. Fedrizzi, An interactive user-friendly decision support system for consensus reaching based on fuzzy logic with linguistic quantifiers, in: *Fuzzy Computing* (M.M. Gupta and T. Yamakawa, Eds.), Elsevier Science Publishers B.V. (North-Holland), Amsterdam (1988), pp.307-321.

A. Kajii, A general equilibrium model with fuzzy preferences, *Fuzzy Sets and Systems* 26

(1988), 131-133.

K.H. Kim, *Boolean Matrix Theory and Applications*, Monographs and Textbooks in Pure and Applied Mathematics, Marcel Dekker, New York - Basel, 1982.

K.H. Kim and F.W. Roush, Generalized fuzzy matrices, *Fuzzy Sets and Systems* 4 (1980), 293-315.

K.H. Kim and F.W. Roush, Fuzzy flows on networks, *Fuzzy Sets and Systems* 8 (1982), 35-38.

K.H. Kim and F.W. Roush, Fuzzy matrix theory, in: *Analysis of Fuzzy Information, Vol.1: Mathematics and Logic* (J.C. Bezdek, Ed.), CRC Press, Boca Raton, FL. (1987), pp.107-129.

W. Kolodziejczyk, On reduction of s-transitive fuzzy matrices, *Proceedings of II joint IFSA-EC EURO-WG Workshop on Progress in Fuzzy Sets in Europe*, Vienna, 6-8 April 1988, 167-170.

W. Kolodziejczyk, On equivalence of two optimization methods for fuzzy discrete programming problem, *European J. Oper. Research* 36 (1988), 85-91.

V.B. Kuzmin and S.V. Ovchinnikov, Group decision I: in arbitrary spaces of fuzzy binary relations, *Fuzzy Sets and Systems* 4 (1980), 53-62.

V.B. Kuzmin and S.V. Ovchinnikov, Design of group decision II: in arbitrary spaces of fuzzy binary relations, *Fuzzy Sets and Systems* 4 (1980), 153-165.

A. Li, Y.F. Li, B.W. Li and J.L. Zha, Systematic structure on fuzzy restricting relation, in: *Proceedings of the Internat. Symposium on Fuzzy Systems and Knowledge Engineering* (X.H. Liu and P.Z. Wang, Eds.), Guangzhou and Guiyang, China, Guangdong Higher Education Publishing House, July 1987, pp.290-292.

H.X. Li, Inverse problem of multifactorial decision, *Fuzzy Math.* 5 (1985), 41-48 (in chinese).

H.X. Li, Multifactorial fuzzy sets and multifactorial degree of nearness, *Fuzzy Sets and Systems* 19 (1986), 291-297.

H.X. Li, Fuzzy clustering methods based on perturbation, *Preprints of II IFSA Congress*, Tokyo, 20-25 July 1987, 637-640.

S.H. Li, A new algorithm for computing the transitive closure of fuzzy proximity relation matrices, *BUSEFAL* 16 (1983).

S.H. Li and Y.S. Li, WK-clustering algorithms, *BUSEFAL* 19 (1984).

S.H. Li, Y.S. Li and P.Z. Wang, A new algorithm for determining the transitive closure of a fuzzy proximity matrix, *Preprints of II IFSA Congress*, Tokyo, 20-25 July 1987, 544-546.

Z.W. Liao, Mutually inverse properties of fuzzy matrix and its inverse, *BUSEFAL* 5 (1980), 13-19.

L.D. Lin, Some properties of a fuzzy system with diagonal superiority, in: *Proceedings of the Internat. Symposium on Fuzzy Systems and Knowledge Engineering* (X.H. Liu and P.Z. Wang, Eds.), Guangzhou and Guiyang, China, Guangdong Higher Education Publishing House, July 1987, pp.780-781.

W.J. Liu, The realizable problem for fuzzy symmetric matrix, *Fuzzy Math.* 1 (1982), 69-76 (in chinese).

S.P. Lou and Z.X. Tong, LSG powers of an L-Fuzzy matrix, in: *Approximate Reasoning in Decision Analysis* (M.M. Gupta and E. Sanchez, Eds.), North-Holland Publ. Co., Amsterdam (1982), pp.51-55.

C.Z. Luo, A method for finding generalized inverses of fuzzy matrices, *Fuzzy Math.* 1 (1981), 31-38 (in chinese).

C.Z. Luo, Generalized inverse fuzzy matrix, in: *Approximate Reasoning in Decision Analysis* (M.M. Gupta and E. Sanchez, Eds.), North-Holland Publ. Co., Amsterdam (1982), pp.57-60.

M. Matloka, On fuzzy vectors and fuzzy matrices, *BUSEFAL* 19 (1984), 92-104.

B. Monjardet, A generalization of probabilistic consistency: linearity conditions for valued preference relations, in: *Non-Conventional Preference Relations in Decision Making* (J.Kacprzyk and M. Roubens, Eds.), Lecture Notes in Economics and Mathematical Systems, Springer-Verlag, Vol. 301, Berlin (1988), 36-53.

F.J. Montero, Measuring the rationality of a fuzzy preference relation, *BUSEFAL* 26 (1986), 75-82.

F.J. Montero, Extended fuzzy preferences (I): randomized extension, *BUSEFAL* 33

(1987), 134-142.

F.J. Montero, Extended fuzzy preferences (II): fuzzified version, *BUSEFAL* 34 (1988), 133-140.

F.J. Montero and J. Tejada, Some problems on the definition of fuzzy preference relations, *Fuzzy Sets and Systems* 20 (1986), 45-53.

F.J. Montero and J. Tejada, A necessary and sufficient condition for the existence of Orlovsky's choice set, *Fuzzy Sets and Systems* 26 (1988), 121-125.

R. Mukherjee, Some observations on fuzzy relations over fuzzy subsets, *Fuzzy Sets and Systems* 15 (1985), 249-254.

V. Murali, Fuzzy equivalence relations, *Preprints of II IFSA Congress*, Tokyo, 20-25 July 1987, 647-649.

K. Nakamura, Preference relations on a set of fuzzy utilities as a basis for decision making, *Fuzzy Sets and Systems* 20 (1986), 147-162.

W.C. Nemitz, Fuzzy relation and fuzzy functions, *Fuzzy Sets and Systems* 19 (1986), 177-191.

H. Nurmi, Approaches to collective decision making with fuzzy preference relations, *Fuzzy Sets and Systems* 6 (1981), 187-198.

W. Ostasiewicz, Multi-valued relations, *BUSEFAL* 27 (1986), 94-104.

S.V. Ovchinnikov, Fuzzy choice functions, in: *Fuzzy Information, Knowledge Representation and Decision Analysis* (M.M. Gupta and E. Sanchez, Eds.), IX IFAC Symposium, Marseille, 19-21 July 1983, Pergamon Press, pp. 359-364.

S.V. Ovchinnikov, A stochastic model of choice, *Stochastica* 9 (1985), 135-152.

S.V. Ovchinnikov, On ordering of fuzzy numbers, in: *Uncertainty and Intelligent Systems* (B. Bouchon, L. Saitta and R.R. Yager, Eds.), Lecture Notes in Computer Science, Springer-Verlag, Vol. 313, Berlin (1988), pp.79-86.

S.V. Ovchinnikov, Preference and choice in a fuzzy environment, in: *Optimization Models Using Fuzzy Sets and Possibility Theory* (J. Kacprzyk and S.A. Orlovsky, Eds.), D. Reidel Publ. Co., Dordrecht (1988), pp.91-98.

S.V. Ovchinnikov and M. Migdal, On ranking fuzzy sets, *Fuzzy Sets and Systems* 24 (1987), 113-116.

S.V. Ovchinnikov and T. Riera, On fuzzy binary relations, *Stochastica* 7, n.3, (1983), 229-241.

K. Piasecki, On any class of fuzzy preference relations in real line. Part 2, *BUSEFAL* 21 (1985), 82-92.

K. Piasecki, On intervals defined by fuzzy preference relation, *BUSEFAL* 22 (1985), 58-65.

X.T. Peng, Solution to a problem of Kim and Roush, *BUSEFAL* 16 (1983), 34-37.

X.T. Peng, A property of strongly transitive matrices over a distributive lattice, *BUSEFAL* 19 (1984), 90-91.

X.T. Peng, A property of matrices over ordered sets, *Fuzzy Sets and Systems* 19 (1986), 47-50.

M. Roubens and P. Vincke, On families of semiorders and interval orders imbedded in a valued structure of preference: a survey, *Inform. Sciences* 34 (1984), 187-198.

M. Roubens and P. Vincke, *Preference Modelling*, Lecture Notes in Economics and Mathematical Systems, N. 250, Springer, Berlin-New York, 1985.

P.A. Rubin, A note on the geometry of reciprocal fuzzy relations, *Fuzzy Sets and Systems* 7 (1982), 307-309.

S. Sarkar and M.K. Chakraborty, A few constants associated to a fuzzy relation, *Proceedings of the Seminar on Fuzzy Systems and Non-Standard Logic*, Calcutta, 14-16 May 1984, 92-105.

R. Spillman and B. Spillman, A survey of some contributions of fuzzy sets to decision theory, in: *Analysis of Fuzzy Information, Vol.2: Artificial Intelligence and Decision Systems* (J.C. Bezdek, Ed.), CRC Press, Boca Raton, FL. (1987), pp.109-118.

Z. Switalski, A concept of fuzzy greatestness and fuzzy maximality for fuzzy preference relations, *BUSEFAL* 26 (1986), 83-91.

Z. Switalski, Choice functions associated with fuzzy preference relations, in: *Non-Conventional Preference Relations in Decision Making* (J. Kacprzyk and M. Roubens

Eds.), Lecture Notes in Economics and Mathematical Systems, Vol. 301, Springer-Verlag, Berlin (1988), pp.106-118.

T. Tanino, Fuzzy preference relations in group decision making, in: *Non-Conventional Preference Relations in Decision Making* (J. Kacprzyk and M. Roubens Eds.), Lecture Notes in Economics and Mathematical Systems, Vol. 301, Springer-Verlag, Berlin (1988), pp.54-71.

M. Togai and P.P. Wang, A study of fuzzy relations and their inverse problem, *Proceedings of 13th Internat. Symposium on Multiple Valued Logic*, Kyoto, 23-25 May 1983, 279-285.

R.M. Tong, A retrospective view of fuzzy control systems, *Fuzzy Sets and Systems* 14 (1984), 199-210.

R.M. Tong, An annotated bibliography of fuzzy control, in: *Industrial Applications of Fuzzy Control* (M. Sugeno, Ed.), North-Holland, Amsterdam, 1985, pp.249-269.

L. Valverde and S.V. Ovchinnikov, Representation of fuzzy symmetric relations, *Proceedings Fall Internat. Seminar on Applied Logic*, Palma de Mallorca (1984), 227-237.

H.X. Wang, One application of a fuzzy cubic matrix, in: *Proceedings of the Internat. Symposium on Fuzzy Systems and Knowledge Engineering* (X.H. Liu and P.Z. Wang, Eds.), Guangzhou and Guiyang, China, July 1987, Guangdong Higher Education Publishing House, pp.586-587.

H.X. Wang, The fuzzy non-singular matrix, *BUSEFAL* 34 (1988), 107-116.

H.X. Wang and Z.X. He, Solution of fuzzy matrix rank, *Fuzzy Math.* 4 (1984), 35-44 (in chinese).

H.X. Wang and C. Wang, Fuzzy cubic matrix algebra, *BUSEFAL* 29 (1987), 12-24.

H.X. Wang and C.X. Yi, The minimum row (column) space of fuzzy matrix, *BUSEFAL* 33 (1987), 38-46.

M. Wang, The realizable conditions for fuzzy matrix and its content, *Fuzzy Math.* 1 (1984), 51-58 (in chinese).

X.Y. Xue, On contents of realizable fuzzy matrices, *Preprints of II IFSA Congress*, Tokyo, 20-25 July 1987, 641-642; *BUSEFAL* 26 (1986), 67-74.

C.X. Yi and H.X. Wang, A table method and computer realization of solving the largest G-inverse of fuzzy matrix, *BUSEFAL* 34 (1988), 117-126.

Y.D. Yu, A note on weakly reflexive and symmetric L-fuzzy relation, *Fuzzy Math.* 5 (1985), 63-64 (in chinese).

S. Zahariev, An approach to group choice with fuzzy preference relations, *Fuzzy Sets and Systems* 22 (1987), 203-213.

J.L. Zha, The Schein rank of fuzzy matrices, *Fuzzy Math.* 2 (1982), 11-19 (in chinese).

C.K. Zhao, Inverses and generalized inverses of L-fuzzy matrix, in: *Proceedings of the Internat. Symposium on Fuzzy Systems and Knowledge Engineering* (X.H. Liu and P.Z. Wang, Eds.), Guangzhou and Guiyang, China, Guangdong Higher Education Publishing House, pp.782-789.

V.E. Zhukovin, F.V. Burshtein and E.S. Korelov, A decision making model with vector fuzzy preference relation, *Fuzzy Sets and Systems* 22 (1987), 71-79.

## Author Index

*Numbers in bold denote the page where a complete reference is given, numbers in italics denote the page where a citation of the author(s) is given only as reference number.*

## Subject Index

**THEORY AND DECISION LIBRARY**

SERIES D: SYSTEM THEORY, KNOWLEDGE ENGINEERING AND PROBLEM SOLVING

*Already published:*

**Topics in the General Theory of Structures**

*Edited by E. R. Caianiello and M. A. Aizerman*

ISBN 90–277–2451–2

**Nature, Cognition and System I**

*Edited by Marc E. Carvallo*

ISBN 90–277–2740–6